Disaster Management: Challenges and Mitigation Techniques

Disaster Management: Challenges and Mitigation Techniques

Editor: Cathy Hogan

Callisto Reference
www.callistoreference.com

Callisto Reference,
118-35 Queens Blvd., Suite 400,
Forest Hills, NY 11375, USA

Visit us on the World Wide Web at:
www.callistoreference.com

ISBN: 978-1-63239-895-6 (Hardback)

The publisher's policy is to use permanent paper from mills that operate a sustainable forestry policy. Furthermore, the publisher ensures that the text paper and cover boards used have met acceptable environmental accreditation standards.

Printed in China.

Cataloging-in-Publication Data

Disaster management : challenges and mitigation techniques / edited by Cathy Hogan.
p. cm.
Includes bibliographical references and index.
ISBN 978-1-63239-895-6
1. Emergency management. 2. Hazard mitigation. 3. Emergency management--Technological innovations. I. Hogan, Cathy.
HV551.2 .D57 2017
363.348--dc23

Table of Contents

Preface

Disaster management is concerned with minimizing the adverse effects of natural as well as manmade disasters. This involves devising strategies and developing programs to mitigate the loss to life and other assets. This book is a compilation of chapters that discuss the most vital concepts and emerging trends in the field of disaster management. The objective of this text is to give a general view of the different areas of this field, and its applications. It will serve as a reference to a broad spectrum of readers including researchers, students and professionals associated with disaster management across various levels.

The researches compiled throughout the book are authentic and of high quality, combining several disciplines and from very diverse regions from around the world. Drawing on the contributions of many researchers from diverse countries, the book's objective is to provide the readers with the latest achievements in the area of research. This book will surely be a source of knowledge to all interested and researching the field.

In the end, I would like to express my deep sense of gratitude to all the authors for meeting the set deadlines in completing and submitting their research chapters. I would also like to thank the publisher for the support offered to us throughout the course of the book. Finally, I extend my sincere thanks to my family for being a constant source of inspiration and encouragement.

Editor

1

Application of a weighted spatial probability model in GIS to analyse landslides in Penang Island, Malaysia

SAMY ISMAIL ELMAHDY†, MAGED MAHMOUD MARGHANY‡* and MOHAMED MOSTAFA MOHAMED†§

†Civil and Environmental Engineering Department, United Arab Emirates University, Al-Ain P.O. Box 15551, United Arab Emirates

‡Faculty of Geoinformation and Real Estate, Institute of Geospatial Science and Technology (INSTeG), Universiti Teknologi Malaysia (UTM), Johor 81310, Malaysia

§Faculty of Engineering, Irrigation and Hydraulics Department, Cairo University, Giza P.O. Box 12211, Egypt

In the current study, Penang Island, which is one of the several mountainous areas in Malaysia that is often subjected to landslide hazard, was chosen for further investigation. A multi-criteria Evaluation and the spatial probability weighted approach and model builder was applied to map and analyse landslides in Penang Island. A set of automated algorithms was used to construct new essential geological and morphometric thematic maps from remote sensing data. The maps were ranked using the weighted probability spatial model based on their contribution to the landslide hazard. Results obtained showed that sites at an elevation of 100–300 m, with steep slopes of 10°–37° and slope direction (aspect) in the E and SE directions were areas of very high and high probability for the landslide occurrence; the total areas were 21.393 km^2 (11.84%) and 58.690 km^2 (32.48%), respectively. The obtained map was verified by comparing variogram models of the mapped and the occurred landslide locations and showed a strong correlation with the locations of occurred landslides, indicating that the proposed method can successfully predict the unpredictable landslide hazard. The method is time and cost effective and can be used as a reference for geological and geotechnical engineers.

1. Introduction

Malaysia is a tropical country with heavy and sometimes with non-stop heavy rainfall. The heavy rainfall creates landslides on steep and cutting slopes in mountainous areas and flash floods at low-lying areas in Penang Island (Jacobson et al. 1993). Landslides and flash floods hit various parts of Penang and have caused heavy traffic, while roads and highways have been cut-off for several days (Chan 1998) (figure 1).

The weathering and urban development has led to decrease the vegetation cover and forest area at the mountainous area (Chan 1998). These have led to the increase in rock erosion and soil saturation by rainwater and occurrence of landslides. In other words, the urban development in the mountainous areas with heavy rainfall and steep slopes and wrong slope designs were found to be essential factors that

*Corresponding author. Email: magedupm@hotmail.com; maged@utm.my

Figure 1. Forms of landslides (a), rock and debris fall (b) and (c) and land creeping (d). These are due to soil saturation by rainfall and wrong slope artificial designs at highly fractured mountainous areas of Malaysia.

cause the occurrence of landslides, while rainwater decelerating at low-lying areas with gentle slopes creates flash floods (Lee & Pradhan 2006; Dahal et al. 2008).

In such mountainous areas, the use of traditional techniques of mapping landslides is time and economic consuming (Chan 1998; Habibah et al. 2009). Thus, the integration of remote sensing and Geographic Information System (GIS) has been particularly proven as successful (Samy & Mohamed 2013).

The weighted probability spatial model (WSPM) and DEM data, particularly, were proven to be powerful tools in mapping geological and geomorphological features that are closely associated with the occurrence of landslides and rockfall, without direct field observations and measurements in remote and inaccessible locations and the knowledge of scientists in the previous similar works on groundwater potentiality and landslide susceptibility mapping (Vijith & Madhu 2007; Ettazarini 2007; Samy & Mohamed 2012).

These geomorphological and geological parameters are essential in landslide prediction and modelling on a regional scale (Nagarajan et al. 1998; Guzzetti et al. 1999; Van Westen 2000; Saha et al. 2005).

Several studies (Guzzetti et al. 1999; Akgün & Bulut 2007; Dahal et al. 2008) have been conducted for landslide susceptibility mapping using GIS and applied probabilistic models. Ohlmacher and Davis (2003) applied the logistic regression model to landslide susceptibility mapping. Shou and Wang (2003) and Zhou et al. (2003) applied geotechnical and safety factor models indicate the landslide hazards. Other techniques (Lee 2007; Vijith & Madhu 2007) have been applied to landslide mapping and modelling using fuzzy logic, fuzzy algebraic operator and artificial neural network (ANN) and cross applied their weight for landslide susceptibility mapping to achieve reasonable prediction accuracy (81.36%).

Previous works (e.g. Pradhan & Lee 2007) conducted in the same study area showed that such landslide mapping was successful and able to accurately highlight the location of landslide occurrence. However, such works were based on a set of subjective model parameters. These parameters were extracted from optical remote sensing using visual interpretation. In addition, these works have been constructed based on geological and geomorphological maps lack to the accuracy and detailed information. Recently, Samy and Mohamed (2012) assessed the natural hazard in Kuala Lumpur, Malaysia using remote sensing and GIS.

The present study considered the accuracy and details of the model parameters using a set of automated and semi-automated algorithms. In particular, the Spectral Angle Mapper (SAM) of Kruse et al. (1993), topographic fabric algorithm of Guth (1999), and the deterministic eight-node (D8) algorithm of Jenson and Domingue (1988) were used to construct a new geospatial database and model parameters based on pixel calculations from different sensors. Then, these parameters were ranked and assessed using weighted spatial probability modelling (WSPM) to predict the occurrence of landslides in Penang Island, Malaysia. Spatial analysis, ranking and weighting, and representation of data are crucial in the natural hazard prediction. This study aimed to map landslide susceptibility and provided a more detailed geotechnical point of view of the major factors of the landslides which would permit better understanding of the natural hazards in Malaysia.

2. Study area

The study area is Penang Island, Malaysia. It is a part of Penang state measuring about 22 km by 14 km (306.842 km^2). It stretches from latitude 5° 15″ N to 5° 31″ N and from longitude 100° 8″ E to 100° 21″ E (figure 2(*a*)). It is located on the north-western coast of Peninsular Malaysia and linked with the Penang state with a 13.5-km long bridge. The mountainous areas in the central, southern and northern parts are composed of weathered and fractured granite, while the low-lying areas in the eastern and western parts are covered with alluvial deposits (figure 2(*b*)).

A tremendous number of geological structures mark the mountainous areas. These features are tensional and wrench, with some low-lying both because of the regional shear stress that affects the study area and the successive tectonic events that occurred during the Silurian and Cretaceous periods SW. Their orientations are predominantly WNW–ESE, NNE–SSW, NW-SE and NE–SW (figure 2(*b*)) .

These features serve as channels for rainwater and increase the erosion of granite. The erosion of granite forms sand (quartz); clay and silt are accumulated in the fault zones and depressions and create landslides.

The relief ranges from complex in the mountainous area in the northern, central and southern parts to low lying and flat in the eastern and western parts of Penang Island (figure 2). The slope ranges from steep slope (>10°) at the mountainous areas to gentle slope (<3°) at low-lying areas (Habibah et al. 2009). The rainfall of the study area ranges from 2670 to 6240 mm. In tropical regions, rainfall is the main reason for landslide occurrence. The annual rainfall during two distinct wet events from September to December and from February to May per annum is ranging from 1500

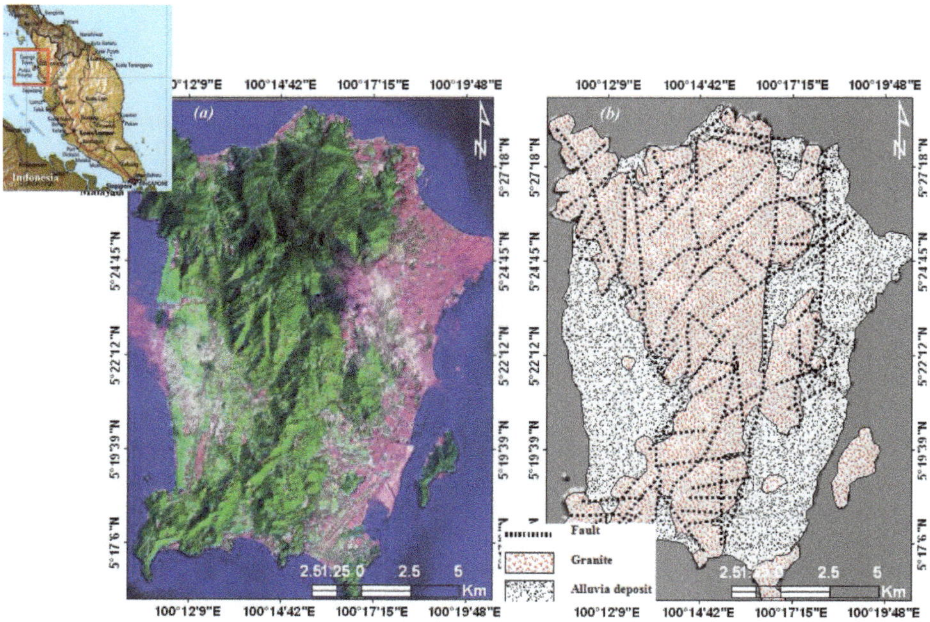

Figure 2. RGB 742 LANDSAT TM image (a) and litho-structural map of the study area (b), showing the major rock in Penang is granite, which is covered by forest.

to 3000 mm/year as reported by the Drainage and Irrigation Department of Malaysia.

The highest rainfall value is reported in the middle part (mountainous area) and considered favourable for landslide occurrence. The heavy rain area is located in the elevated areas in the middle part and received rainwater much more than the lowland area in the coastal area (east and west). For example, in the elevated areas rainfall rate reaches 3000 mm on mountainous areas, while it does exceed 1500 mm in the coastal area. At a higher elevation and steep slope, the rainwater accelerates and rock erosion increases, while at lower elevation the rainwater decelerates and sediment accumulation on steep slopes increases. The sediment accumulation on steep slopes increases the probability of landslide occurrence.

3. Data and methods

3.1. Data and preprocessing

The digital elevation mode (DEM) used in this study was resampled and fused with the Shuttle Radar Topographic Mission (SRTM) DEM (http://srtm.csi.cgiar.org/Index.asp) of ~90 m spatial resolution and an Advanced Spaceborne Thermal Emission and Reflection Radiometer (ASTER) DEM (http://www.gdem.aster.ersdac.or.jp/search.jsp) of 30 m spatial resolution.

To enhance the quality of DEM, four steps were applied. In the first step, the original cell size (0.000834 in decimal degrees) of SRTM DEM should be 0.000278 and the bilinear interpolation had higher accuracy and was quite effective. In the second step, elevation data from topographic map (of scale 1:25,000) were used to provide a correction layer to reduce the height offset of the ASTER and SRTM DEM. In the third step, the artifacts in the DEMs were reduced by applying a 3 × 3 smooth filter (Hengl et al. 2003). In the fourth step, the enhanced DEMs were then merged using merge and filter DEMs tool that implemented in ArcGIS v.10 software.

3.2. *Methods*

3.2.1. GIS database and input layer construction. Prior to the weighted spatial probability modelling, it is important to construct accurate landslides causative parameters. A set of the model parameters and/or factors (that are very closely associated with landslide occurrence) was taken into consideration for the landslide analysis. In the landslide susceptibility mapping, the potential event and its probability of occurrence were combined. These factors were constructed from ancillary and remote sensing data using visual interpretation and automated algorithms (figure 3). The SAM algorithm (supervised classifier) was employed to map the urban development on mountainous areas and cutting a slope from 30 m spatial resolution Landsat (figure 4(*a*)). The deterministic eight-node (D8) algorithm of Jenson and Domingue (1988), which is implemented in the ArcGIS v. 9.1 software was used to delineate stream network from DEM (figure 4(*b*)).

The maps of topographic slope (figure 5(*a*)) and aspect (figure 5(*b*)) were calculated from DEM. The maps of topographic slope, aspect and topographic relief (figure 6 (*a*)) were used to extract fractures and fault zones based on the slope breaks and slope directions (figure 6(*b*)). Finally, the topographic relief was calculated from DEM using the ArcGIS v. 9.2 software. All landslide causative factors were converted to raster layers for the weighted spatial probability modelling as seen in the flowchart of the landslides susceptibility mapping (figure 3).

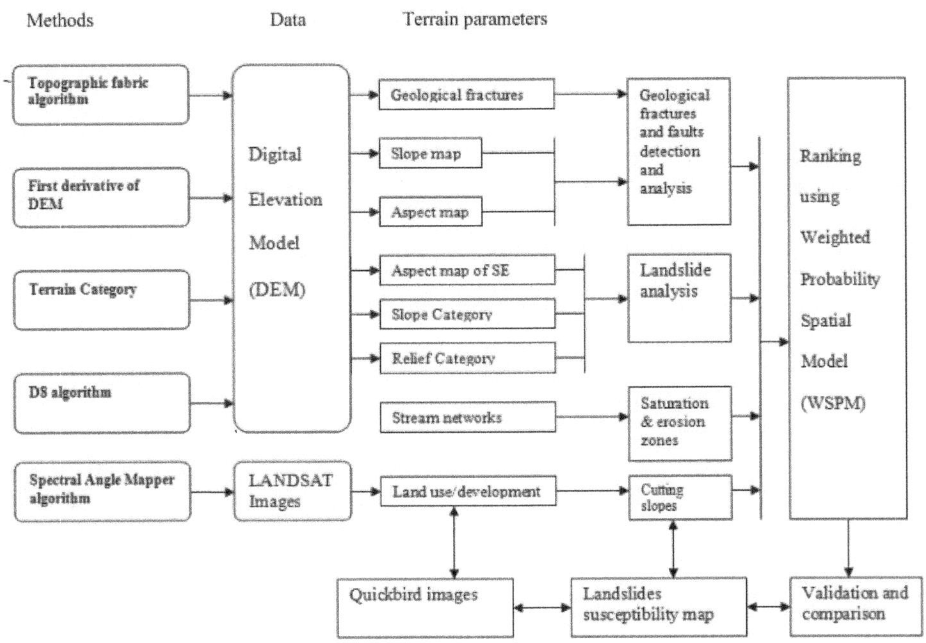

Figure 3. Flowchart of methodology used for landslide susceptibility mapping

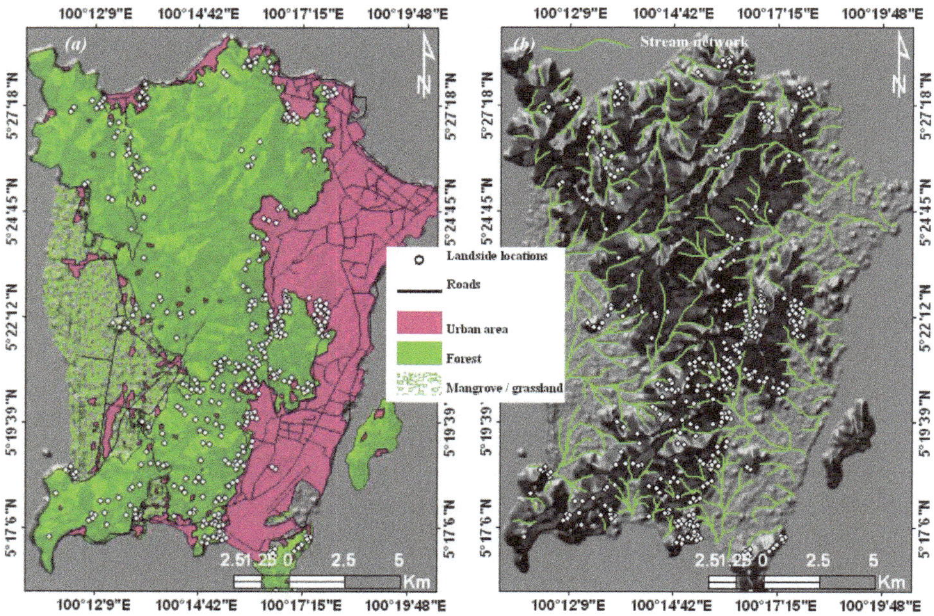

Figure 4. Land cover/use map generated from LANDSAT images (a), stream network drains Penang (b), showing the spatial association between the occurred landside locations (the white points) and the granite erosion by forest roots and stream network.

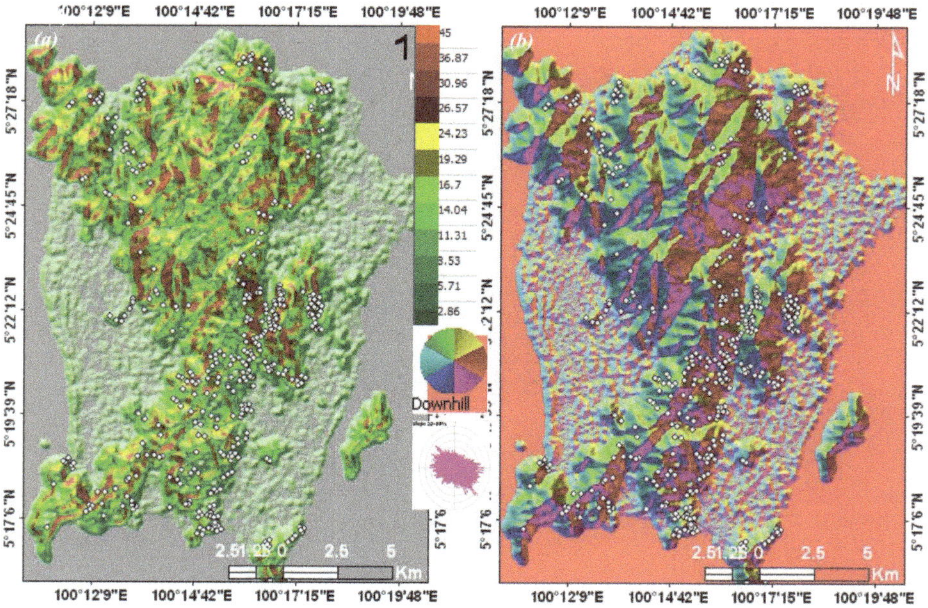

Figure 5. Slope (a), and aspect (b) maps calculated from DEM, showing the positive relationship between the slopes and the occurred landslide locations. Rose diagram shows that the common direction of escarpments and the occurred landslides is in NW–SE direction.

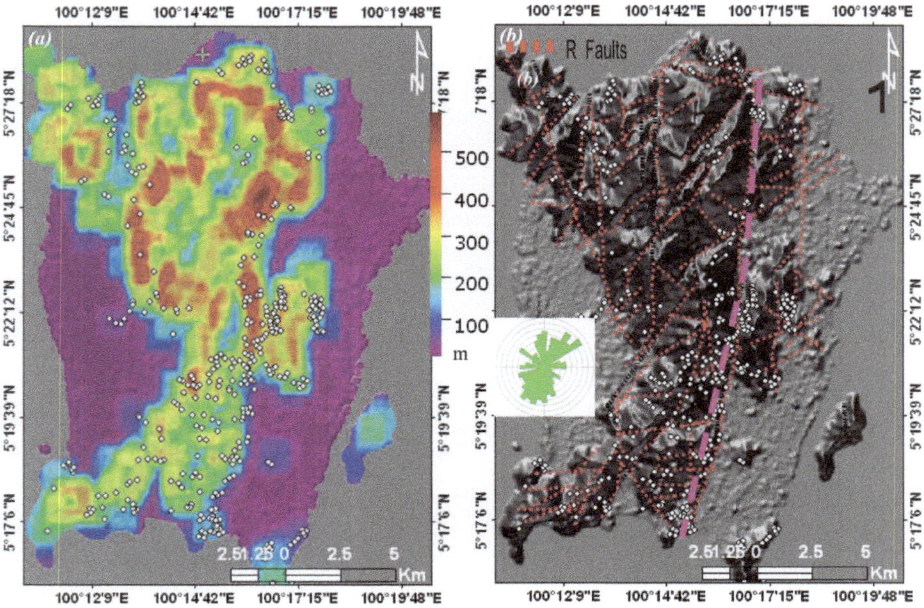

Figure 6. Relief map (a), and fault zones extracted from slope and aspect maps based on slope breaks and slope directions (b). The pink colour line highlights the regional fault zones that divide Penang into two parts and controls the direction of landslides. Rose diagram highlights the common directions of fault trends crosscutting the study area. White points highlight the known landslide locations.

3.2.2. Weight calculation. Not all landslide-related factors have the same influence on the landslide occurrence. For example, cutting slopes along hilly roads have much more effect than topographic relief and slope. In other talks, locations where landslides have not occurred and slope is flat were assigned a very low value of weight and rate of effectiveness, and locations where landslides have occurred were assigned a very high value for weight and rate of effectiveness (table 1). Thus, each layer was assigned a weight and rate of effectiveness based on its level of contribution in landslide occurrence. Each landslide-related factor was classified into five classes from very high to very low.

Suppose 100% as a maximum value for the rank; therefore, for the five class ranks will be categorized as 100%–80%, 80%–60%, 60%–40%, 40%–20% and 20%–0%, respectively. The average ranks will be as 90%, 70%, 50%, 30% and 10% in the WSPM. Cutting slopes along hilly roads were given the highest relative weight of 22% (0.22), while topographic slope was given the least weight of 18% (0.18).

To calculate the rate of effectiveness (rt) for each layer, the weight was multiplied by the rank (Wf × Rf). For instance, if the weight of land use and cutting slopes equals to 22%, and this is multiplied by the average rank of 90 (class I), the rate of effectiveness will be

$$\mathrm{rt} = \mathrm{Wf} \times \mathrm{Rf} = 0.22 \times 90 = 19.8 \tag{1}$$

In the present approach, the landslide susceptibility index was derived as the sum of the rate of effectiveness calculated to the landslide related factors and

Table 1. Ranks and weights for factors and their influencing classes used for landslides susceptibility mapping.

Hazard factor	Probability classes	Average rank (Rf)	Weight (Wt)	Rate of effectiveness (rt)
Land use and cutting slopes	I (very high)	90	22% (0.22)	19.8
	II (high)	70		15.4
	III (moderate)	50		11
	IV (low)	30		6.6
	V (very low)	10		2.2
Stream network	I (very high)	90	21% (0.21)	18.9
	II (high)	70		14.7
	III (moderate)	50		10.5
	IV (low)	30		6.3
	V (very low)	10		2.1
Geological structures	I (very high)	90	20% (0.20)	18
	II (high)	70		14
	III (moderate)	50		10
	IV (low)	30		6
	V (very low)	10		2
Slope	I (very high)	90	19% (0.19)	17.1
	II (high)	70		13.3
	III (moderate)	50		9.5
	IV (low)	30		5.9
	V (very low)	10		1.9
Relief	I (very high)	90	18% (0.18)	16.2
	II (high)	70		12.6
	III (moderate)	50		9
	IV (low)	30		5.4
	V (very low)	10		1.8

represented as follows:

$$\text{LSI} = \text{LCrt} + \text{Frt} + \text{Drt} + \text{Srt} + \text{Rrt} \tag{2}$$

where LSI is landslide susceptibility, LCrt, Frt, Drt, Srt and Rrt are the rate of effectiveness of land cover, fracture intersections, drainage pattern, slope and relief, respectively.

The landslide-related factors were imported into the weight sum overlay that implemented in ArcGIS v. 9.2 software to produce a landslide susceptibility map.

4. Landslide susceptibility mapping and spatial analysis

Through WSPM, it was possible to observe that the high probability of the occurrences of landslide is mostly located in the areas where the slope is of wrong design and the density of stream networks and geological fractures intersected is higher. These areas are suited in areas at an elevation of 100–300 m with slope angle of 10°–37° and face mostly to the E and SE directions. Furthermore, most of the

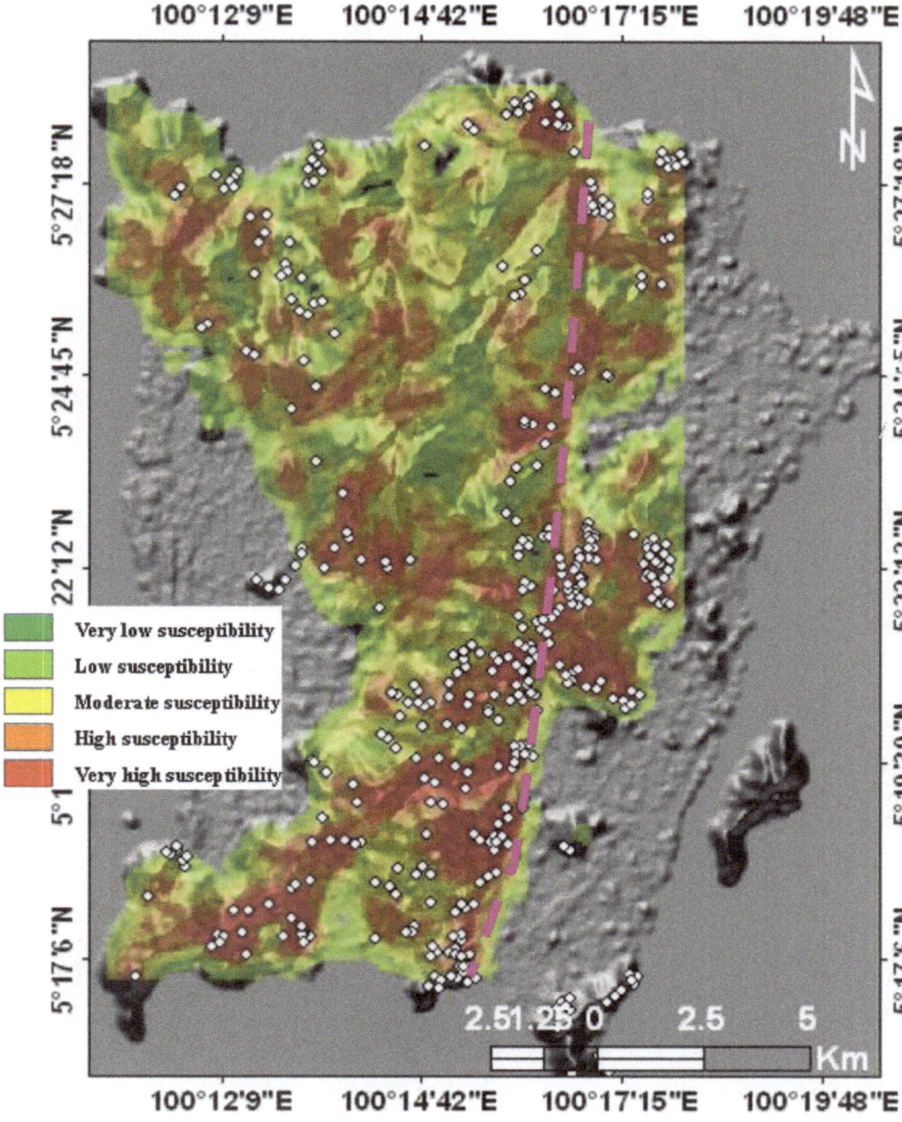

Figure 7. Landslide susceptibility map constructed using weighted spatial probability modelling. The map is draped over DEM. The white points highlight the known landslide locations. Most of the landslides are clustered along the NNE–SSW trending regional fault zone (pink colour).

Table 2. Areas of landslide hazard classes (km^2). Total granite area: 180.650 km^2.

Landslides	Very high susceptibility	High susceptibility	Moderate susceptibility	Low susceptibility	Very low susceptibility
Granite area (km^2)	21.393	58.690	53.57	30.98	16.017
Area% of granite area	11.84	32.48	29.65	17.14	8.86
No. landslides occurred	100	140	125	30	12

Note: Total number of reported landslides: 407.

reported landslides were clustered along the NNE–SSW and NW–SE trending regional fault zones (valleys) crosscutting the southern part of Penang Island (figure 7).

The areas having very high probability of landslide occurrence show an area of 21.393 km^2 (11.84%), whereas the areas characterized by low susceptibility for the occurrence of natural hazard are approximately 16.017 km^2 (8.86%) of the total Penang Island. However, the high landslides and rockfall susceptibility class occupies 58.690 km^2 (32.48%), while the moderate landslides and rockfall susceptibility class occupies 53.57 km^2 (29.65%) of the map, indicating the overall high and moderate landslide susceptibility of the Penang Island (figure 7 and table 2).

The zones of very high susceptibility for the occurrence of landslides are marked by the presence of intensive intersected fractures, wrong slope design and saturated bedrocks, while the zones of low susceptibility for the occurrence of landslides are marked by high stability of slope design, massive rocks.

Through the data processing and analysis, it was observed that the landslide occurrences in Penang Island are very strongly controlled by four parameters: (1) geological parameters such as major fault zones and geological fracture intersections, (2) hydrological parameters including intensive rainfall that exceeds 2700 mm, intensity stream network and shape and size of hydrological basins, (3) morphometric parameters including slope, aspect and relief shape, and (4) land use development on mountainous areas with wrong slope design.

The dentritic stream network reflects areas with steep slopes, low precipitation and high rate of rock erosion, while longitudinal streams reflect areas with gentle slopes, high precipitation and accumulation. These stream networks are often connected with fractured zones, which can serve as channels for rainwater downward movement and increase rock saturation. This rock saturation increases granite erosion and probability occurrence of rockfall and landslides, especially when the slope is greater than 10° (Gerasimov & Korzhuev 1979; Ollier 1981). Out of 407 landslides, 300 occurred within 200 m of the stream network.

On a steep slope, surface water accelerates and rock erosion increases, while on a lower slope surface flow decreases and sediment accumulation increases, and therefore, probability of landslide occurrence increases (Samy et al. 2012). The relationship between landslide occurrence and topographic slope and topographic relief shows that the areas with topographic slope above 11° and topographic relief of 400 m have a high frequency of landslide occurrence, indicating a higher probability of landslide occurrence. The spatial correlation also shows that there is an inverse relationship between distance to cutting slopes and farming areas and probability of landslide occurrence.

5. Verification

The landslide susceptibility map was validated using the known landslide locations. The validation was performed by constructing two grids (raster maps) of the known landslide locations and the landslide susceptibility map. The constructed grids were

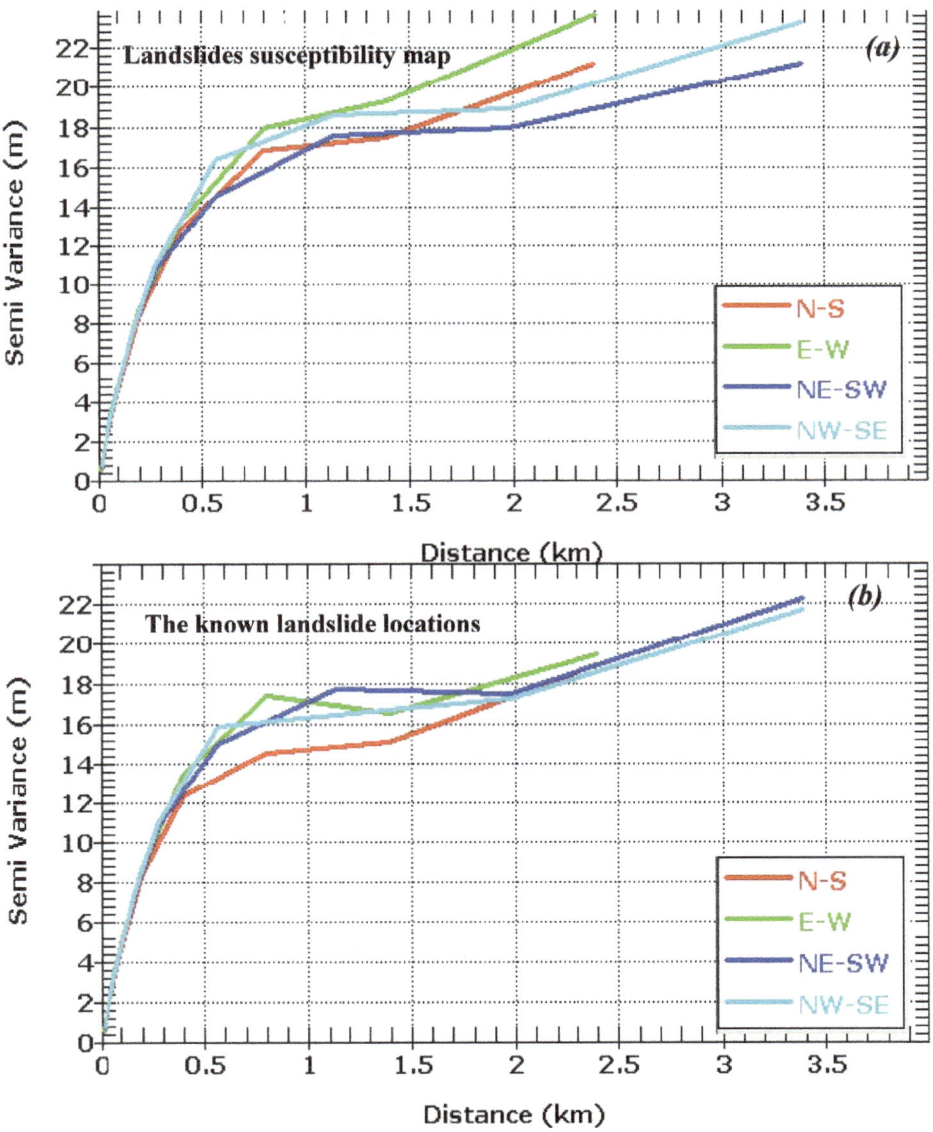

Figure 8. Semivariograms models in N–S, E–W, NE–SW and NW–SE directions. These were generated from the two grids; the first grid constructed from the landslide susceptibility map (a) with the second one of the known (b) landslide locations.

used to generate two semivariogram models (that are the key functions in geostatistics as it used to fit a model of the temporal/spatial correlation of the observed phenomenon). The resultant semivariogram models were used to correlate the landslide map with the occurred landslide locations. The four-directional semivariogram models for the occurred and the constructed landslides map show reasonable consistency (figure 8). Furthermore, a sill value, which defines the maximum value of the range, did not differ greatly between locations of the constructed map and the occurred landslides. All directional semivariogram models arise from an original point and continue to change at about 0.5 km with not much variation or dip in the curve with increasing distance. The NE–SW semivariogram rises a little more steeply in the known landslide, while the NW–SE values are comparatively reduced in the constructed landslide susceptibility map (figure 8). The variogram models for the constructed landslide map and the occurred landslide locations show positive correlation in the NW–SE, NE–SW and N–S directions with a little more steeply in the known landslides and difference in the E–W direction with root mean square errors (r^2) varying from 0.75 to 0.81 (table 3). The little difference between two grids due to the difference in the spatial resolution between aerial photographs that used to map the known landslide locations and the remote sensing data used to map the susceptibility of landslides.

6. Discussion

Prior to the result completed in this study, the precise mapping of landslide model parameters and the practical role of terrain parameters (e.g. slope, aspect and relief) on landslides occurrence in Malaysia had not been exercised in detail. Studies beyond the year 2005 (e.g. Pradhan & Lee 2007; Habibah et al. 2009) relied on the slide susceptibility mapping using parameters extracted visually. However, the influence of terrain parameters on landslides occurrence in these later studies is of limited investigation. The landslides in the Penang occur at highly fractured granite, zones with steep slopes, high relief and weathered granite. Weathering of granite rock by intensive rainwater on the wrong slope design appears to be a significant reason of landslide occurrence. The large amount of debris (sand and clay), creeping from the mountainous areas suggests chemical and mechanical weathering of granite ($SiO_2Al_2O_3K_2ONa_2O$) by rainwater in the following reaction:

$$\underset{\text{Granite}}{SiO_2\,Al_2O_3\,K_2ONa_2O} + \underset{\text{Water}}{H_2O} + H = \underset{\text{Quartz}}{SiO_2} + \underset{\text{Potassium ion}}{K + Al_4Si_4\,O_{10}(OH)_8} + \underset{\text{Clay mineral}}{SiO_2} \quad (3)$$

Table 3. Linear fit of constructed known and predicted semivariogram models in the four directions.

	Variogram of occurred landslides			Variograms of predicted landslides		
Directions	Intersect	Slope	r^2	Intercept	Slope	r^2
N–S	4.75	6.99	0.75	4.86	8.20	0.77
E–W	5.27	7.37	0.69	4.94	9.25	0.81
NE–SW	6.28	5.56	0.72	6.30	5.36	0.70
NW–SE	6.38	5.32	0.70	6.51	5.93	0.71

The proposed method uses essential geological, hydrological and morphometric maps such as stream network, geological linear structures that can be deduced from digital elevation models. Higher landslide hazard areas can be simulated and detected by studying the magnitude and contribution of essential factors to the landslide hazard.

The results of the study have permitted a better understanding of the spatial relationship between the occurrence of landslides and natural and anthropogenic factors. The analysis presented in the study shows the complexity of the landslides and the role of stream network and wrong slope design and terrain parameters in landslide occurrence in Malaysia for the first time. The produced map was tested against the occurred landslide locations to understand if the latter parameters in terms of topographical and structural control.

The proposed model based on a statistical criterion showed high precision with time and cost effectiveness. The resultant map is of great help for geotechnical engineers and urban planners for selecting suitable locations to employ developmental activities.

Future work will involve field measurement and the application of high resolution of SAR data (e.g. Phased Array L-band Synthetic Aperture Radar [PALSAR]) in order to extract more accurate model parameters. These data may enable the construction of more landslides hazard modelling of Penang Island.

7. Conclusions

In the current study, the integration of remote sensing, GIS and WSPM takes benefits from the computer-assisted computation of morphometric parameters and features (e.g. geological fractures, slope, aspect and stream patterns) to predict areas where landslide occurrences are higher. The landslides were mapped not only on the basis of WSPM but also on the spatial association between landslide hazard and geological, hydrological and morphometric parameters. The landslide susceptibility map is considered significant as it shows that the areas with higher landslide occurrences have strong agreement with the reported landslide locations, proving the ability of the proposed WSPM to predict the higher probability landslide hazard areas in Penang Island, Malaysia.

The study demonstrated that most of the landslide locations are clustered in the south-eastern and eastern parts of Penang Island due to tilting in elevation and slope under the influence of fault zone displacements. The study also demonstrated that the landslide response is controlled by five parameters: (1) geological structures on regional and small scale, (2) morphometric parameters such as slope, aspect and relief, (3) cut slope and land development such as wrong slope designs, (4) hydrological parameters in terms of intensive rainfall and stream networks and shape and size of drainage basins and (5) relief and topography. Sites of highly intensive geological fractures, heavy rainfall and wrong slope designs were very closely associated with landslide occurrences.

Acknowledgements

The authors are grateful to the anonymous reviewers and the Earth Remote Sensing Data Analysis Centre (ERSDAC) Japan, the NASA Land Processes Distributed Active Archive Center, the User Services USGS Earth Resources Observation and Science (EROS), and the

Consortium for Spatial Information (CSI) of the Consultative Group for International Agricultural Research (CGIAR) for providing and distributing remotely sensed data.

References

Akgün A, Bulut F. 2007. GIS-based landslide susceptibility for Arsin-Yomra (Trabzon, North Turkey) region. Environ Geol. 51:1377–1387.

Chan NW. 1998. Responding to landslide hazards in rapidly developing Malaysia: a case of economics versus environmental protection. Disaster Prev Manag. 7:14–27.

Dahal RK, Hasegawa S, Nonomura S, Yamanaka M, Masuda T, Nishino K. 2008. GIS-based weights-of-evidence modelling of rainfall-induced landslides in small catchments for landslide susceptibility mapping. Environ Geol. 54:314–324.

Elewa HH, Qaddah AA. 2010. Groundwater potentiality mapping in the Sinai Peninsula, Egypt, using remote sensing and GIS-watershed-based modeling. Hydrogeol J. 19:613–628.

Ettazarini S. 2007. Groundwater potentiality index: a strategically conceived tool for water research in fractured aquifers. Environ Geol. 52:477–487.

Gerasimov IP, Korzhuev SS, editors. 1979. Morphostructural analysis of the drainage network of the USSR. Moscow: Nauka. Russian.

Guth PL. 1999. Quantifying topographic fabric: eigenvector analysis using digital elevation models. In: Merisko RJ, editor. 27th Applied Imagery Pattern Recognition (AIPR) Workshop: Advances In Computer-assisted Recognition; 1988 Oct 14–16; Washington (DC); p. 233–243.

Guzzetti F, Carrarra A, Cardinali M, Reichenbach P. 1999. Landslide hazard evaluation: a review of current techniques and their application in a multi-scale study, Central Italy. Geomorphology. 31:181–216.

Habibah LK, Younus AK, Fauziah A. 2009. Mechanism of shallow landslide at 3.9 Km Tun Sardon Road of Penang Island. Proceedings of the South China Sea Tsunami Workshop held at Universiti Sains Malaysia; Malaysia.

Hengl T, David G, Rossiter F. 2003. Supervised landform classification to enhance and replace photo-interpretation in semi-detailed soil survey. Soil Sci Soc Am J. 67:1810–1822.

Jacobson RB, Mcgeehin JP, Cron ED, Carr CE, Harper JM, Howard AD. 1993. Landslides triggered by the storm of November 3–5, 1985, Willis Mountain anticline, West Virginia and Virginia. US Geol Surv Bull. C1–C33.

Jenson SK, Domingue JO. 1988. Extracting topographic structure from digital elevation model data for geographic information system analysis. Photogramm Eng Remote Sens. 54:1593–1600.

Kruse FA, Lefkoff AB, Boardman JB, Heidebrecht KB, Shapiro AT, Barloon PJ, Goetz AFH. 1993. The Spectral Image Processing System (SIPS) - interactive visualization and analysis of imaging spectrometer data. Remote Sensing Environ. 44:145–163.

Lee S, Pradhan B. 2006. Probabilistic landslide hazards and risk mapping on Penang Island, Malaysia. J Earth Syst Sci. 115:661–672.

Lee S. 2007. Application and verification of fuzzy algebraic operators to landslide susceptibility mapping. Environ Geol. 52:615–623.

Nagarajan R, Mukherjee A, Roy A, Khire MV. 1998. Temporal remote sensing data and GIS application in landslide hazard zonation of part of Western Ghat, India. Intern J Remote Sensing. 19:573–585.

Ohlmacher GC, Davis JC. 2003. Using multiple logistic regression and GIS technology to predict landslide hazard in northeast Kansa, USA. Eng Geol. 69:331–343.

Ollier C. 1981. Tectonics and landforms. London: Longman.

Pradhan B, Lee S. 2007. Utilization of optical remote sensing data and GIS tools for regional landslide hazard analysis by using an artificial neural network model at Selangor, Malaysia. Earth Sci Front. 14:143–152.

Saha AK, Gupta RP, Irene SI, Arora MK, Csaplovics E. 2005. An approach for GIS-based statistical landslide susceptibility zonation with a case study in the Himalayas. Landslides. 2:61–69.

Shou KJ, Wang CF. 2003. Analysis of the Chiufengershan landslide triggered by the 1999 Chi-Chi earthquake in Taiwan. Eng Geol. 68:237–250.

Samy IE, Shattri M, Bujang BK, Ahmad RM. 2012. Application of terrain analysis to the mapping and spatial pattern analysis of subsurface geological fractures of Kuala Lumpur Limestone Bedrock, Malaysia. Intern J Remote Sensing. 33:20.

Samy IE, Mohamed M. 2012. Natural hazards susceptibility mapping in the Kuala Lumpur, Malaysia: an assessment using remote sensing and Geographic Information System (GIS). Geomatics Nat Hazards Risk J. 4:71–91.

Samy IE, Mohamed M. 2013. Relationship between geological structures and groundwater flow and groundwater salinity in Al Jaww Plain, UAE; mapping and analysis by means of remote sensing and GIS. Arab J Geosci. 7:1249–1259.

Van Westen CJ. 2000. The modelling of landslide hazards using GIS. Surv Geophys. 21: 241–255.

Vijith H, Madhu G. 2007. Application of Gis and frequency ratio model in mapping the potential surface failure sites in the Poonjar subwatershed of Meenachil river in western ghats of Kerala. J Indian Soc Remote Sensing. 35:275–285.

Zhou G, Esaki T, Mitani Y, Xie M, Mori J. 2003. Spatial probabilistic modeling of slope failure using an integrated GIS Monte Carlo simulation approach. Eng Geol. 68:373–386.

2

Impact assessment of rainfall-vegetation on sedimentation and predicting erosion-prone region by GIS and RS

MAHBOOB ALAM†, RAJA RIZWAN HUSSAIN‡ and
A.B.M. SAIFUL ISLAM*§
†Department of Meteorology, COMSATS Institute of Information Technology, Islamabad 44000, Pakistan
‡Civil Engineering Department, College of Engineering, King Saud University, Riyadh 11421, Saudi Arabia
§Department of Civil Engineering, Faculty of Engineering, University of Malaya, Kuala Lumpur 50603, Malaysia

Water reservoirs are facing universal sedimentation problems worldwide. Land covers, whether natural or manmade, eventually change, and the vegetation cover and rainfall have a great effect on the sediment load. Traditional techniques for analysing this problem are time-consuming and spatially limited. Remote sensing (RS) provides a convenient way to observe land cover changes, and geographic information system (GIS) provides tools for geographic analysis. This study demonstrates a GIS-based methodology for calculating the impact of vegetation and rainfall on the sediment load using remotely sensed data. Moderate resolution imaging spectroradiometer data were used to observe temporal changes in the vegetation-cover area of the watershed surface. The total drainage area for the reservoir was calculated from shuttle radar topographic mission data. The annual rainfall amount was used to compute the annual available rainwater for the watershed, and the impact of the annual available rainwater on the vegetation-covered area was determined. In addition, areas that were adding sedimentation to the reservoir were identified. An inverse relationship between the rainfall and vegetation cover was observed, clearly showing the triggering of erosion.

1. Introduction

Water is an important resource for any country, especially for agricultural countries, such as Pakistan. It is a resource that cannot be generated but can be preserved. Water resources are affected by many natural hazards, such as sedimentation, earthquakes and floods. Anthropogenic activities such as land clearance and dam construction have led to an increase in sedimentation (Walling 2008). The study revealed that the accelerating pace of human impact in various regions of the world indicated the changes in the sediment loads of its rivers likely to be intensifying. The world's 13 largest rivers carry 5.8 billion tons of sediment into reservoirs every year (Nasir et al. 2000; Islam, Jameel, & Jumaat 2012).

*Corresponding author. Email: abm.saiful@um.edu.my

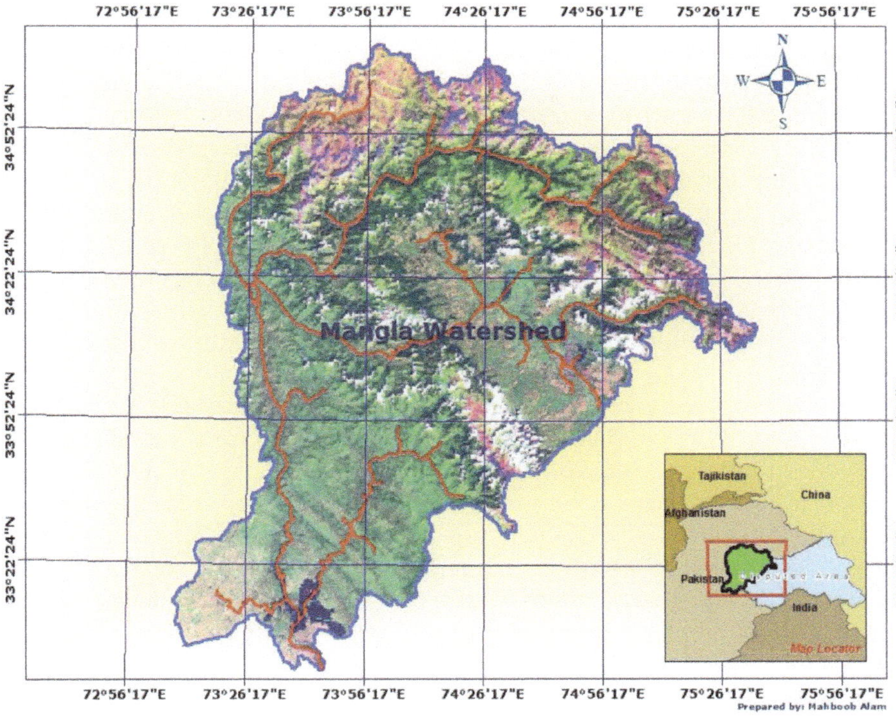

Figure 1. Study area/Mangla watershed.

This study was performed in the Mangla domestic watershed (Mangla dam 2010), shown in figure 1. The Mangla dam (Mangla dam 2010) was constructed in 1967 across the Jhelum River and is situated 60 km south-east of Islamabad, Pakistan. The storage capacity of the Mangla dam was 725,004 million cubic metres (mcm) when it was constructed, but now the capacity is only 585,675 mcm. The sedimentation rate is so high that the reservoir has lost 19.2% of its capacity since 1967. The precious Mangla and Tarbela reservoirs are becoming filled with silt at a rate of 51,806,238 and 51,806,238 cubic metre per annum, respectively, because of poor practices in the catchment areas (Khan 1985).

Erosion around the catchment area of the reservoir contributes a sediment load that reduces the capacity of the dam. Recently, several studies have been performed to uncover the causes of sedimentation (Tanaka & Yagisawa 2010; Day et al. 2011; Fritz et al. 2012; Islam, Jameel, Jumaat, Shirazi, & Salman 2012; Noe 2013). The vegetation parameter for modelling soil erosion has been estimated using a linear spectral mixture analysis of Landsat Enhanced Thematic Mapper (ETM) data in some studies (de Asis & Omasa 2007; Islam, Hussain, et al. 2013; Islam, Jameel, et al. 2013). In several studies, the landscape vegetation and rainfall influence have been explored. Nouri et al. (2012) have reviewed evapotranspiration measurement techniques for estimating the water requirements of urban landscape vegetation. Sandoval et al. (2013) evaluated rainfall influence over water quality effluent of an urban catchment incorporating a data-driven approach. Bookhagen (2010) presented the appearance of extreme monsoonal rainfall events and their impact on erosion in the Himalayas.

The remote sensing (RS) as well as geographic information system (GIS) based assessment has been carried out in a few studies. Koehn et al. (2011) dealt with quantification of storm water runoff using a combined GIS and curve number approach as a case study for an urban watershed in the Ozark Highlands, USA. Goldshleger et al. (2012) reported generalizing urban runoff and street network density relationship with hydrological and remote-sensing case study in Israel. Dahal et al. (2012) reported the replication of landslide hazard mapping at catchment scale. Pourghasemi et al. (2012) performed a comparative assessment of prediction capabilities of Dempster–Shafer and Weights-of-evidence models in landslide susceptibility mapping using GIS. Khosrokhani and Pradhan (2013) carried out spatio-temporal assessment of soil erosion at Kuala Lumpur metropolitan city using remote sensing data and GIS.

Most previous work has been accomplished via traditional methods, such as physical surveying and sampling. The Revised Universal Soil Loss Equation (RUSLE) has been used for three decades in combination with physical sampling and is known to work well in erosion management (Tiwari et al. 2000). However, the traditional methods of land management are time-consuming and spatially limited. The factors causing sedimentation have only been analysed on a small scale and on an experimental basis. A large-scale analysis and implementation of recommendations are required. This problem of sedimentation prompted the initiation of the Mangla Watershed Management Project at an early stage of the dam project, and biological and engineering measures were introduced to prevent erosion and to reduce sedimentation in the reservoirs (Haq & Ahmad 2004). Jat et al. (2009) conducted remote sensing and GIS-based assessment of urbanization and degradation of watershed health. Differently to the antecedent studies, the present case promotes the use of the geographic information system (GIS) and has reduced the effort involved in performing surveys and sampling and also increases the scale of analysis and requires less time than traditional methods.

Therefore, the aim of the study is to use simple GIS-based and remote sensing-based methodologies to identify the effects of vegetation and rainfall on reservoir capacity throughout the entire area providing water to the Mangla dam (the watershed). The annual rainfall is spatially interpolated in the GIS environment to compute the annual available rainwater. Remote sensing technology is well suited to assessing the land cover and to acquiring topographic information. For geo-visualization, shuttle radar topographic mission (SRTM) elevation data were used. Moderate resolution imaging spectroradiometer (MODIS) sensor data were used to determine changes in the vegetation cover area of the watershed. The effect of both vegetation and rainfall parameters on the sediment load was observed, and erosion-prone areas were identified in this research.

2. Materials and methods

The study considers the collection of data through different approach and categories. Four types of data were acquired which are stated as follows. These have been used for the detail estimation of essential parameters.

(1) Satellite data.
(2) SRTM data.
(3) Annual rainfall data.
(4) Sediment load data.

MODIS images were used to classify the land cover features in the study area, and temporal MODIS images were used to detect changes. Six images acquired at one-year intervals from 2000 to 2005 were selected for this study. The SRTM data were used to generate the watershed area of the Mangla dam. Eight tiles of SRTM data were used to generate the watershed. The total annual rainfall data from 2000 to 2005 were used in this study to compute the available water for the watershed and to identify areas of heavy rains. Fourteen meteorological observatories were selected to acquire rainfall data. The sediment load data for the Mangla dam (Mangla dam 2008) were used to observe the effect of vegetation and rainfall on the sediment load. Figure 2 illustrates the general methodology of this research.

2.1. *Watershed area*

The SRTM data were taken in height files that contain topographic information about the region. To generate a digital elevation model (DEM), the SRTM data require processing. Using the Landserf software, the height tiles are first converted into Landserf format (".srf"), and then voids in the data (due to sensor defects) are removed with the help of the software. The eight Landserf files are then converted to text files.

These text files contain latitude, longitude and elevation information for each pixel. A triangular unregulated network is generated from the text files in ArcGIS 9.1. Triangulated irregular network raster files are converted to DEM raster files and then converted into a single DEM file using the Mosaic tool.

ESRI's ArcHydro tool is used to extract the watershed area from the DEM. The first step in this process is to *fill sinks*, sudden changes in pixel height values in the DEM. Using hydrological modelling, the *flow direction* raster is generated from the DEM. The flow direction raster shows the actual direction of water flow. Each pixel in the flow direction raster is assigned a slope value.

The next process is the *flow accumulation* step. The flow accumulation is computed from the flow direction raster and contains the accumulated number of cells upstream of a particular cell.

By using this flow accumulation raster, the stream definition raster is generated. The *stream definition* function takes a flow accumulation raster as the input and creates a stream raster for a user-defined threshold. This threshold is defined either as a number of cells (default: 1%) or as a drainage area in square kilometres. In this study, the 1% default is selected.

To obtain a stream network for the study area, the *stream segmentation* raster is generated from the stream definition raster. This function links the stream definition raster to create stream networks for the area.

The catchments delineation raster is computed from the stream segmentation. The catchments raster delineation function creates a raster in which each cell carries a value that indicates which cells belong to the catchments. This value corresponds to the value carried by the stream segment that drains a particular area, as defined in the stream segment link raster.

To extract a polygon shape file of the catchments, the *catchments polygon processing* step is performed on the catchments delineation raster. Adjacent cells in the raster with the same raster code are combined into a single area, whose boundary is then vectorized. Single-cell polygons and "orphan" polygons generated as artefacts of the

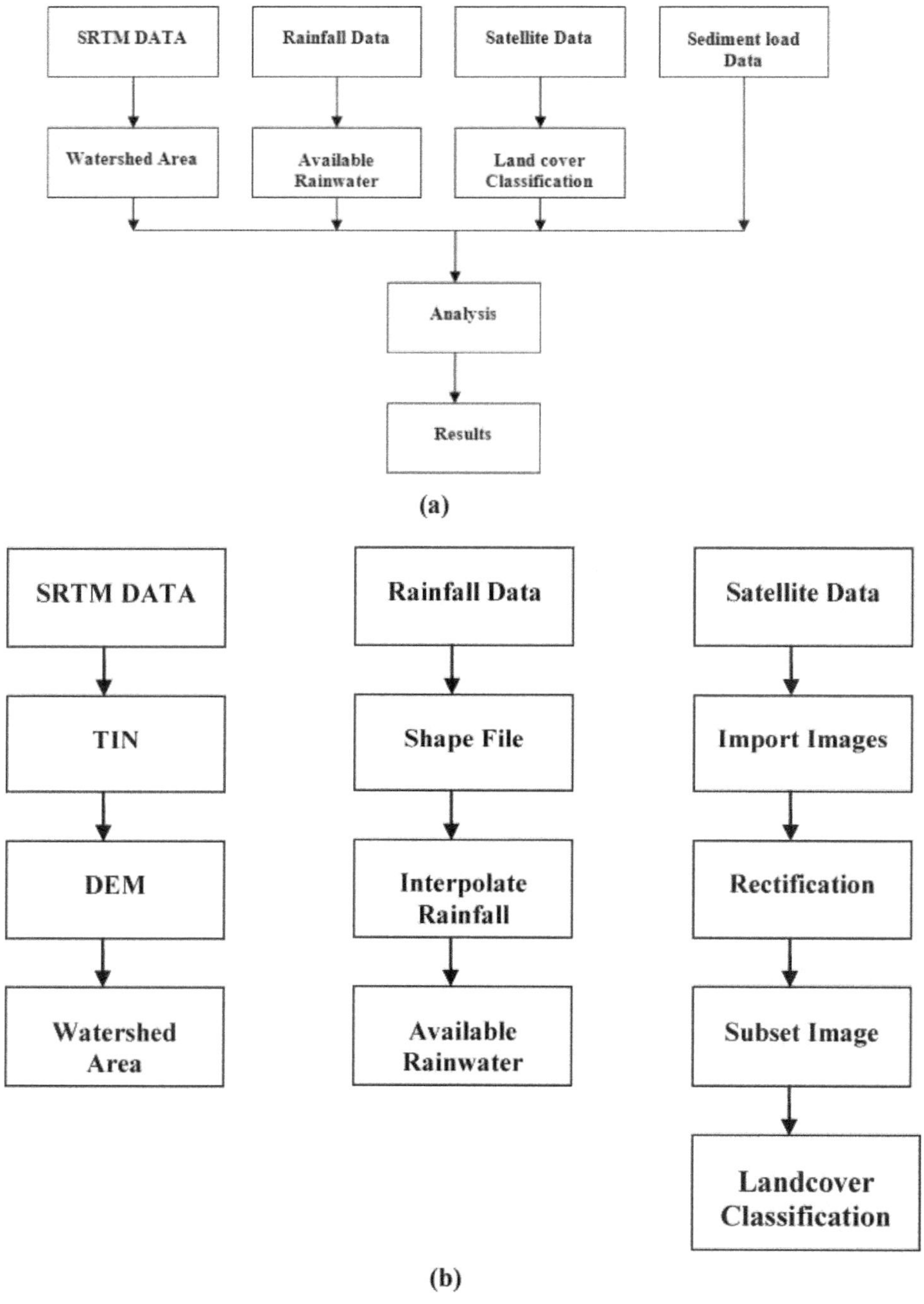

Figure 2. Methodology.

vectorization process are automatically dissolved; thus, at the end of the process, there is only one polygon per catchment.

The drainage lines are generated from the stream definition raster and the flow accumulation raster using the Archydro tool *drainage line processing*. The drainage line processing function converts the input stream link raster into a drainage line feature class. Each line in the feature class carries an identifier of the catchment in which it resides.

Finally, the watershed extraction is performed by the *adjoint catchment processing* function. The adjoint catchment processing function generates the aggregated upstream catchments from the "Catchment" feature class. For each catchment that is not a head catchment, a polygon representing the entire upstream area draining to its inlet point (reservoir) is constructed and stored in a feature class with an "Adjoint Catchment," which represents the watershed of the Mangla dam (Indus Basin 2008; Mangla dam 2008). The Mangla watershed and its stream network are shown in figure 1.

2.2. *Available rainwater*

The annual available rainwater for the watershed was calculated from the annual rainfall data. Total rainfall data from 14 stations for 2000 to 2005 were acquired. Six raster files for each year from 2000 to 2005 were generated from the annual rainfall data with a projected coordinate system, "WGS 84 North UTM zone 43."

To calculate the available rainwater, the watershed area was separated from the rainfall-interpolated raster files. Therefore, the rainwater was calculated using the "Area and Volume" function from the rainfall raster files. The available rainwater for the watershed for each year from 2000 to 2005 in million cubic meters is given in table 1.

2.3. *Land cover classification*

The land cover classification used to monitor the vegetative cover in the watershed area was performed using MODIS data. Seven MODIS images were used to classify the land cover.

To observe the land cover changes over the six-year study period, a supervised classification technique was adopted. Six images, as discussed previously, were used for classification. The band combination used for the interpretation was 7, 5, 3 of the MODIS data. All of the images were classified using the supervised classification technique. In this classification, the user must select the training area to specify a land cover feature; then, four classes are addressed, i.e., vegetation, snow, water body and bare land. All six MODIS images were classified according to these four classes. In the first step, spectral signatures were created for each image by selecting 10–15 training areas for each land cover feature, and then, using these signatures, the images were classified using ERDAS Imagine 8.7. The classification results of the

Table 1. Computed annual available rainwater.

Year	Available rainwater (mcm)
2000	22,853
2001	19,851
2002	18,533
2003	21,983
2004	22,697
2005	21,742

Table 2. Land cover grades.

Land cover practice	2000 Area (km^2)	2001 Area (km^2)	2002 Area (km^2)	2003 Area (km^2)	2004 Area (km^2)	2005 Area (km^2)
Vegetation	12,543	18,126	18,228	17,062	17,828	16,915
Bare land	7230	6939	6578	5139	8095	9099
Water body	1554	808	1009	487	768	494
Snow	7932	3391	3439	6555	2569	2754

vegetation cover area composed by the four land cover classes in square kilometres (km^2) for all six images are given in table 2.

3. Results and discussion

The relationship between rainfall and vegetation has immense impact on soil erosion (de Asis & Omasa 2007). The exposed surface behaves differently than covered land by the rain drops and this behaviour rate of the surface is very much dependent on the intensity of rain. This study is performed on a mountainous watershed where the amount of rainfall also effects the vegetation, especially if mountains have light vegetation like grass and bushes.

3.1. Influence of rainfall on vegetation

The vegetation trend is observed inversely proportional to rainfall water. The impact of rainfall on vegetation is shown in figure 3. It is observed that the vegetation-covered area in 2000 was 12,543 km^2, and the annual available rainwater for the same year was 22,853 mcm. The amount of rain fall water available in the watershed in 2001 was 19,850 mcm, and the vegetation-covered area was 18,125 km^2 which is greater than that of year 2000. In 2002, the vegetation trend remained the same as in previous years, with a decrease in the available rainwater to 18,533 mcm. Furthermore, there was an increase in the vegetation area to 18,228 km^2. Again in 2003, the available rainwater increased to 21,983 mcm, and the vegetation decreased

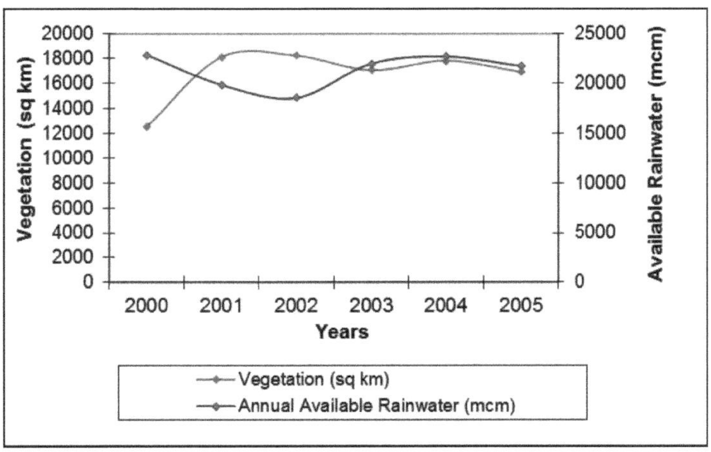

Figure 3. Impact of rainfall on vegetation.

to 17,062 km^2. The vegetation trend got changed slightly in 2004, with an increase in the amount of rainwater from 21,983 to 22,697 mcm. An upturn in vegetation was observed from 17,062 km^2 for 2003 to 17,828 km^2 for 2004. Besides, in 2005, the vegetation and rainwater both decreased, to 21,742 mcm (rainwater) and 16,915 km^2 (vegetation).

3.2. *Influence of rainfall and vegetation on reservoir capacity*

The comparison between sediment loads is performed differently because the data for sediment load is recorded as total load for two to three years gap rather than for single year gap by the Water and Power Development Authority (WAPDA). The relationship is made here by averaging the recorded sediments load data.

The sediment load for 2000 was 35 mcm with 12,543 km^2 of vegetation-covered area, and the rainwater was 22,853 mcm. The sediment load, vegetation-covered area and rainfall are given in table 3. The average vegetation-covered area for the years 2001 and 2002 was 18,177 km^2, and the average rainwater for the same two years was 19,192 mcm, with total a sediment amount of 116 mcm. The average rainwater available from 2003 to 2005 was 22,141 mcm, and the average vegetation-covered area was 17,268 km^2, while the total sediment load was 74 mcm. The sediment load has been observed greater with less in rainwater and with larger area covered by vegetation. Sediment load 116 mcm for year 2001–2002 is higher than other years with having maximum average vegetation covered area and minimum average rainwater available for watershed. The sediment loads that were calculated in the last three years (2003–2005) are almost same which was seen around 25 mcm each.

Therefore, the results of the study assist to build an important relationship as follows:

$$\text{More rainfall} = \text{less vegetation} = \text{more erosion}$$

During high rainfall in a low vegetation covered area, high soil erosion rates were registered due to the high surface runoff. Such imperative mechanisms of soil erosion ensure indicative messages for the prediction of sediments landing.

3.3. *Identification of erosion-prone regions*

The erosion-prone areas had been identified individually for each year from 2000 to 2005. Figure 4 shows the erosion-prone areas and rainfall distribution maps. Six sub-figures included in the figure 4 legend sequentially illustrate total yearly rainfall over and above the identified erosion-prone regions.

Table 3. Rain water and vegetation impact of sediment load.

Name	2000	(2001–2002)	(2003–2005)
Vegetation (km^2)	12,543	Average 18,177	Average 17,268
Annual available rainwater (mcm)	22,853	Average 19,192	Average 22,141
Sediment load (mcm)	35	Total 116	Total 74

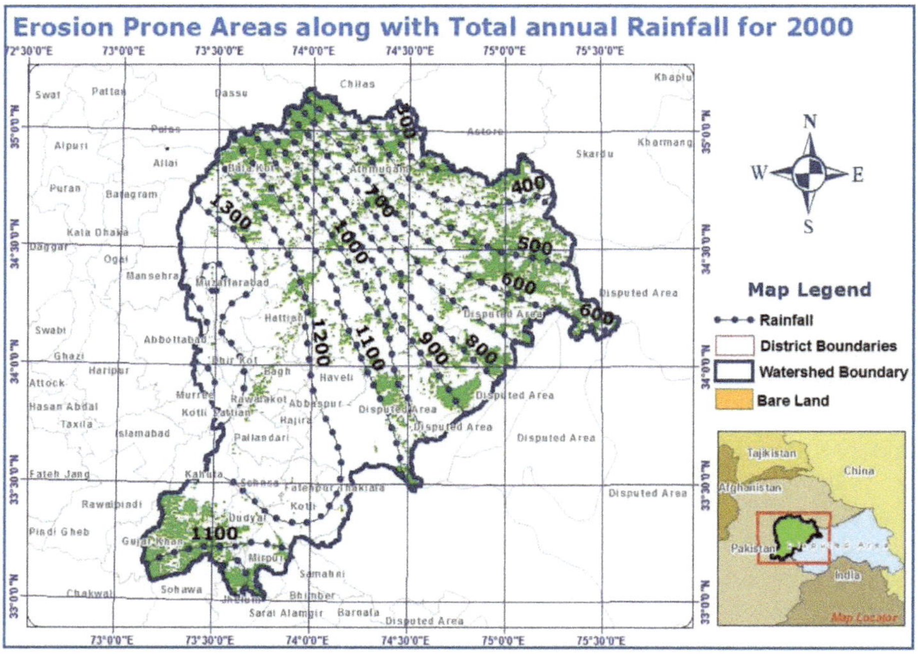

(a)

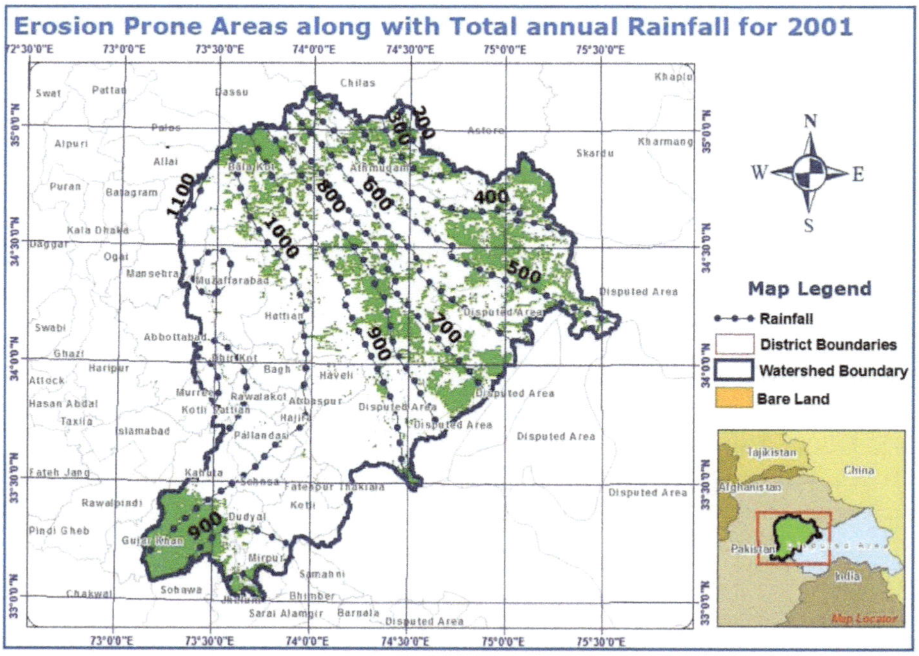

(b)

Figure 4. (*a*) Map of erosion-prone areas in 2000. (*b*) Map of erosion-prone areas in 2001. (*c*) Map of erosion-prone areas in 2002. (*d*) Map of erosion-prone areas in 2003. (*e*) Map of erosion-prone areas in 2004. (*f*) Map of erosion-prone areas in 2005.

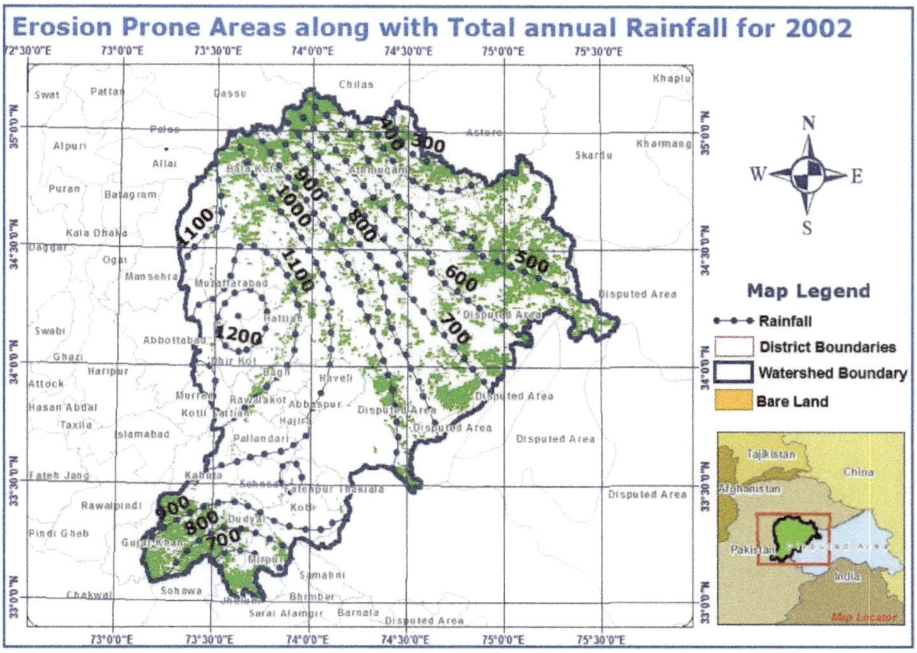

(c)

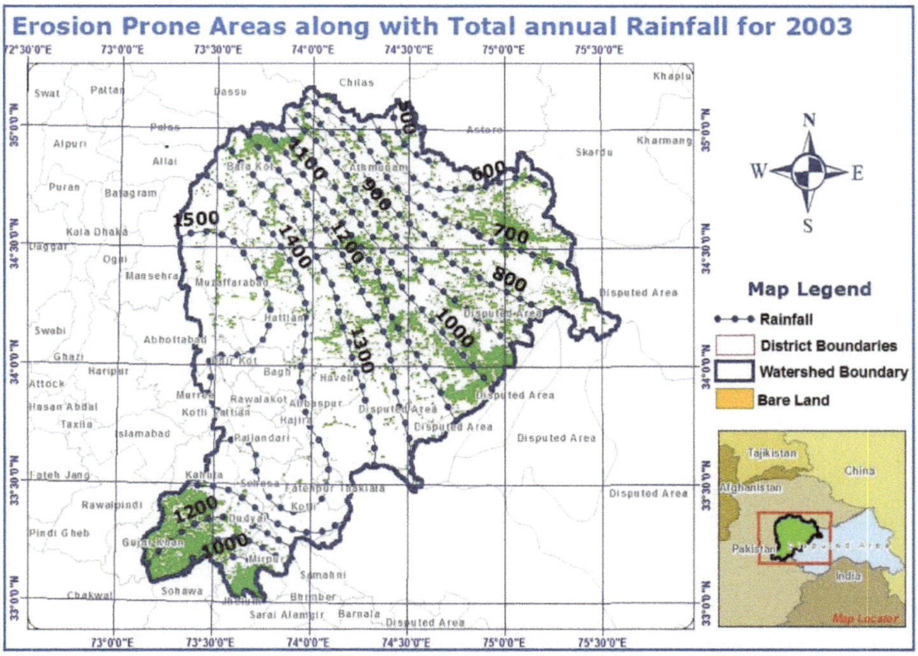

(d)

Figure 4. (Continued)

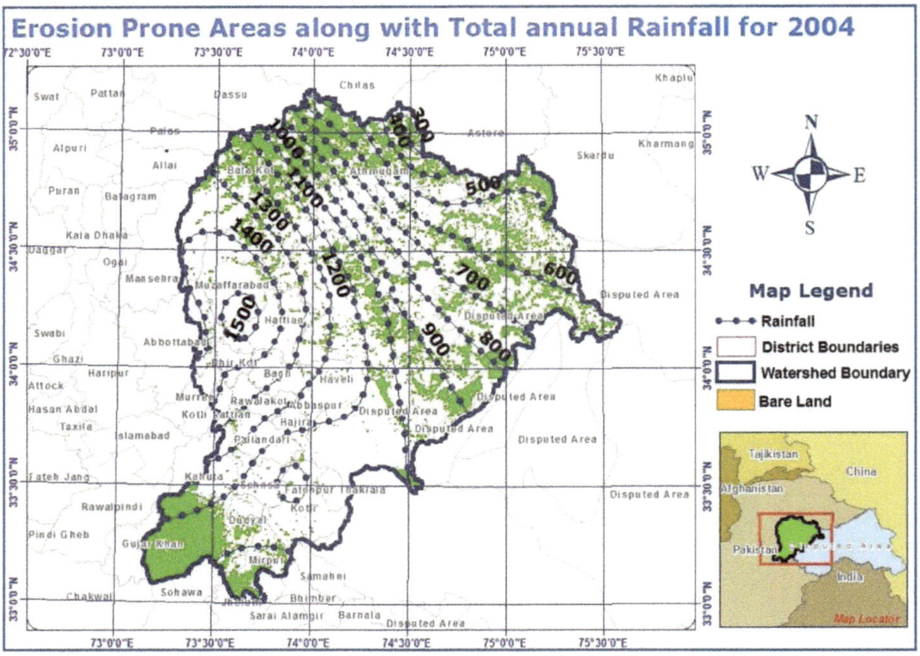

(e)

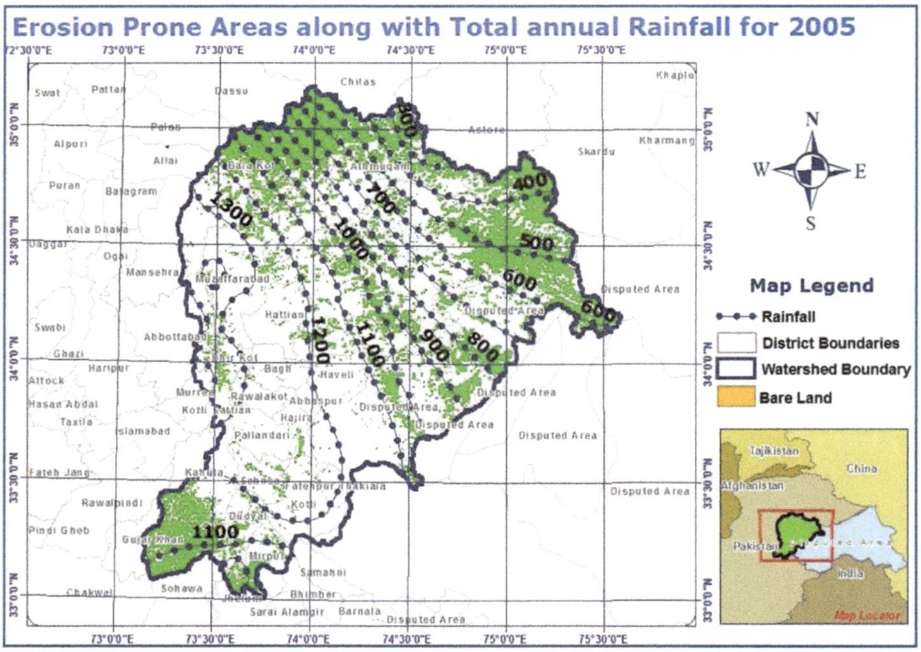

(f)

Figure 4. (Continued)

4. Conclusions

Resource conservation has always been a critical issue for efficient preservation and utilization. Through this study, the effects of different sedimentation-related parameters in a local water reservoir have been shown. The incorporation of rainfall interpolation procedures has revealed the complex relationship between vegetation and available rainwater, demonstrating that a decrease in vegetation-covered area, added to high rainfall rates, can trigger the erosion process. Furthermore, usually, areas of high rainfall intensities always have a large surface of vegetation. The surface terrain presents dramatic changes in the research zone. The catchment area is covered by scattered vegetation cover (bushes and grass), which cannot stop runoff flow living out high levels of soil loss rates due to high anthropogenic impact.

With an increase in rainfall, the vegetation-covered area decreases dramatically, causing high runoff events and soil loss rates. Some of the main soil-producing areas were also identified along with the rainfall patterns. In the future, this work will be useful in term of decision-making in land management of new reservoirs, with similar conditions and constraints, applying the appropriate land preservation against the different forms of soil erosion in water shed.

Acknowledgements

The authors gratefully acknowledge the support by King Saud University, Deanship of Scientific Research, College of Engineering Research Center.

References

Bookhagen B. 2010. Appearance of extreme monsoonal rainfall events and their impact on erosion in the Himalaya. *Geomatics Nat Hazards Risk*. 1:37–50.

Dahal RK, Hasegawa S, Bhandary NP, Poudel PP, Nonomura A, Yatabe R. 2012. A replication of landslide hazard mapping at catchment scale. *Geomatics Nat Hazards Risk*. 3:161–192.

Day JW, Kemp GP, Reed DJ, Cahoon DR, Boumans RM, Suhayda JM, Gambrell R. 2011. Vegetation death and rapid loss of surface elevation in two contrasting Mississippi delta salt marshes: the role of sedimentation, autocompaction and sea-level rise. *Ecol Eng*. 37:229–240.

de Asis AM, Omasa K. 2007. Estimation of vegetation parameter for modeling soil erosion using linear Spectral Mixture Analysis of Landsat ETM data. *ISPRS J Photogramm Remote Sens*. 62:309–324.

Fritz M, Herzschuh U, Wetterich S, Lantuit H, De Pascale GP, Pollard WH, Schirrmeister L. 2012. Late glacial and Holocene sedimentation, vegetation, and climate history from easternmost Beringia (northern Yukon territory, Canada). *Quat Res*. 78:549–560.

Goldshleger N, Karnibad L, Shoshany M, Asaf, L. 2012. Generalising urban runoff and street network density relationship: a hydrological and remote-sensing case study in Israel. *Urban Water J*. 9:189–197.

Haq IU, Ahmad F. 2004. Experience of the Mangla Dam watershed management project in Pakistan. In: Barker DH, Watson AJ, Sombatpanit S, Northcutt B, Maglinao AR, editors. *Ground and water bioengineering for control and slope stabilization*. Enfield, USA: Science Publishers, Inc.; p. 253–262.

Indus Basin. 2008. Pakistan water and power development authority. [cited March 15]. Available from: http://www.waterinfo.net.pk/sites/default/files/knowledge/Indus%20Basin.PDF

Islam ABMS, Hussain RR, Jumaat MZ, Rahman MA. 2013. Nonlinear dynamically automated excursions for rubber-steel bearing isolation in multi-storey construction. *Autom Constr*. 30:265–275.

Islam ABMS, Jameel M, Jumaat MZ. 2012. Oil and gas energy potential at Malaysian seabed and Spar platform for deepwater installation. *Int J Green Energy*. 9:111–120.

Islam ABMS, Jameel M, Jumaat MZ, Rahman MM. 2013. Optimization in structural altitude for seismic base isolation at medium risk earthquake disaster region. *Disaster Adv*. 6:23–34.

Islam ABMS, Jameel M, Jumaat MZ, Shirazi SM, Salman FA. 2012. Review of offshore energy in Malaysia and floating Spar platform for sustainable exploration. *Renew Sustain Energy Rev*. 16:6268–6284.

Jat MK, Khare D, Garg PK, Shankar V. 2009. Remote sensing and GIS-based assessment of urbanisation and degradation of watershed health. *Urban Water J*. 6:251–263.

Khan SM. 1985. Management of river and reservoir sedimentation in Pakistan. *Water Int*. 10:18–21.

Khosrokhani M, Pradhan B. 2013. Spatio-temporal assessment of soil erosion at Kuala Lumpur metropolitan city using remote sensing data and GIS. *Geomatics Nat Hazards Risk*. 5:252–270.

Koehn K, Brye KR, Scarlat C. 2011. Quantification of stormwater runoff using a combined GIS and curve number approach: a case study for an urban watershed in the Ozark Highlands, USA. *Urban Water J*. 8:255–265.

Mangla dam. 2008. Pakistan Water and Power Development Authority. [cited February 10]. Available from: http://www.waterinfo.net.pk/sites/default/files/knowledge/Mangla%20Dam.pdf

Mangla dam. 2010. Pakistan water and power development authority, Tarbela, Pakistan [cited January 10]. Available from: http://en.wikipedia.org/wiki/Mangla_Dam

Nasir A, Uchida K, Ashraf M. 2000. Estimation of soil erosion by using RUSLE and GIS for small mountainous watershed in Pakistan. *Pak J Water Resour*. 10:11–21.

Noe GB. 2013. 12.21 Interactions among hydrogeomorphology, vegetation, and nutrient biogeochemistry in floodplain ecosystems. In: Shroder JF, editor. *Treatise on geomorphology*. San Diego (CA): Academic Press; p. 307–321.

Nouri H, Beecham S, Kazemi F, Hassanli AM. 2012. A review of ET measurement techniques for estimating the water requirements of urban landscape vegetation. *Urban Water J*. 10:247–259.

Pourghasemi H, Pradhan B, Gokceoglu C, Moezzi KD. 2012. A comparative assessment of prediction capabilities of Dempster–Shafer and weights-of-evidence models in landslide susceptibility mapping using GIS. *Geomatics Nat Hazards Risk*. 4:93–118.

Sandoval S, Torres, A, Duarte M, Velasco A. 2013. Assessment of rainfall influence over water quality effluent of an urban catchment: a data driven approach. *Urban Water J*. 11:116–126.

Tanaka N, Yagisawa J. 2010. Flow structures and sedimentation characteristics around clump-type vegetation. *J Hydro-environ Res*. 4:15–25.

Tiwari AK, Risse LM, Nearing MA. 2000. Evaluation of WEPP and its comparison with USLE and RUSLE. *Trans ASAE 2000 Am Soc Agric Eng*. 45:1129–1135.

Walling DE. 2008. The changing sediment loads of the world's rivers. *Land Reclamation*. 39:3–20.

3

Debris flow impact assessment along the Al-Raith Road, Kingdom of Saudi Arabia, using remote sensing data and field investigations

AHMED M. YOUSSEF†‡, MOHAMED AL-KATHERY‡, BISWAJEET PRADHAN §* and TURKI EL-SAHLY‡

†Geology Department, Faculty of Science, Sohag University, Sohag 82524, Egypt
‡Geological Hazards Department, Applied Geology Sector, Saudi Geological Survey, Jeddah 21514, Kingdom of Saudi Arabia
§Department of Civil Engineering, Faculty of Engineering, Geospatial Information Science Research Center (GISRC), University Putra Malaysia (UPM), Serdang 43400, Malaysia

Jizan mountainous areas in Kingdom of Saudi Arabia are suffering from a variety of slope failures. Most of these failures happen due to heavy rainfalls from time to time. These landslides include rock topples, rockslides, debris flow, and some combination of these which affected many roads, highways, and buildings. The Al-Raith Road is one of these roads connecting Red Sea coast cities with Asir and Al-Hasher areas. The length of this road reaches about 45 km and it has been exposed to landslides during each heavy rain storm. One of these events happened in 24 August 2013, which caused huge debris flows that cut and damaged the road. The current research aims to evaluate the debris flow assessment along this highway using remote sensing data and field studies. According to the detailed analysis of geological and geomorphological maps, as well as field investigation, it is evident that the debris flow materials are mainly related to the different types of landslides. These landslides included rock topples which are frequently observed along the side walls of the channels (flexture which occur in foliated rocks and block which occurs in massive rocks), rock sliding (planner failures) where many rock joints and shear zones dip towards the channel, and rockfalls. Debris range in their size from up to 2 m in diameter to fine materials less than 2 mm. These materials can be easily moved with water causing a risk to vehicles, roads, and housing in the area. Field study indicated that these debris channels especially at the lower part have been reactivated several times in the past. Finally, suitable solutions have been suggested to these critical sites to minimize and/or avoid the debris flow hazards in the future.

1. Introduction

Western and southern regions of the Kingdom of Saudi Arabia are affected by various natural disasters including earthquakes, flooding, earth fissures, and landslides (Youssef, Maerz, et al. 2012; Youssef, Pradhan, et al. 2012; Youssef & Maerz 2013; Youssef, Sabtan, et al. 2014). Landslides are the most catastrophic natural hazard all over the world among the different types of geomorphological hazards (land

*Corresponding author. Email: biswajeet24@gmail.com; biswajeet@lycos.com

degradation processes) causing billions of dollars in damaging the infrastructures and thousands of deaths each year (Aleotti & Chowdhury 1999). Landslides represent a type of mass movements that happened due to a variety and combination of different processes including falls, topples, avalanches, slides, and flows (Shroder & Bishop 1998; Regmi, Devkota, et al. 2014; Regmi, Yoshida, et al. 2013). Different factors such as seismic activity, high groundwater pressures (after heavy rainfall), geological factors, and human activities can trigger large rock/soil blocks or even larger assemblages of rock to crash down on to the road surface below. The California Department of Transportation (CADOT) (McCauley et al. 1985; VanDine 1985; Church & Miles 1987; Guzzetti et al. 2008; Baum & Godt 2010; Iverson et al. 2011) determined different factors that cause landslides. These factors include rainfall intensity, freeze–thaw, fractured rock, wind, snowmelt, channel run-off, channel profile, adverse planner fracture, burrowing animals, differential erosion, tree roots, springs or seeps, wild animals, truck vibration, debris availability in streams, and soil decomposition.

Swanston (1974) identified the composition of debris according to the texture and found that debris is a mixture of sand, gravel, cobbles, and boulders with different proportions of silt and clay, and sometimes it contains a significant amount of organic materials such as logs and tree stumps. Debris flow occurs when masses of poorly sorted sediment (different sizes) move downslopes due to the effect of water. Many events identified as debris slides, debris torrents, debris floods, mud flows, mudslides, mud spates, and lahars may be regarded as debris flow (Varnes 1978; Johnson 1984; Pierson & Costa 1987; Youssef, Pradhan, et al. 2013). Many authors studied the debris flows, their types, and mechanisms among them are Evans (1982), O'Connor et al. (Forthcoming), Johnson (1984), Hungr et al. (2001), VanDine (1985), and Pierson (1986).

In addition, due to the high density and mobility of debris flows, they represent a serious hazard, which impose serious problems for people, properties vehicles, and infrastructure in mountainous regions. Different authors indicated the hazard impact of the debris flows (e.g. Hungr et al. 1987; Prochaska et al. 2008). They indicated that these problems are due to the indirect impact lower energy of coarse-grained and fine-grained debris that can bury structures; and flood water that are forced from the normal channel by debris deposits and have the potential to erode unprotected surfaces and cause flood damage.

Materials collected in the ravines, gullies, and streams are related to different types of landslides along the sides of the networks. These slope failures can be classified into one of the four categories depending on the geometrical and mechanical nature of the discontinuities and the conditions of the rock masses which include circular, planar, wedge, and toppling failures. In many areas, the discontinuities are oriented in a way that contributes to create wedge, planar, or toppling failures. The dip/dip direction measurements at any area can be measured to determine the rock sliding/toppling potentiality. Landslides such as rockfalls, rockslides, and rock toppling have been studied and described by many authors, e.g. Aydan & Kawamoto (1992), Evans (1981), Farrokhnia et al. (2010), Goodman & Bray (1976), Ishida et al. (1987), and Varnes (1978). Rock toppling usually develops in the slope of foliated rock mass and can occur in cut slope in massive rock with regularly spaced joints, which strike parallel to the slope and dip towards or away from the slope. Whereas, planner and wedge failures can happen along structures such as shear zones, faults, and/or discontinuities that dip towards the highways. They can be analysed using limiting equilibrium analysis (Watts 2003). Other types of landslides are called rock failure as ravelling mechanism cannot be analysed using limiting equilibrium analysis Piteau (1979). This type of landslide is caused by many factors including adverse groundwater, excavation methods (poor blasting practices during original construction or reconstruction), climatic conditions, weathering, and tree levering (Brawner 1994). Franklin and Senior (1997) analysed 415 rock-slope failures along highways in Northern Ontario. They found that 33% of those failures involved toppling or planar blocks and wedges. While 67% of the rockslide incidents were identified to be involved in these complex mechanisms.

Debris flow mitigation structures may be required to minimize their risks which have been applied in many research areas such as DeNatale et al. (1997), Frenez et al. (2004), Rickenmann (1999), and Rimbock and Strobl (2002).

Few landslide hazard studies were carried out in the Kingdom of Saudi Arabia along the road and highway sections. With the help of remote sensing and GIS techniques, landslide studies such as susceptibility mapping become more easier and efficient (Youssef et al. 2009; Pradhan et al. 2011; Akgun et al. 2012; Althuwaynee et al. 2012; Tien Bui et al. 2012; Pourghasemi et al. 2012). This paper intends to describe the debris flows that caused a serious hazard along the Al-Raith Road from time to time. This research aims to determine various types of landslides occurring along the sides of each debris channel and causing accumulation of debris later moved with water; to recognize the structurally controlled landslides types and non-structural types; detailed geomorphologic characteristics of the different types of landslides; and the rock types that are most affected by landslides and forming the debris along the channels. In addition, it is aimed to detect the impact of the anthropogenic activities on the formation of the debris.

2. Study area and problem evaluation

2.1. *Study area*

Al-Raith Road section is one of the most landslide-affected roads and highways in Jizan Region. The road section is about 46 km long, and it passes through areas that are prone to debris flows. It is located in Al-Raith Governorate of the Jizan Region, southwest of the Kingdom of Saudi Arabia (figure 1). It connects the Red Sea coastal plain with Al-Hasher and Asir areas. It represents an important road, as it offers private vehicles and light-duty trucks convenient access between these cities. The study area is located at latitude between 17°35′8.8″ N and 17°36′1.7″ N, and a longitude between 42°52′37″ E and 42°53′29″ E. Debris flows are the most common landslides along the Al-Raith Road and they possess very high damaging effect. Many of the debris flow channels crossing the road were not remediated effectively. They are observed in relating to rock mass failures (natural phenomena) and man-made (due to dumping materials in old channels) due to excavated slopes. Consequently, the road is commonly closed from time to time due to landslides (debris flows).

Geologically, the study area represents part of the Wadi Baysh quadrangle (GM-77c) that covers 17,550 km^2 in the Asir and Tihamah provinces (Fairer 1985). It is composed of the Baish group that consists of volcanic and volcaniclastic rocks formed during the development of the Arabian ensimatic island arc (greenstone,

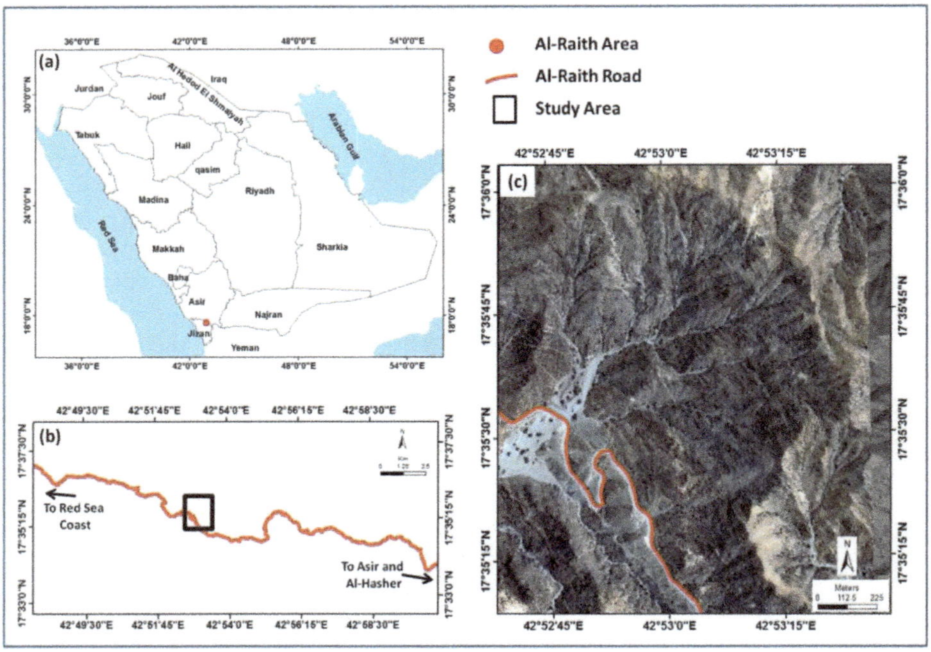

Figure 1. Location of the study area in the KSA map: (a) Kingdom of Saudi Arabia; (b) along Al-Raith Road; and (c) road section affected by debris flow problem.

metabasalt and minor metagraywacke, metachert, and marble) (figure 2). Later they were intruded by mafic plutonic rocks that range in age from about 1000 to 760 Ma (Fairer 1985).

Extreme rainfall events were reported within the historical records according to the data of rain gauge (SA145) that located about 9 km west of the study area. This rain gauge is operated by the Ministry of Water and Electricity in the form of daily data. The data in the rain gauge cover a time span from 1966 to 2013. The maximum daily precipitation in a day noticed as amount of 99 mm on 8 December 1972, 99 mm on 13 January 1973, and 93 mm on 9 March 1999. In addition, the average annual precipitation is reported as about 290.4 mm/year, while the maximum sum of rainfall value of 1441.9 mm in year 1972 and a minimum rainfall sum of 8 mm reported in 1966. The seasonal average precipitation for the whole period is reported as about 290.4 mm in autumn and about 52.7 mm in summer.

2.2. *Problem evaluation*

Al-Raith Road encounters debris flow from time to time after rainfall storm event. One of these debris flow events happened during the day of 24 August 2013 due to a heavy rainfall that occurred for few hours along the Al-Raith area. The rainfall caused huge amounts of debris to be flowed along different locations in the road section causing serious hazard to the area (figure 3). This debris covered the entire road section and led to close in both directions for few days. These debris were related to two sources: one is related to the debris that accumulate inside the channels (natural

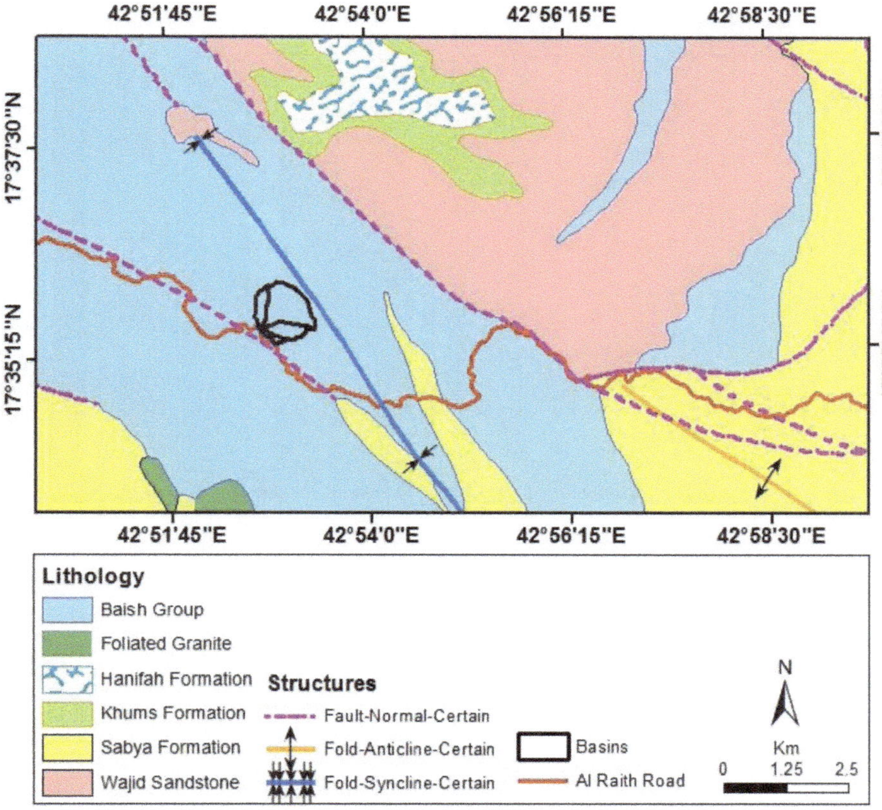

Figure 2. Geological map of the study area.

Figure 3. Panorama view showing Al-Raith landslide (debris flow). Several scarps at the upper part, thick debris in the channels, and serious road damage at the lower part of the landslide can be seen in the image.

Figure 4. Photos taken at the time of debris flow that cut and accumulated above the road, where some of the road sections have been damaged.

materials) and the other source is according to anthropogenic activities (dumping materials) that are related to road widening and modification. Figure 4 shows different photos taken at the time of the debris flow occurrence along this section of the road showing different features in the study area. It is also obvious from field studies that the road and houses nearby are in critical hazard due to the debris that come from these channels from time to time. In addition, several gabion walls are seen at the mouth of the channels to control the debris; however, they are destroyed and the debris come over them (figure 4). After getting these preliminary views of the landslide, further study was carried out at outcrop scale.

3. Methodology

Lithological, morphometrical, hydrological, and structural, in addition to anthropogenic activities, might have influenced the formation of debris flows. Distribution of debris flows and the landslides that cause mass movements in the study area were collected using standard geological and geomorphic field techniques. General field techniques were employed to identify and map different types of debris flows, as well as to determine different types of landslides and to collect the rock and soil samples for laboratory analysis.

In the current study, debris flow channels have been mapped using different data types including (1) digital elevation model (DEM 5 m resolution) which was extracted from a topographic map (1:10,000 scale), (2) high-resolution satellite images including Geo-Eye and QuickBird imagery with ground resolution of 2.5 m /pixel (after resampling) and ~61 cm /pixel, respectively, and (3) by field investigations where two field trips were done to investigate the study area and collect the data. The remote sensing images were obtained from the King Abdulaziz City for Science and Technology. All the data used in the current study were geo-referrenced to UTM coordinate system, WGS84 datum, and zone 38N.

Different software were used in the current study including watershed modelling system (WMS 8.1) to extract different catchments and their morphometric parameters, Global mapper 15 to prepare the three-dimensional model, and Arc GIS 10 to compile different data types.

Detailed field investigations (large and small scales) were carried out in the study area in order to understand and analyse the detailed characteristics of the debris flow of these channels and the sources of these debris. Additionally, the geomorphic situation of the channels was studied in detail in order to define the characteristics of the different types of landslides causing the debris to be accumulated in these channels. Finally, laboratory investigation was carried out for the collected rock samples from the study area to determine the friction angles. Based on the properties of the intact rock samples and rock masses characteristics (discontinuities, filling materials, and rock types), friction angles in this study were measured using RockData software.

4. Results and discussions

4.1. Mapping debris flow locations using high-resolution images

The debris flow channels have been mapped using digital elevation model (DEM 5 m resolution) and high-resolution satellite images (figures 5 and 6), and the results have been verified during the field investigations. According to the field data collected and remote sensing analysis, debris flows from the basins may have entrained material along their travel paths.

Four catchments have been extracted using Arc-Hydro tools in ArcGIS 10. These catchments were responsible for these debris flows at the Al-Raith Road section (figure 5). Sinotech Engineering Consultants INC (2008) indicated that the topography and characteristic of the catchment play essential factors in accelerating the debris flow.

In the current study, the morphometric factors for these four basins are summarized in table 1 including catchment characteristics area, length, slope, perimeter, average elevation, shape factor, and sinuosity. Other factors are related to stream

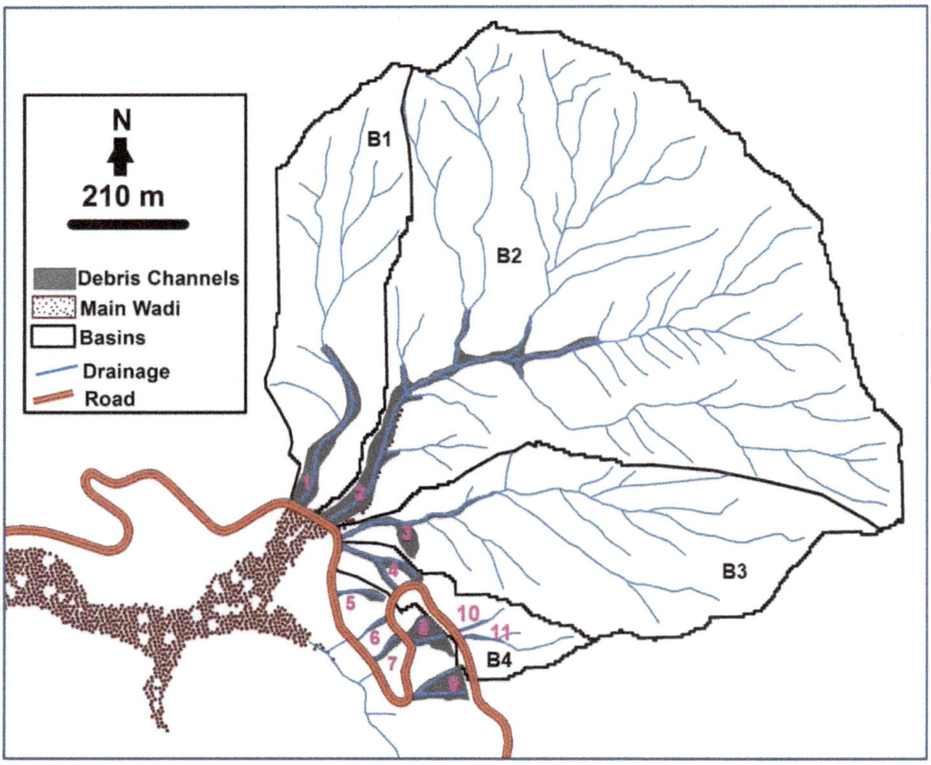

Figure 5. Eleven debris channels, four basins, and drainage networks have been determined and mapped in the study area.

characteristics which include main flow length and slope, main stream length and slope, and centroid out distance and slope. Other important factors including profile shape and availability of the debris were determined. It was found that parameters that are related to catchment characteristics, channel characteristics, and debris availability inside these channels have essential impacts on the formation of debris flow in the study area. The catchment slope ranges from 0.735 m/m for catchment 1 to 0.873 m/m for catchment 3, main stream distance slope ranges from 0.4306 m/m for catchment 2 to 0.544 m/m for catchment 3 (table 1). The study indicates that the basin slope and the main stream distance slope are very high by representing the most important factor to increase the water velocity. According to the Manning equation, the velocity of the water was calculated as follows:

$$V = \frac{R^{2/3} \times S^{1/2}}{n} \tag{1}$$

where V = bottom slope of channel (m/s), R = hydraulic radius = A/P (m), S = bottom slope of channel (m/m), n = Manning roughness coefficient (empirical constant), A = cross-sectional area of flow perpendicular to the flow direction (m^2), and P = wetted perimeter of cross-sectional flow area (m).

According to the field survey and remote sensing analysis of the images, the average width of each channel was calculated and the average erosion height was found

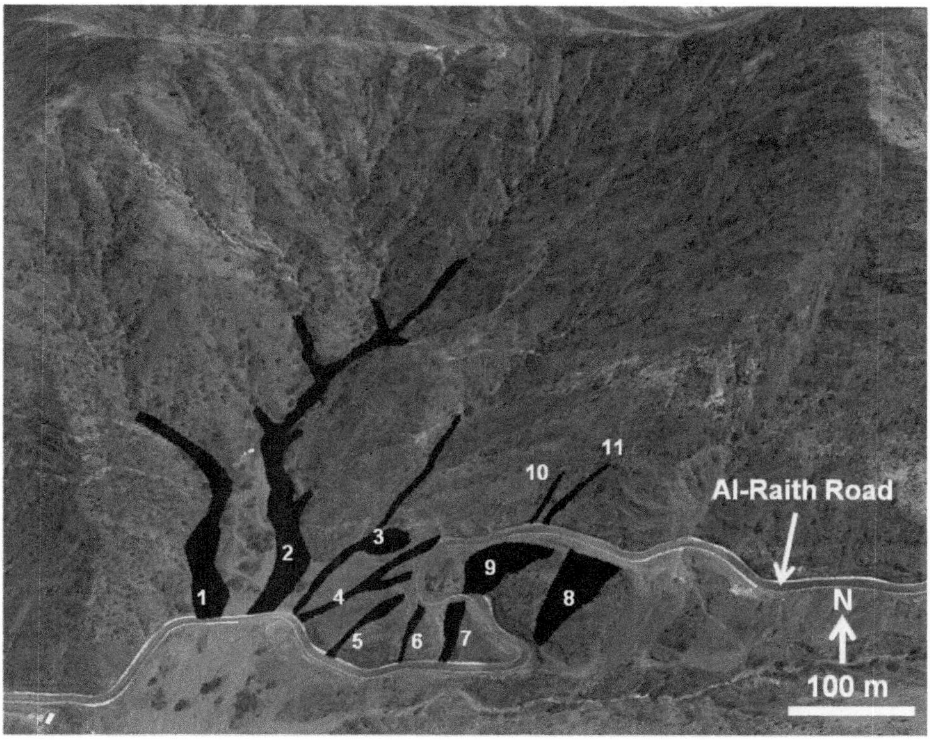

Figure 6. Three-dimensional image showing debris flows channels that intersect Al-Raith Road.

to be 1.5 m, the bottom of these channels was ill-sorted materials and we assumed the roughness value to be 0.03, and the main stream distance slope was used to be bottom slope of the channel. Accordingly, the water velocity was calculated as shown in table 1. The velocity ranges from 25.9 m/s for channel 4 to about 28 m/s for channel 1 (table 1). This velocity of water could carry any materials in its way and for that reason

Table 1. Main characteristics of the basins and drainage in the study area.

Basin ID /basin characteristics	B1	B2	B3	B4
Catchment area (km^2)	0.15	0.63	0.20	0.03
Catchment slope (m /m)	0.7351	0.7985	0.8733	0.7852
Catchment length (m)	843.4	1060.0	850.9	462.9
Catchment perimeter (m)	2647.2	4751.6	2803.4	1541.1
Mean elevation of catchment (m)	1389.1	1511.9	1452.9	1356.7
Main stream distance (m)	751.1	1195.9	715.8	344.7
Main stream distance slope (m /m)	0.4939	0.4306	0.5440	0.4508
Average channel width (m)	17	30	10	14
Velocity (m /s)	28	27.2	27.9	25.9
Material availability	High	High	High	High

the availability of debris along the channels is a very important factor. The study indicated that the larger the basin area, basin slope, and main stream distance slope, the larger the water that can carry debris, and cause a serious problem to the infrastructure.

4.2. General characteristics of the debris flow materials

This area of landslide is one of the active landslides (debris flow) along this highway (figure 6). The rock debris accumulated in the channels that were generated by different types of landslide mobilized into a disastrous debris flow in many events and creating many problems. This landslide happened in August 2013 and it was a huge one about few hundred metres length and width. It came from different channels as shown in figures 5 and 6. These channels were filled by deposits which some were due to natural processes due to the combination of rock toppling, sliding, ravelling, as well as colluvium materials, and others were due to the anthropogenic effects by dumping crashed rock materials that came from rock cuts along the roads. The depth of the debris ranges from 3 to 20 m. The landslide is located in the rock formations belonging to Baish group in which the rock units include greenstone, tholeiitic metabasalt (local pillow structures), and minor metagraywacke, metachert, and marble (Fairer 1985).

Field study indicated that most of the lower and central parts of the channels are covered by thick debris. The thickness of these debris ranges from 3 to 20 m, and they consist of different materials ranging in size from boulders up to 2 m to fine materials (size less than 2 mm) (figures 7(*a*)–(*d*)). Some of the materials that accumulate inside the channels and along their sides are related to landslides processes and these are characterized by ill-sorted materials and include large blocks that reach sometimes to 2 m in diameter with irregular shape (figures 7(*a*)–(*d*)). Some of these debris are coming from the colluviums that accumulate at the top of the mountains (forming a soil layer) due to weathering, trees, and erosional processes along these soil materials. There are discrete trees along the valley wall where they are grown along the fractured that filled with soil materials. Field study indicated that the weathering degree of the rocks along the sides of the channels ranges from highly weathered especially for the upper parts to slightly moderately weathered rocks for lower part rocks; however, fresh rocks are encountered for dykes. Furthermore, degree of fracturing ranges from highly fractured rocks as in foliated meta-volcanics to low fracturing in the rocks range from highly fractured rocks and foliated especially in meta-volcanics and greenstones and close to the major structures to less fractured rocks especially for marble and massive rocks.

Figures 7(*a*), (*e*), and (*f*) represent other debris that related to anthropogenic work but these materials are characterized by their grey colour and its sizes range from few 30 cm to less than 2 cm with few boulders. These materials accumulate along the sides of the channels and some of them are compactor.

4.3. Large- and small-scale investigation of one of the channels

4.3.1. Large-scale investigation. Detailed field investigation for one of the channels (channel of basin 3) has been carried out in order to understand different processes that cause debris to accumulate and prepare the different sources of the debris that

Figure 7. (*a*) Large boulders up to 0.5 m deposited in the bottom of debris channel (note that along the left side of the debris channel are dumping materials and along the right side are colluvium materials); (*b*) colluviums (ill-sorted) along the channel side slopes; (*c*) debris channels cut through old debris coherent materials; (*d*) debris channel with different material sizes along steep gully; (*e*) dumping materials in the channels were eroded and flown due to running water along the channel; (*f*) dumping materials accumulated in the channels (notice the height of the dimming materials more than 10 m).

accumulate in the different channels (figure 8). It was found that the debris accumulated in the gullies and channels are due to a combination of different landslides including rock topples, rockslides, ravelling (rockfalls), small debris slides, and anthropogenic work (figure 9). It was found that the rock exposure exists along the sides of the channels as we move towards the upstream part where as at the downstream part there are thick debris that are located along the channel base and sides. Rockslides are observed in the valley walls at the central and upstream parts of the

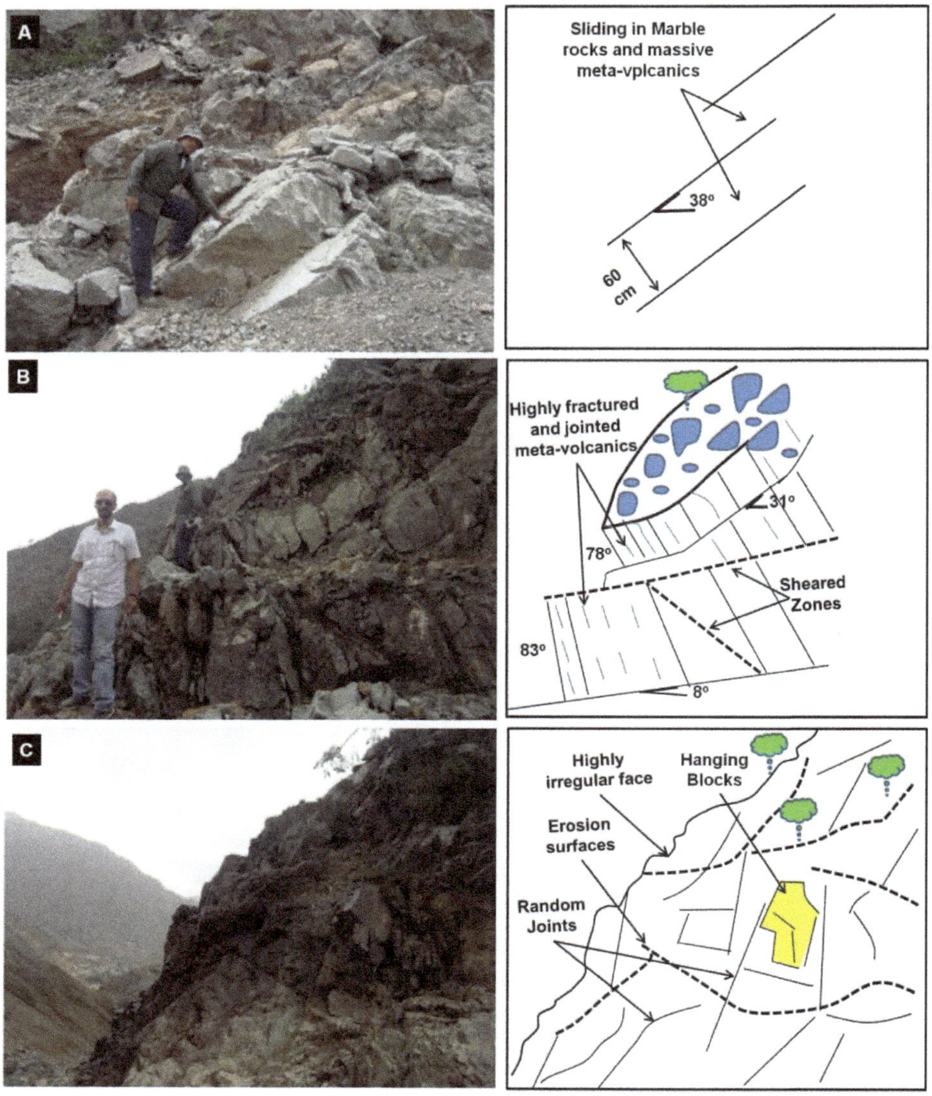

Figure 8. (*a*) Photographs showing plane failure dips towards the channel and some fallen blocks are located underneath these planes; (b) different types of toppling (flexture and block toppling); and (*c*) falling blocks in debris channels due to ravelling and trees' effect.

channel (figure 8). Also, it is seen that the slope of the channel gradually increases from 21.5° in the lower part to the maximum of 28° in the middle section and increases towards the upper reaches of the channel (figure 9).

4.3.2. Small-scale investigation. Most slope failures can be classified into one of the four categories depending on the geometrical and mechanical nature of the discontinuities and the conditions of the rock masses which includes circular, planar, wedge, and toppling failures. In many areas, the discontinuities are oriented in such a way that they contribute to create wedge, planar, or toppling failures. The dip/dip

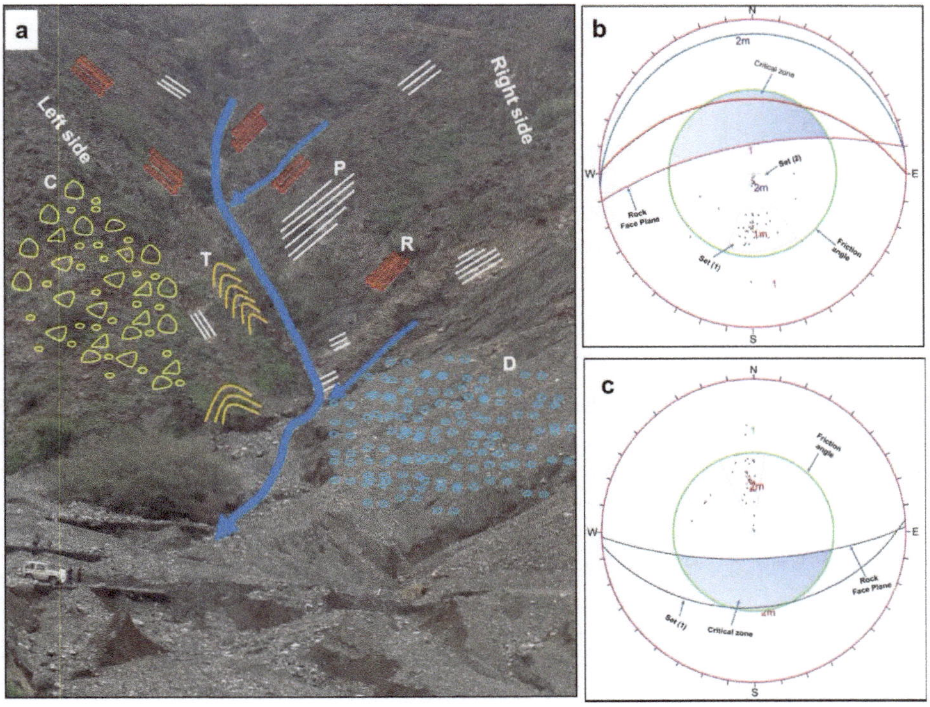

Figure 9. (*a*) An example of different types of landslides and materials on both sides of the debris channel (P = planner failure; R = rockfalls by ravelling; T = toppling failure; C = colluviums covering the mountain; and D = dumping materials by contractors working on the roads). (*b*) and (*c*) Markland stereonet test for the right side and left side, respectively.

direction and friction angle measurements at all rock stations were plotted on stereonets using ROCKPACK III (Watts 2003) and DIPs 5.1 software. Rock-slope stability analysis utilizing the Markland Test Plot method was applied to determine the potentiality for planar and toppling failures along the identified discontinuities in the study area.

4.3.2.1. Rock sliding. Different rocks have planar failures in which the planes have dip direction towards the channel. The dip/dip direction values of the main joint sets for both sides of the rock slopes from the channel were calculated and was found as follows: (1) for the right side it was found that there are two main joint sets with a dip/dip direction of 38°/359° and 06°/353° (figures 9(*a*) and (*b*)); and (2) for the left side there is only one main set of joints with a dip/dip direction of 38°/175° (figures 9(*a*) and (*c*)). The joint vertical spacing is about 30 and 50 cm for the right and left sides of the channel, respectively, and sometimes reaches 1 m for massive rocks figure 8(*a*). The friction angle of the right-side samples ranges from 33° to 39° with an average value of 36°, whereas the friction angle of the samples collected from the left side ranges from 35° to 45° with an average value of 40°. To be more conservative, the lowest value of the friction angle for each side was used to test the planner failure for both sides. Markland Test Plots for both sides showed that there is a potential for planar failure as shown in Figures 9(*b*) and (*c*). Figure 8(*a*) shows one of these examples where clear surfaces appear after sliding.

4.3.2.2. Rock toppling. This type is located at the foliated rocks and meta-volcanics at the central and lower reaches of the channel (figure 9). Rock topples are observed in most of the rock exposure especially along the outer part of rock out crops. Figure 6(*b*) is the section of the lower part of the channel with a width of 40 m. The lower part covered partially with debris in the middle and from both sides there was a large thickness. The debris is the product of the landslides along the wadi walls. There are two main types of the rock topple that were observed and detected in the study area (figure 8(*b*)).

4.3.2.3. Rockfalls (rock ravelling). In most of the rock cuts, rockfalls are not simple blocks and wedges, and are more difficult to analyse. In the current study, the "Modified Colorado Rockfall Rating System" was applied for the study area (Russell et al. 2008). The system indicated that many areas along both sides of the channel fall from time to time and these sides are unstable (figure 8(*c*)). Rockfalls (ravelling) in the study area happened in both crystalline rocks and colluvial sediments (boulders with fines at the top of the mountains) due to overhanging, undercutting, erosions, and the impact of trees as shown in figure 8(*c*).

4.4. Anthropogenic activities' impact in accelerating the debris flow

Anthropogenic activities act as part of the causative factor of the problem. The impact of the anthropogenic activities in this debris flow is obvious. Field investigation showed that a construction work has been done in widening the road, and in this section, the road switches back for two times. The construction company used these old debris channels to accumulate the materials that are related to widening the road. These channels that have many dumping materials are shown in figure 10. Some of these materials are loose and ready to flow down with rain and others are

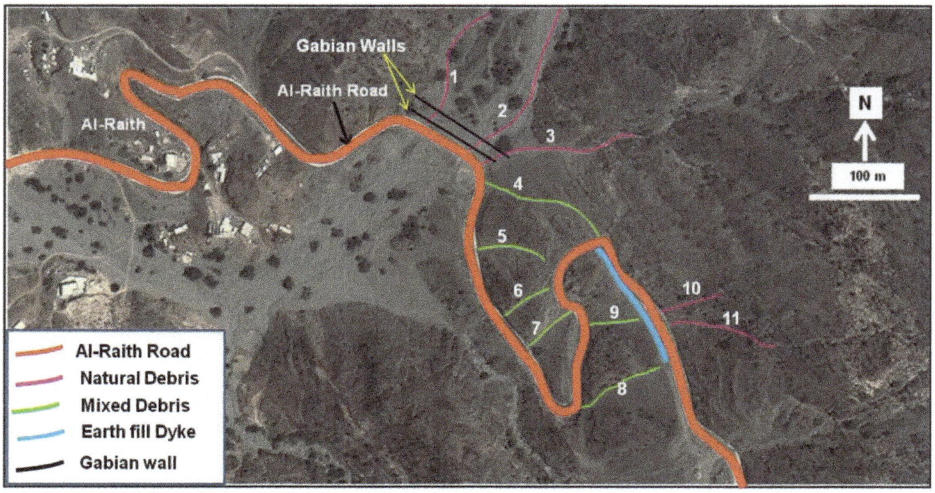

Figure 10. The location of the debris channels influenced by natural and anthropogenic processes.

compacted due to time. The study indicated that debris in channels 1, 2, 3, 10, and 11 were due to natural processes and debris in channels 4, 5, 6, 7, 8, and 9 were related to natural and anthropogenic processes (mixed debris). And also, the company created an earth-fill dyke that converts water and debris from channels 10 and 11 at the event time and run by the road where the road slope is towards the channel 4 and that increased the problem (figures 5, 6, and 10). Another anthropogenic impact is related to the establishing gabian walls in front of channels 1–3 (figures 5, 6, and 10). However, these channels bring debris from time to time and the space behind these gabians was filled by debris, and at this event time, most of the debris come from these channels over these gabian walls and destroy them as well and close the road.

5. Mitigation methodologies

VanDine (1996) determined the design consideration parameters for the debris flows including debris flow volume, flow paths, run-out distance, impact forces, run-up, and probable storage angle. Different types of measures can be used to reduce the impact of debris flows including decreasing run-off and erosion by land management techniques through run-off diversion or channel bed alterations; controlling water discharge by water management through run-off diversion; and controlling debris by engineering the movement of the flow.

Many authors are interested in debris flow remedial works. Among them, Eisbacher and Clague (1984), Government of Japan (1984), Hollingsworth and Kovacs (1981), Hungr et al. (1984), Lo DOK (2000), and Huebl and Fiebiger (2005) are interested in deflection and terminal walls, berms, and barriers, which could be constructed across the debris flow path to encourage deposition by presenting a physical obstruction to flow or to deflect dams which can be built downslope of the debris channels. These structures can be used to protect infrastructures by deflecting the flow to another area, or by increasing the length of the flow path, decreasing the overall gradient, encouraging deposition, and decreasing the angle of impact on a structure. The deflection walls can be constructed of reinforced concrete, local materials, or composite.

Other methods are debris racks, grizzlies, or other types of straining structures which can be used to separate the coarse-grained debris from the fine-grained debris (Thurber Consultants 1984; VanDine 1985; Hungr et al. 1987). These methods are used to prevent culvert openings and bridge clearances from becoming blocked with debris. At the same time, to remain effectiveness of these remedial structures, the coarse-grained debris must be removed from behind of these structures regularly.

In arid areas such as some of the mentioned methods could be worked especially if the debris channels are not steep and there is a good space to build these structures. However, for very steep and high-volume debris flows, land management techniques typically revolve around vegetation and reforestation, a practical impossibility in arid climates (Youssef, Pradhan, et al. 2013). For this particular environment, design options are considered to stop the debris flows, and stabilizing slopes included starving the potential flow of the water and/or starving the flow of the solid elements, intercepting the flow using barriers, or alternatively allowing the flow to proceed under the highways. The most effective and permanent solution is to raise the highway above the debris flow channels. This would involve either bridges or extremely large culverts with structural protection between culverts. This would ideally allow flows to pass harmlessly below the road. The problem with this kind of solution in a mountainous terrain is where the road switches back below, the debris flow problems are merely passed down the side of the mountain to the next road where they still leaves the highway at risk for overtopping or damage from high-volume debris flow events.

Another method is by starving the potential debris flow of water which is normally an effective solution. This would require interception and diversion of surface flow. This method will never work in high-steep slopes with significant rainfall events, because the water has many sources that have to go somewhere, and ultimately may find its way back to the debris flow channel or into another channel where it can be equally as destructive.

Intercepting the debris flows, using structural barriers as the flow approaches the highway is a solution that was originally implemented here during highway construction. Freestanding retaining walls and gabions extending the height of the structure have been used to protect the highway from debris flows. This is probably the most cost-effective solution, but the design fell short on these steep slopes with high volumes of debris flow resulting in overflow of the physical barriers.

For this study area, our recommended solutions are as follows:

(1) For channels 1–3, walls and berms can be good solutions especially as they can be built from local materials, as well as for gabians located in front of channels 1–3. It is recommended to increase the volume capacity of the bulkhead-type barriers by implementing a method of removing the debris from behind the barriers. Therefore, that debris can be removed following flow events, creating more catchment space for future flow events, and removing the materials behind these barriers.
(2) For moderately sloped areas, diversion of water could be a good solution as for channels 10 and 11 especially along the road section.
(3) It is recommended to clean all debris from the high-steep debris channels such as channels 4–9.
(4) Increase and develop a proactive maintenance programme and incur an indefinite maintenance liability.

The use of techniques to block debris emanating above the highway will also serve to starve the debris flow channels below the highway from mobilizing and affecting the switch back below.

6. Conclusions

Heavy rainfalls in Al-Raith area caused different debris to flow downward and block the road in different locations. There are two types of debris that have been recognized in the study area: one is related to natural processes (natural debris) and the other one is related to anthropogenic activities that damp the products of the road widening along the old debris channels (mixed debris). Different factors affect the mobilization of these different types of debris including intense rainfall, steepness of the channels, area of watershed, and the presence of materials.

The current study indicated that no method has been used to stabilize the debris along the different channels or to make suitable remedial work along the large catchment areas. Furthermore, no any attempt has been done in order to establish any drainage system to divert the water away from the channels that contain debris. Detailed analysis has been done in the study area to determine the different sources of debris and to recognize the main influential factors that can cause debris flow. The findings of this research showed that the natural debris are formed due to different landslides along the sides of the channels including rock sliding, toppling, and ravelling, as well as related to fall of the colluviums that located at the top of the mountain slopes. Finally, suitable mitigation techniques have been suggested to minimize and/or prevent the impact of these debris channels on the infrastructure in the study area.

References

Akgun A, Sezer EA, Nefeslioglu HA, Gokceoglu C, Pradhan B. 2012. An easy-to-use MATLAB program (MamLand) for the assessment of landslide susceptibility using a Mamdani fuzzy algorithm. Comput Geosci. 38:23–34.

Aleotti P, Chowdhury R. 1999. Landslide hazard assessment: summary review and new perspectives. Bull Eng Geol Environ. 58:21–44.

Althuwaynee OF, Pradhan B, Lee S. 2012. Application of an evidential belief function model in landslide susceptibility mapping. Comput Geosci. 44:120–135

Aydan O, Kawamoto T. 1992. Stability of slopes and underground openings against flexural toppling and their stabilisation. Rock Mech Rock Eng. 25:143–165.

Baum RL, Godt JW. 2010. Early warning of rainfall- induced shallow landslides and debris flows in the USA. Landslides. 7:259–272.

Brawner CO. 1994. Rockfall Hazard Mitigation Methods, Participant Workbook, NHI Course No. 13219. Publication No. FHWA SA-93-085. Washington (DC): U.S. Department of Transportation, Federal Highway Administration.

Church M, Miles M. 1987. Meteorological antecedents to debris flow in southwestern British Columbia: some case histories. In: Costa JE, Wieczorek GF, editors. Debris flows/avalanches: process, recognition and mitigation. Reviews in Engineering Geology. Boulder (CO): Geological Society of America; p. 63–79.

DeNatale JS, Fiegel GL, Iverson RM, Major JJ, LaHusen RG, Duffy JD, Fisher GD. 1997. Response of flexible wire rope barriers to debris-flow loading. Paper presented at: Proceedings of the First International Conference on Debris-Flow Hazards Mitigation; San Francisco, CA.

Eisbacher GH, Clague JJ. 1984. Destructive mass movements in high mountains: hazard and management, Geological Survey of Canada Paper 84-16. Ottawa: Geological Survey of Canada.

Evans RS. 1981. An analysis of secondary toppling rock failures – the stress redistribution method. J Eng Geol. 14:77–86.

Evans SG. 1982. Landslides and surficial deposits in urban areas of British Columbia: a review. Can Geotech J. 19:269–288.

Fairer GM. 1985. Geologic map of the Wadi Baysh quadrangle, Kingdom of Saudi Arabia. Ministry of Petroleum and Mineral Resources, Deputy Ministry for Mineral Resources. Sheet 17 F, GM-77, scale 1:250,000.

Farrokhnia A, Pirasteh S, Pradhan B, Pourkerman M, Arian M. 2010. A recent scenario of mass wasting and its impact on the transportation in Alborz Mountains, Iran: contribution from Geo information technology. Arab J Geosci. 4:1337–1349.

Franklin JA, Senior SA. 1997. Rockfall Hazards – strategies for detection, assessment, and remediation. Paper presented at: Proceedings International Symposium on Engineering Geology and the Environment; Athens, Greece; p. 657–663.

Frenez T, Roth A, Kaestli A. 2004. Debris flow mitigation by means of flexible barriers. Paper presented at: Proceedings of the 10th Congress Interpraevent 2004; Trento, Italy.

Goodman RE, Bray JW. 1976. Toppling of rock slopes. In: Proceedings of a Specialty Conference on Rock Engineering for Foundation and Slope; August 15–18; Boulder (CO). American Society of Civil Engineering, Vol. 2, p. 201–234.

Government of Japan. 1984. Basics of planning the measures against debris flows and planning countermeasure facilities against debris flow. Kyoto (Japan): Mininstry of Construction; p. 39.

Guzzetti F, Peruccacci S, Rossi M, Stark CP. 2008. The rainfall intensity duration control of shallow landslides and debris flows: an update. Landslides. 5:3–17.

Hollingsworth R, Kovacs GS. 1981. Soil slumps and debris flows: prediction and protection. Bull Assoc Eng Geol. 18:17–28.

Huebl J, Fiebiger G. 2005. Debris flow mitigation measures. In: Jakob M, Hungr O, editors. Debris-flow hazards and related phenomena. Berlin: Springer; p. 445–488.

Hungr O, Evans SG, Bovis MJ, Hutchinson JN. 2001. A review of the classification of landslides in the flow type. Environ Eng Geosci. VII:221–228.

Hungr O, Morgan GC, Kellerhals R. 1984. Quantitative analysis of debris torrent hazards for design of remedial measures. Can Geotech J. 21:663–677.

Hungr O, Morgan GC, VanDine DF, Lister DR. 1987. Debris flow defences in British Columbia. In: Costa E, Wieczorek GF, editors. Debris flows/avalanches: process, recognition and mitigation. Reviews in Engineering Geology. Boulder (CO) Geological Society of America; p. 201–222.

Ishida T, Chigira M, Hibino S. 1987. Application of the distinct element method for analysis of toppling observed on a fissured slope. Rock Mech Rock Eng. 20:277–283.

Iverson RM, Reid ME, Logan M, LaHusen RG, Godt JW, Griswold JP. 2011. Positive feedback and momentum growth during debris-flow entrainment of wet bed sediment. Nat Geosci. 4:116–121.

Johnson AM. 1984. Debris flow. In: Brunsden D, Prior DB, editors. Slope instability. New York (NY): Wiley; p. 257–361.

Lo DOK. 2000. Review of natural terrain landslide debris-resisting barrier design. GEO Report No. 104. Hong Kong: Civil Engineering Department, Geotechnical Engineering Office, The Government of Hong Kong Special Administrative Region.

McCauley ML, Works BW, Naramore SA. 1985. Rockfall mitigation. Report FHWA/CA/TL-85/12. FHWA, U.S. Department of Transportation.

O'Connor JE, Hardison JH, Costa JE. Forthcoming. Debris flows from moraine-dammed lakes in the Three Sisters and Mt. Jefferson Wilderness areas, Oregon. Oregon: USGS Water Supply Paper.

Pierson TC. 1986. Flow behavior of channelized debris flows, Mount St. Helens, Washington. In: Abra-hams AD, editor. Hill slope processes. Winchester: Allen and Unwin; p. 269–296.

Pierson TC, Costa JE. 1987. A rheologic classification of subaerial sediment-water flows. In: Costa JE, Wieczorek GF, editors. Debris flows/avalanches: process, recognition and mitigation. Reviews in Engineering Geology. Washington (DC): Geological Society of America; p. 1–12.

Piteau DR. 1979. Engineering geology considerations and basic approach to rock slope stability analysis for highways. Part A. Rock Slope Engineering Reference Manual: FHWA Report – FHWA-TS-79-208. Washington (DC); p. 78.

Pourghasemi HR, Pradhan B, Gokceoglu C. 2012. Application of fuzzy logic and analytical hierarchy process (AHP) to landslide susceptibility mapping at Haraz watershed. Iran Nat Hazards. 63:965–996.

Pradhan B, Mansor S, Pirasteh S, Buchroithner M. 2011. Landslide hazard and risk analyses at a landslide prone catchment area using statistical based geospatial model. Int J Remote Sens. 32:4075–4087.

Prochaska AB, Santi PM, Higgins JD, Cannon SH. 2008. A study of methods to estimate debris flow velocity. Landslides. 5:431–444.

Regmi AD, Devkota KC, Yoshida K, Pradhan B, Pourgasemi HR, Kumamoto T, Akgun K. 2014. Application of frequency ratio, statistical index, and weights-of evidence models and their comparison in landslide susceptibility mapping in Central Nepal Himalaya. Arab J Geosci. 7:725–742.

Regmi AD, Yoshida K, Dhital MR, Devkota K. 2013. Effect of rock weathering, clay mineralogy, and geological structures in the formation of large landslide, a case study from Dumre Besei landslide, Lesser Himalaya Nepal. Landslides. 10:1–13.

Rickenmann D. 1999. Empirical relationships for debris flows. Nat Hazards. 19:47–77.

Rimbock A, Strobl T. 2002. Rope nets for woody debris entrapmentin torrents. Technical Document. Rimböck: Technische Universitat Muchen.

Russell CP, Santi P, Humphrey JD. 2008. Modification and statistical analysis of the Colorado Rockfall Hazard Rating System: Report No. CDOT-2008-7. Colorado; p. 139.

Shroder JF, Bishop MP. 1998. Mass movement in the Himalaya: new insights and research directions. Geomorphology. 26:13–35.

Sinotech Engineering Consultants INC. 2008. [The investigation of vulnerability factors and risk analysis, risk management of debris flows]. Report to the Soil and Water Conservation Bureau. Council of Agriculture, Executive Yuan. Chinese.

Swanston DN. 1974. Slope stability problems associated with timber harvesting in mountainous regions of the western United States. General Technical Report PNW-021. Portland (OR): Forest Service, U.S. Department of Agriculture; p. 14.

Thurber Consultants Ltd. 1984. Debris Torrents: a review of mitigative measures. A report to Ministry of Transportation and Highways, British Columbia. Saanich: Thurber Consultants Limited.

Tien Bui D, Pradhan B, Lofman O, Revhaug I, Dick OB. 2012. Landslide susceptibility assessment in the Hoa Binh province of Vietnam: a comparison of the Levenberg–Marquardt and Bayesian regularized neural networks. Geomorphology. 171:12–29.

VanDine DF. 1985. Debris flows and debris torrents in the southern Canadian Cordillera. Can Geotech J. 22:44–68.

VanDine DF. 1996. Debris flow control structures for forest engineering. Working Paper 08/1996. Victoria: Research Board, British Columbia Ministry of Forests.

Varnes DJ. 1978. Slope movement types and processes. In: Schuster RL, Krizek RJ, editors. Landslides – analysis and control, transportation. Special Report 176. Washington (DC): Transport Research Board, National Research Council; p. 11–33.

Watts CF. 2003. User's Manual Rockpack III for Windows. Rock Slope Stability Computerized Analysis Package, Part One - Stereonet Analyses. North Carolina: C.F. Watts & Associates; p. 48.

Youssef AM, Maerz NH. 2013. Overview of some geological hazards in the Saudi Arabia. Environ Earth Sci. 70:3115–3130.

Youssef AM, Maerz HN, Al-Otaibi AA. 2012. Stability of rock slopes along Raidah escarpment road, Asir Area, Kingdom of Saudi Arabia. J Geogr Geol. doi:10.5539/jgg.v4n2p48

Youssef AM, Maerz NH, Hassan AM. 2009. Remote sensing applications to geological problems in Egypt: case study, slope instability investigation, Sharm El-Sheikh/Ras-Nasrani area, Southern Sinai. Landslides. 6:353–360.

Youssef AM, Pradhan B, Maerz NH. 2013. Debris flow impact assessment caused by 14 April 2012 rainfall along the Al-Hada Highway, Kingdom of Saudi Arabia using high-resolution satellite imagery. Arab J Geosci. doi:10.1007/s12517-013-0935-0

Youssef AM, Pradhan B, Sabtan AA, El-Harbi HM. 2012. Coupling of remote sensing data aided with field investigations for geological hazards assessment in Jazan area, Kingdom of Saudi Arabia. Environ Earth Sci. 65:119–130.

Youssef AM, Sabtan AA, Maerz NH, Zabramawi YA. 2014. Earth fissures in Wadi Najran, Kingdom of Saudi Arabia. Nat. Hazards 71:2013–2027.

4

Deterministic slope failure hazard assessment in a model catchment and its replication in neighbourhood terrain

KIRAN PRASAD ACHARYA*†, RYUICHI YATABE‡,
NETRA PRAKASH BHANDARY‡ and RANJAN KUMAR DAHAL§

†Graduate School of Science and Engineering, Ehime University, 3 Bunkyo-cho, Matsuyama 790-8577, Japan

‡Department of Civil and Environmental Engineering, Graduate School of Science and Engineering, Ehime University, 3 Bunkyo-cho, Matsuyama 790-8577, Japan

§Department of Geology, Trichandra Campus, Tribhuvan University, Ghantaghar, Kathmandu, Nepal

In this work, we prepare and replicate a deterministic slope failure hazard model in small-scale catchments of tertiary sedimentary terrain of Niihama city in western Japan. It is generally difficult to replicate a deterministic model from one catchment to another due to lack of exactly similar geo-mechanical and hydrological parameters. To overcome this problem, discriminant function modelling was done with the deterministic slope failure hazard model and the DEM-based causal factors of slope failure, which yielded an empirical parametric relationship or a discriminant function equation. This parametric relationship was used to predict the slope failure hazard index in a total of 40 target catchments in the study area. From ROC plots, the prediction rate between 0.719–0.814 and 0.704–0.805 was obtained with inventories of September and October slope failures, respectively. This means September slope failures were better predicted than October slope failures by approximately 1%. The results show that the prediction of the slope failure hazard index is possible, even in a small catchment scale, in similar geophysical settings. Moreover, the replication of the deterministic model through discriminant function modelling was found to be successful in predicting typhoon rainfall-induced slope failures with moderate to good accuracy without any use of geo-mechanical and hydrological parameters.

1. Introduction

Shallow rainfall-induced landslides or simple slope failures in topographic hollows generally occur due to a rapid increase in porewater pressure during heavy rainfalls. The process of an increase in porewater pressure depends on a number of factors, e.g. a topographic hollow area, soil type, soil depth, slope morphology, slope gradient, vegetation pattern, microclimate, lithology, geological history, and rainfall pattern. (Crosta 1998; Crozier 1999; Van Asch et al. 1999; D'Odorico and Fagherazzi 2003; Wieczorek and Glade 2005). Rainfall events are temporal and their impacts are localized, so they make slope failure process a stochastic phenomenon (Talebi

*Corresponding author. Email: kiran_acharya60@yahoo.com

et al. 2008). It is often hard to identify 'when' and 'where' hillslopes fail and 'how' widespread a potential slope failure event could be during a rainfall event (Godt et al. 2008). Moreover, rainfall-triggered slope failure is a recurring problem in the hillslopes of loose colluvium. So, slope failure hazard analysis and its management in catchment scales has been a primary concern of geoscientists and geoengineers (Montgomery and Dietrich 1994; Guzzetti et al. 2005; Zolfaghari and Health 2008; Hadmoko et al. 2010; Arnone et al. 2011; Ching et al. 2011; Ghimire 2011; Dahal et al. 2012; Segoni et al. 2012). Slope failure hazard maps in catchment scales can be used as a cost-effective tool for slope failure hazard mitigation planning and risk analysis. With the availability of advanced mapping tools, such as ILWIS, GIS, Remote sensing, etc., and high-resolution digital topographical data, a number of methods have been developed mainly for the spatial prediction of potential slope failures, which include heuristic methods, statistical methods, and deterministic methods (Soeters and van Westen 1996; Guzzetti et al. 2006).

Heuristic methods (such as in Barredo et al. 2000, Dai et al. 2002; Saha et al. 2002; Pavel et al. 2008; Bijukchhen et al. 2013) are based on expert opinion. Therefore, these methods can not be easily replicated in other areas. In addition, these methods do not include the influence of geo-mechanics and porewater pressure, which are deemed necessary in deterministic methods (Jia et al. 2012).

In statistical methods, a relationship between slope failure location and the independent causal factors of the slope failure is established in the form of an empirical parametric function, which is then used to obtain a slope failure hazard map of the target catchment area (Carrara et al. 1999; Cannon et al. 2004; Bell 2007; Dahal et al. 2008a; Pradhan et al. 2010), and also to replicate the hazard index of one catchment to another with a similar set of parameters (Ghimire 2011; Dahal et al. 2012, etc.). However, as with heuristic methods, the statistical methods do not incorporate the influence of geo-mechanical processes and hillslope hydrology.

In the deterministic methods, on the other hand, the causal factors are expressed in algebraic terms to determine the factor of safety (Terlien et al. 1995; Wu and Sidle 1995; Borga et al. 1998; D'Odorico and Fagherazzi 2003; Godt et al. 2008; Salciarini et al. 2008; Harp et al. 2009). These approaches utilize a large amount of detailed geo-mechanical and hydrological parameters derived from field and laboratory experiments, and computer tool-based modelling. On most occasions, however, the geo-mechanical and hydrological parameters are not available for a large portion of the target area (Aleotti and Chowdhury 1999; Wang et al. 2013). Therefore, the application of a deterministic method in a larger area or its replication from one catchment to another through the same set of geo-mechanical and hydrological parameters is difficult due to limitations in the availability of these parameters.

In this study, we employ the deterministic hazard modelling technique to prepare a slope failure hazard map in a tertiary sedimentary terrain in the Shikoku Region in west Japan that suffered a large number of slope failures and debris flows leading to a heavy loss of life and property in 2004. The study area is relatively small, so the spatial variations in the causal factors, such as the amount of precipitation, geology, vegetation, aspect, etc. are negligible. This leads to a situation in which if the number of causal factors considered in the hazard analysis is low, it results in an extremely low prediction rate. It is inevitable that such an area needs a deterministic approach for the hazard analysis. So, the prime objective of this study is to prepare a deterministic slope failure hazard model for a catchment, and replicate it for other catchments of the study area so as to generate slope failure hazard maps. For this, we selected a small catchment in the study area as the model site, which was severely damaged during the extreme typhoon rainfall events of 2004, and 40 other catchments in the same terrain in order to replicate the hazard model.

2. Study area

The study area is situated in the northeastern hills of Niihama city in Ehime Prefecture, Japan (figure 1). In 2004, this area suffered extensive slope failure damage due to extreme typhoon rainfalls of September and October. The total study area covers 30.5 sq. km, with ground elevation ranging from 2.5 m to 285 m. It is surrounded by river flood plains in all directions except for the northeast corner where it faces the Seto Inland Sea. Two rivers in Niihama, the Kawahigashi and the Kawanishi, divide the study area hills into two main regions; Kita-yama and Nishi-no-yama. The study area is well forested with short-height evergreen and deciduous plants, such as Japanese red pine (*Pinus deniflora*), camphor (*Cinanamomum camphora*), and Japanese oak (*Quercus serrata* and *Quercus variabilis*), and Baby rosa (*Rosa multiflora*), and China root (*Smilax china*). The upper sections of the hillslopes consist of loose, thin colluvial soil over impervious bedrock, which is subject to frequent slope failures. The bases of the slopes have debris flow deposits. Also, densely populated residence areas are found close to the base of the hills.

The geology of the Shikoku region is characterized by three distinct units, namely the Ryoke, Sambagawa-Chichibu, and Shimanto Belts. Two northerly dipping major faults, the Median Tectonic Line (MTL) in the north and the Butsuzo Tectonic Line (BTL) in the south, separate the three geological units (Hashimoto 1991; Dahal et al. 2011). The selected study area falls in the Ryoke Belt, and consists of tertiary shales and sandstones of the late Cretaceous age often known as the Izumi group. Typical feature of this geological formation is that piles of intercalated sandstone and shale run in an east–west direction.

Japan often experiences heavy rainfalls during periods of extreme climate such as a typhoon. June to October is usually the typhoon season in Japan. In 2004, a series of nine typhoons hit Shikoku Island causing extreme typhoon rainfalls of various intensities, which resulted in a massive loss of life and property. Typhoon 0423 and 0421 severely impacted Ehime, Kochi, and Kagawa Prefectures in the region. Niihama city (about 140,000 residents) in Ehime Prefecture was the hardest hit area where many debris flows near the hill bases caused 25 deaths and 40 billion yen worth of property damage (Bhandary and Yatabe 2005). A total of 424 slope failures in September and 1396 slope failures in October were identified to have occurred in the selected area (figure 2). A higher number of slope failures was found in the Kita-yama area than the Nishi-no-yama area. It was later understood that the maximum hourly rainfall exceeded 50 mm and the maximum one-day rainfall exceeded 300 mm, and were the main causes of slope failures (Dahal et al. 2006; Bhandary et al. 2013). Altogether 41 catchments were delineated in the topographical map (figure 2) in the study area. The catchments were selected on the basis of an abundance of both September and October slope failures. Figure 3 shows the selected catchments in detail (indicated by MC and W1-W40; here, MC is the model catchment and others are the test catchments). In this study, both September and October slope failures were utilized separately to assess the prediction accuracy of slope failure hazard maps in the selected test catchments.

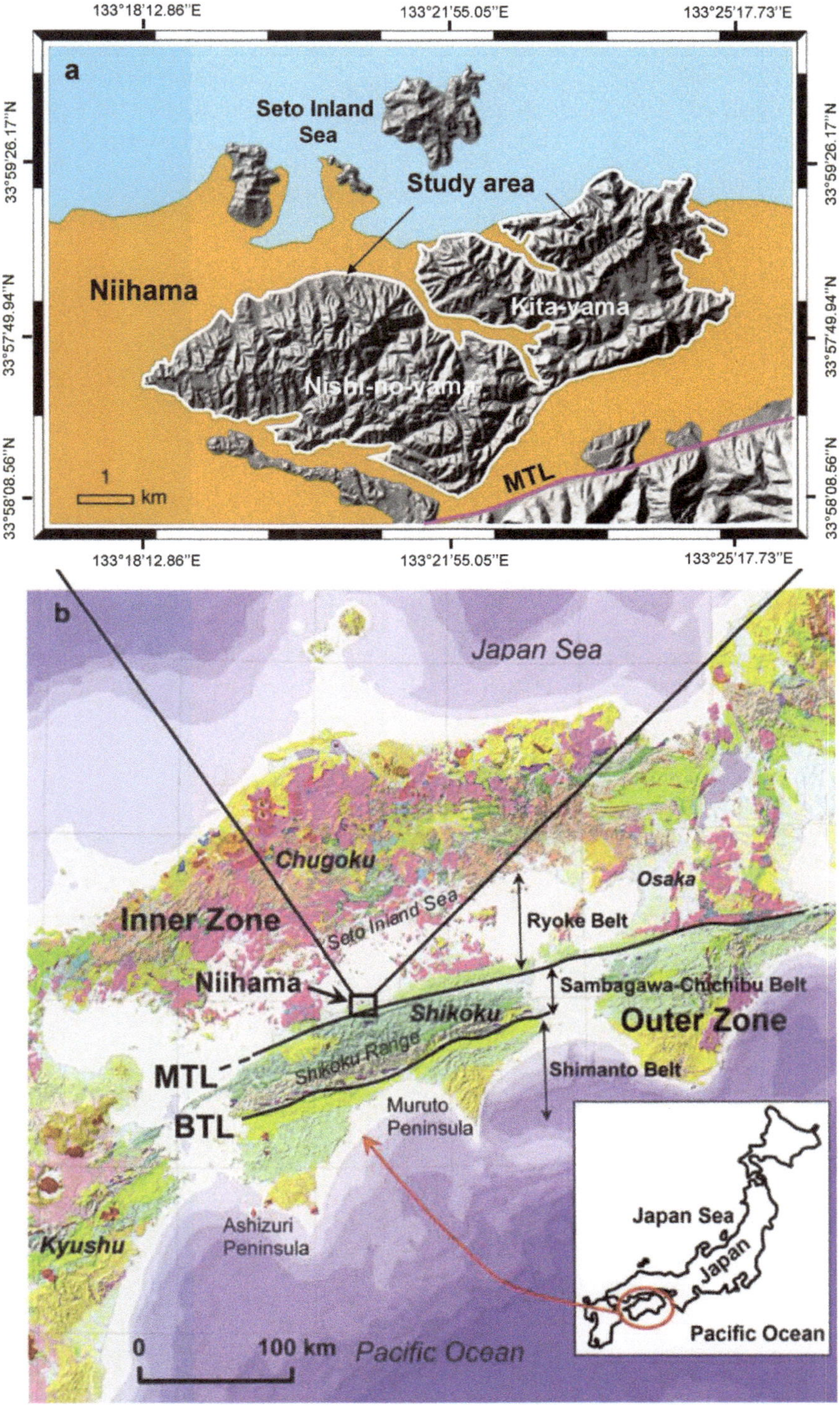

Figure 1. (*a*) Location map of the study area; and (*b*) geological outline of Shikoku Island (modified after Bhandary et al. 2013).

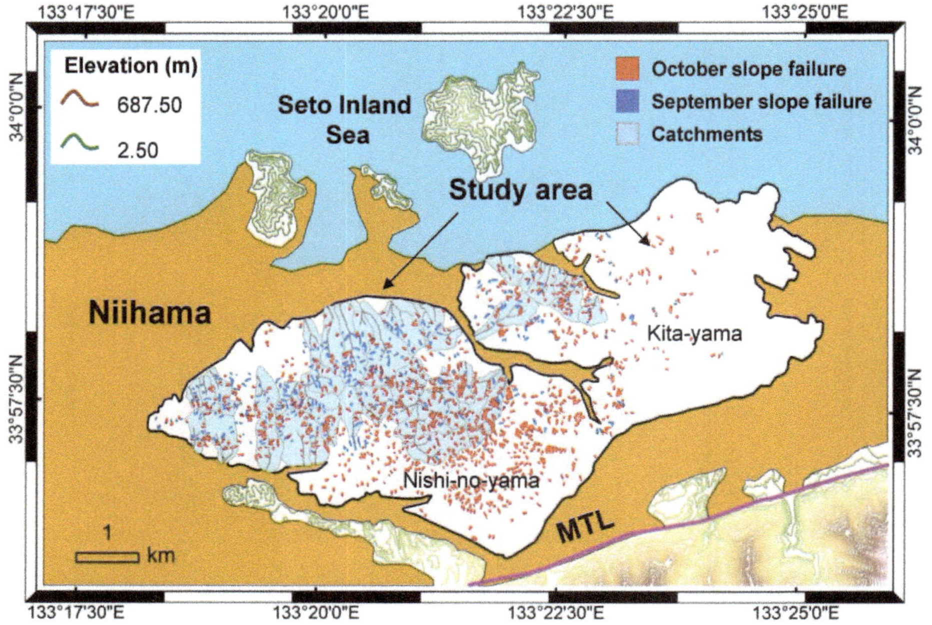

Figure 2. Distribution of slope failures of September and October of 2004 in selected catchments of Niihama.

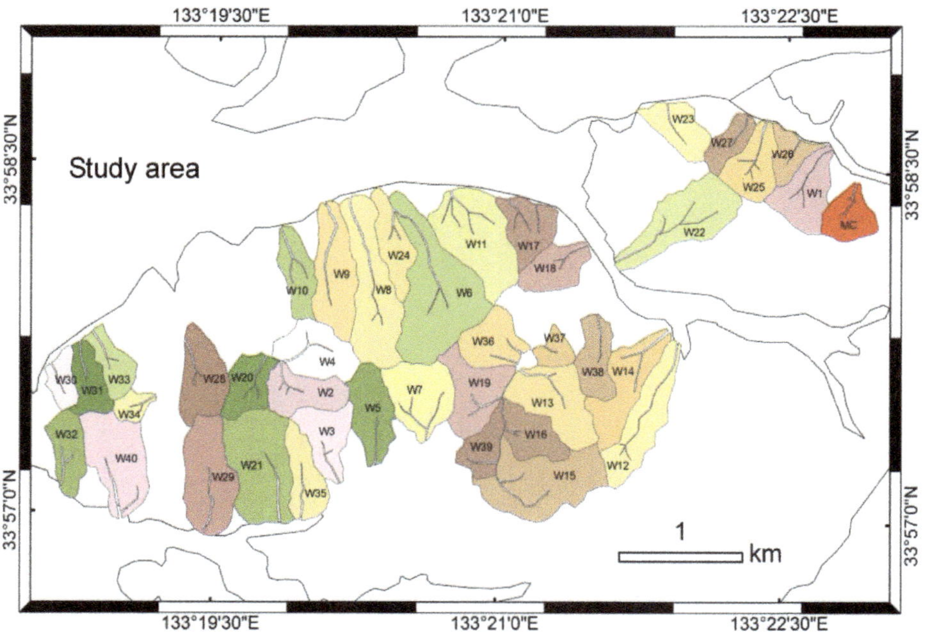

Figure 3. The selected model catchment and test catchments in detail.

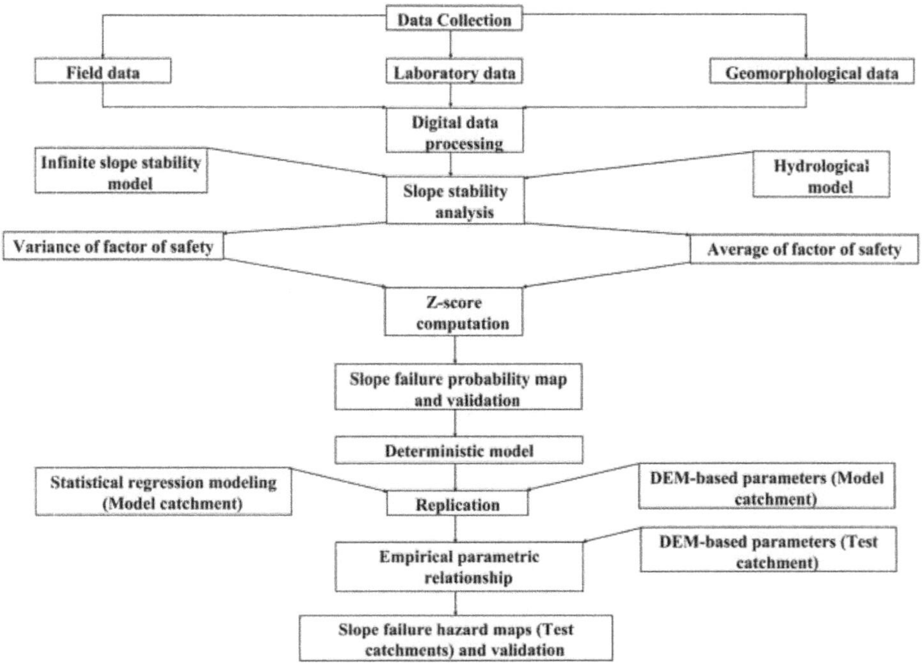

Figure 4. Research flow in this study.

3. Methodology

The research framework of this study is illustrated in figure 4. The Higashifukubegawa catchment, indicated by MC in figure 3, was used for the deterministic slope failure hazard assessment. For this, basic geo-mechanical and hydrological properties of soil obtained from various sources were utilized. Both hydrological and stability models were considered in the hazard assessment. Hydrological modelling was done at a hillslope scale. The catchment scale hydrology was derived from the sub-surface hydrology of individual hillslope profiles. The results obtained from the hydrological modelling were then used in stability modelling in pixel basis for the entire model catchment so as to obtain a deterministic slope failure hazard model. The deterministic model thus obtained was then replicated to the test catchments through statistical regression modelling using DEM-derived parameters. The following subsections describe the details of the methods and materials used in the analysis.

3.1. Model catchment

The Higashifukubegawa catchment has a tentative oval shape (figure 5). It has a spatial extension of 142,000 sq. m and the elevation varies from 2.5 m to 285 m from the mean sea level. The catchment is characterized by steep slopes with shallow colluvium. The slope varies between 0 and 60.57° with a mean value of 32.37°. Almost 2/3 of the catchment has slope between 20° and 40°. A detailed slope failure inventory of the catchment was prepared through field checks, Google Earth Image interpretation and inventories of September and October slope failures of 2004. After this, there

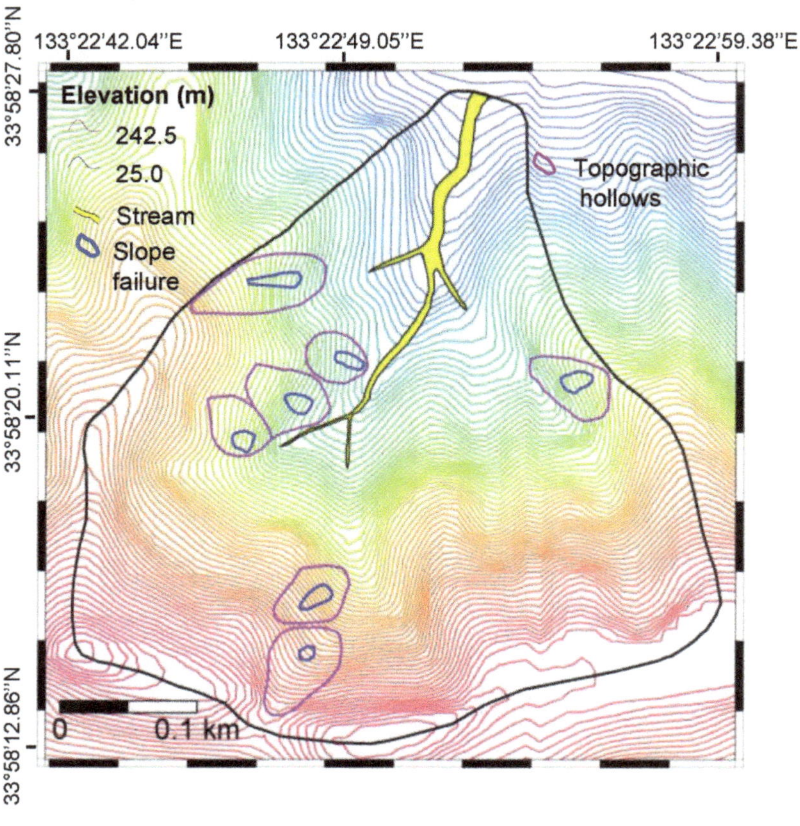

Figure 5. Higashifukubegawa catchment with detailed slope failure inventory, showing slope failures lying in topographic hollows.

were no inventories of September and October slope failures for model catchment. The detailed slope failure inventory contains seven slope failures (figure 5), each lying on separate topographic hollows or zero order basins.

3.2. Geo-mechanical and hydrological data collection

Field visits were made in October, November, December, and April of 2011 as well as November of 2012 in the model catchment to observe the response of catchment slopes during various rainfall events in these months. To obtain detailed in-situ/laboratory data, the model catchment was divided into 25 sq. m blocks. The total number of blocks was 261 (figure 6). The in-situ tests, such as the dynamic cone penetration test to measure the soil thickness above weathered bedrock and the Hasegawa in-situ permeability test (Daitou Techno Green 2009) to measure soil permeability within the unsaturated zone, were conducted in each square block. These tests revealed that soil thickness varied from 0.1 m to 1.91 m and soil permeability from 10^{-5} m/s to 10^{-8} m/s in the model catchment. At least one soil sampling was done at each block to determine the basic geo-mechanical properties of soil and to classify the soil type. Mohr-Coulomb parameters of soil strength (C', ϕ') were determined from direct shear test results. Root cohesion (C_r') was estimated from the findings of Neupane (2005) and

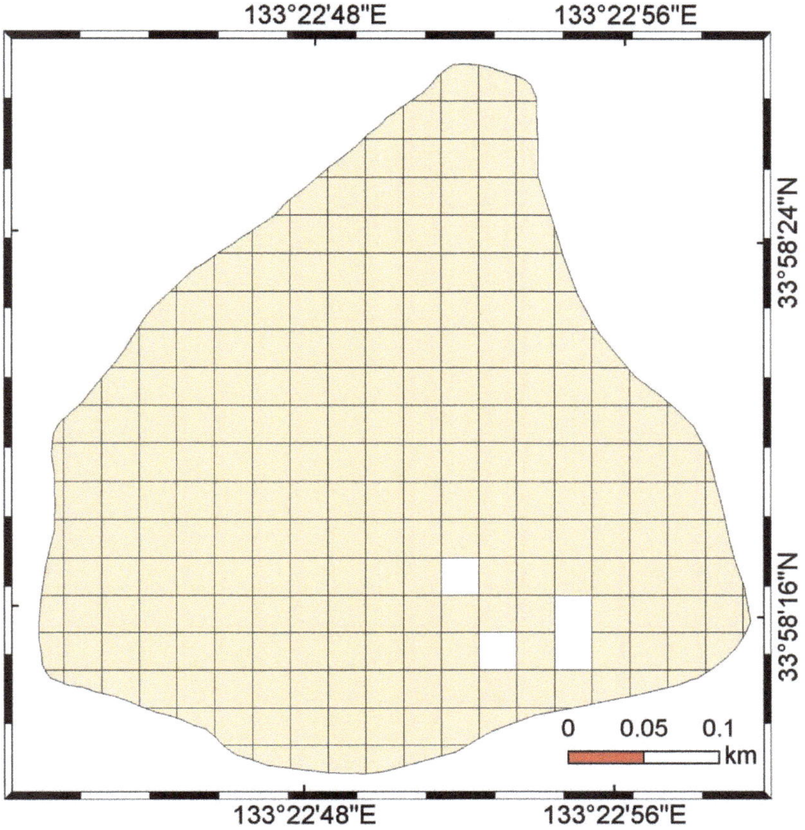

Figure 6. 25 sq. m blocks in model catchment. The hollow blocks representing bedrock were excluded from the analysis.

information provided by Sidle (1991) for similar types of plants in this study. Also, the laboratory tests for saturated unit weight (γ_{sat}), bulk unit weight (γ_t), porosity or volumetric water content (n), and particle size distribution of soil were conducted following the specifications of ASTM. Based on the results of particle size distribution (after referring to USCS and ASTM standards), the soils were classified into five major soil domains, namely SM (fine), SM (coarse), SM (medium), GM, and M [SM is silty sand, GM is silty gravel, and M is silt]. The extreme values of geo-mechanical and hydrological properties in five major soil domains are given in table 1. Based on soil domains, the model area was partitioned into five zones (figure 7) for deterministic slope failure hazard modelling (see in Section 3.5).

3.3. Model selection

The reliability of a hazard map depends not only on quality/amount of the incorporated data and working scale but also on appropriate methodology (Baeza and Corominas 2001). The models should be chosen also with climatic realization. For this study, we chose SEEP/W (GeoStudio 2005) as a hydrological modelling tool and considered infinite slopes for the stability modelling. Likewise, we used a

Table 1. Geo-mechanical and hydrological properties of soil domains

	Major soil domains									
	SM (fine)		SM (medium)		SM (coarse)		GM (medium)		M	
Soil properties	MIN	MAX	MIN	MAX	MIN	MAX	MIN	MAX	MIN	MAX
C' (kN/m^2)	0.31	10.76	1.62	14.11	0.31	12.50	4.49	10.85	9.48	12.50
ϕ' (°)	23.91	43.59	20.92	40.59	22.15	44.59	21.91	33.14	24.33	34.64
C'_r (kN/m^2)	1.92	4.79	0.00	4.79	1.50	4.48	1.92	4.31	1.40	3.35
γ_t (kN/m^3)	10.58	15.04	11.74	16.74	9.58	14.07	14.44	15.07	11.59	13.94
γ_{sat} (kN/m^3)	15.43	18.90	16.33	19.90	15.33	17.39	16.79	17.19	17.43	18.29
n	0.48	0.51	0.45	0.55	0.45	0.52	0.42	0.47	0.41	0.51
k (m/s)	6.12×10^{-8}	2.66×10^{-5}	7.18×10^{-8}	4.67×10^{-5}	8.01×10^{-8}	3.69×10^{-5}	7.78×10^{-8}	1.48×10^{-5}	6.16×10^{-8}	2.17×10^{-5}

Note: Saturation depth (h) and soil depth (z) data are not included in this table. The spatial distribution of these two properties will be prepared in raster format for the whole model area (see in Section 3.5).

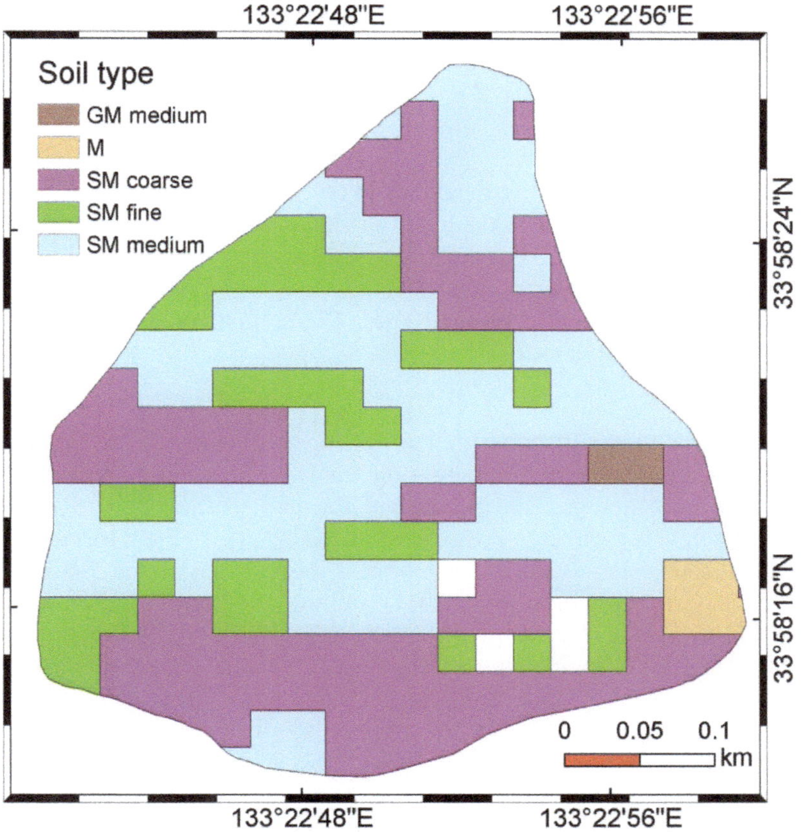

Figure 7. Major soil domains in the model catchment after soil classification.

discriminant function model (multivariate regression model) for the replication purpose. These models have been acknowledged worldwide in recent landslide research such as Gasmo et al. 2000; Acharya et al. 2006; Godt et al. 2008; Lee et al. 2008; Trandafir et al. 2008; Eeckhaut et al. 2009; Cascini et al. 2010; Tarolli et al. 2011; Ghosh et al. 2012; Harris et al. 2012; and Jagielko et al. 2012. The SEEP/W (GeoStudio 2005) hydrological model is first used in slope failure hazard assessment in this study. The following subsections briefly describe the models.

3.3.1. Hydrological model. As also stated earlier, rainfall is a temporal stochastic phenomenon. The amount of rainfall varies with time (Ng and Shi 1998), and with time varying infiltration, the porewater pressure rises/falls above impervious bedrock. This rise/fall (transient behaviour) of porewater pressure often controls hillslope instability (Reid and Iverson 1992), but it is difficult to estimate the transient porewater pressure information required for the slope stability analysis of hillslopes (Harp et al. 2009). SEEP/W (GeoStudio 2005) solves this problem. It is a finite element mesh model of non-linearized Richard's equation. Richard's equation explains both saturated and unsaturated Darcian flows through soil layers. Richards's equation to compute 2-dimensional seepage in SEEP/W has the following form:

$$\frac{\partial}{\partial x}\left(k_x\frac{\partial H}{\partial x}\right)+\frac{\partial}{\partial y}\left(k_y\frac{\partial H}{\partial y}\right)+q=m_w^2\gamma_w\frac{\partial H}{\partial t}, \tag{1}$$

where k_x is the coefficient of permeability in the x direction; k_y is the coefficient of permeability in y direction; H is the hydraulic head or total head; q is applied flux at the boundary; ${m^2}_w$ is the slope of soil–water characteristics curve; and γ_w is the unit weight of water. With these inputs, this study uses SEEP/W to compute transient porewater pressure distribution in hillslopes. The computed seepage is then used in the infinite slope stability model.

3.3.2. Infinite slope stability model. The hillslopes with loose soil colluviums can be recognized as infinite slopes when the soil thickness is limited compared to slope length and width, slope gradient is constant throughout the length, and underlying ground water flow is parallel to the ground surface (Taylor 1948). In such slopes, the failure mass is analyzed as the movement of a single block of soil, disregarding the head and toe portions (Arnone et al. 2011). Infinite slopes are easily destabilized due to a rapid rise in positive porewater pressure or a loss of soil suction during rainfall (Iverson et al. 1997). The slope mass fails typically at or near the contact between the soil colluvium and impervious bedrock. Generally, the nature of failure is translational. The shear strength acting along the slip surface is given by the Mohr–Coulomb criterion (Terzaghi and peck 1967).

$$\tau = C' + (\sigma_n - u_w)\tan\phi', \tag{2}$$

where τ is the shear strength of unsaturated soil; C' is the effective cohesion (kN/m^2); $(\sigma_n - u_w)$ is the net normal stress; σ_n is the total normal stress (kN/m^2); u_w is porewater pressure (kN/m^2); and ϕ' is the angle of shearing resistance (°). During failure, shear stress (T) exceeds shear strength (τ). The ratio $F = \tau/T$ is called the factor of safety. For an infinite slope, the factor of safety can be expressed as below (Hammond et al. 1992).

$$FS = \frac{C' + C_r' + A\cos^2\beta\tan\phi'}{B\sin\beta\cos\beta} \tag{3}$$

In which

$$B = h\gamma_{sat} + \gamma_t(D-h);$$
$$A = h(\gamma_{sat} - \gamma_w) + \gamma_t(D-h),$$

where C_r' is the effective root strength; β is the slope inclination (°); $\gamma_{\rm sat}$ (kN/m^3) is the saturated unit weight of soil; $\gamma_{\rm t}$ (kN/m^3) is the bulk unit weight of soil; h is the vertical saturation depth (m); and D is the vertical soil depth (m) (as expressed in Section 3.3.1; h is estimated from hydrological modelling). The effect of surcharge is disregarded in this study. When the factor of safety is greater than 1, the hillslope is stable and when it is equal to 1, the slope mass is on the verge of failure (limit equilibrium state). An infinitesimal perturbation can result in sliding.

The Mohr–coulomb strength parameters (i.e. C', ϕ') computed in the laboratory and the root strength (C_r') estimated from past studies may vary from the actual field values. So, these have some degree of uncertainty. The erroneous strength parameters either should not be used (van Westen and Terlien 1996) or the factor of safety computed by using such parameters should be checked for error. Various methods exist in the literature for analyzing the propagation of error (Ward et al. 1981; Hammond et al. 1992; Terlien 1996; Burrough and McDonnell 1998). Calculating the expected value and the variance of the factor of safety and incorporating these into a stability analysis to compute a z-score is a reliable method (Dahal et al. 2008b). Based on these, equation 4 is converted in the following linear form.

$$FS = \alpha_1 C' + \alpha_1 C_r' + \alpha_2 \tan\phi', \tag{4}$$

where

$$\alpha_2 = \frac{A\cos\beta}{B\sin\beta};$$

$$\alpha_1 = \frac{1}{B\sin\beta\cos\beta}.$$

In this study, the expected value and variance of the factor of safety, and the z-score are ascertained by using the following equations (Ward et al. 1981; Ross 2004).

$$E[FS] = \alpha_1 E[C'] + \alpha_1 E[C_r'] + \alpha_2 E[\tan\phi'] \tag{5}$$

$$V[FS] = \alpha_1{}^2 V[C'] + \alpha_1{}^2 V[C_r'] + \alpha_2^2 V[\tan\phi'] \tag{6}$$

$$Z = \frac{1 - E(FS)}{\sqrt{V(FS)}}, \tag{7}$$

where, $E[.]$ is the expected value; $V[.]$ is the variance, and Z is the z-score. The expected values and variance of the Mohr–Coulomb strength parameters are computed based on their nature of distribution (see in Section 3.5). $E[FS]$ is the average value of the factor of safety. $V[FS]$ gives the total propagation of error due to the variation of C', C_r', and $\tan\phi'$. The z-score is equivalent to the corrected factor of the safety score or the factor of safety score after adjusting for error. It is considered as the susceptibility/hazard score (Guzzetti et al. 2005; Dahal et al. 2008b). In this study, the failure probability score is computed from the z-score that is then implemented into ILWIS 3.8.1 to prepare the slope failure probability map. Thus, with the availability of geo-mechanical (C', C_r, ϕ', γ_{sat}, γ_t), hydrological (h), and geometric properties (β), the evaluation of hillslope instability or slope failure hazard is possible as carried out in this study.

3.3.3. Discriminant function model. The discriminant function model is used to determine the relationship between the outcome or the dependent variable (e.g. slope failure) and the independent or predictor variables (causal factors of slope failure)

(Nagarajan et al. 2000; Carrara et al. 2003; Ghosh et al. 2012; Jagielko et al. 2012). This model is required when there are multiple (i.e. >2) categories of dependent variables (e.g. types of failures). The model determines multiple combinations of employed independent variables (which are the discriminant functions) and selects the most significant combination (it is the combination that best discriminates categories of dependent variables) (Baeza et al. 2010; Dong et al. 2009). For example, if there are *m* categories of dependent variables, the possible combination of independent variables or discriminant functions will be *m*-1. Finally, the selected combination of variables is included in the discriminant function equation. The discriminant function equation is a linear combination of weighted independent variables as given below.

$$DFs = b_0 + b_1 \times X_1 + b_2 \times X_2 + b_3 \times X_3 + \cdots\cdot b_n \times X_n \quad (8)$$

Where *DFs* is the discriminant function score; b_o is the coefficient that maximizes the variability between categories of dependent variables; b_n is the discriminant weight value or discriminant coefficient. The negative/positive sign of the discriminant weight values determine the contribution of employed independent variables to discriminant categories of dependent variables (or these determine the significance of independent variables to cause instability); X_n is the most significant combination of independent variable (slope failure causal factor). The goodness of fit of the discriminant function model was tested using the Wilk's lambda test (Baeza et al. 2010; Jamaludin et al. 2006; Ghosh et al. 2012; He et al. 2012). The Wilk's Lambda score measures the discriminating potential of a combination of parameters. The smaller the value of Wilk's Lambda, the higher the discriminating capability. Hence, a small value of Wilk's Lambda (<1) is always preferred in discriminant function analysis. In the Wilk's Lambda test, if the significance of the Chi-square value is less than 0.05, the combination of parameters is significant and the discriminant function model can be accepted (Baeza et al. 2010, Ghosh et al. 2012). In this study, different low and high hazard classes of the deterministic slope failure hazard model were used as categories of dependent variables and slope failure causal factors (DEM-based parameters) as independent variables in discriminant function modelling.

3.4. Seepage modelling

Seepage modelling was performed in SEEP/W (GeoStudio 2005) to obtain porewater pressure information required for the slope stability analysis in the model catchment. For this, 55 slope profiles were drawn in the model area parallel to the direction of maximum subsurface flow. Their locations are represented by slope profile lines in figure 8. The profile geometry of these lines was prepared using DEM (5 m resolution) of the study area and the soil thickness data considering the topographical break in slope (figure 9). All the profile continuums were discretized into a mesh of fine squares of four nodes and a nine-integration order. The dimensions of all the elements were the same with a side length of <0.2 m and unit thickness. The soil water characteristics curve (SWCC) and soil permeability were the two major input parameters required for the seepage simulation of each profile unit. For this, the mean values of volumetric water content (porosity) and soil permeability of all the square

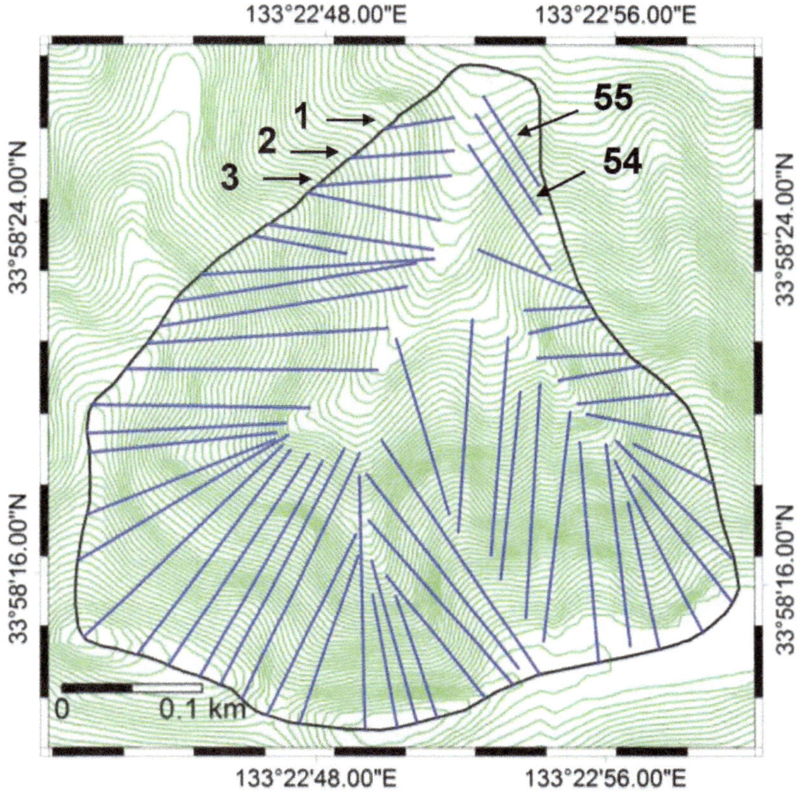

Figure 8. Slope profile lines constructed in the entire model catchment.

blocks lying below slope profile were computed for each slope profile. The computed mean values were integrated with SWCC and the soil permeability functions available in the SEEP/W library. SEEP/W is highly sensitive to initial ground water table conditions. To prevent an unnecessary negative porewater pressure regime, a limiting value of −20 kPa suction was applied as the initial condition. The initial ground water table was defined along the boundary between the soil colluvium and the bedrock. Boundary conditions control porewater pressure distribution along the profile length. A null flux boundary condition was applied on the left vertical edge, the right vertical edge, and along the bedrock (figure 10). The phreatic line itself acts as the upper boundary along the exposed sloping face. The potential seepage face was considered to be on the right vertical edge and along the exposed sloping face to avoid ponding on the slope.

There was considerable rainfall almost every day in October 2004 and slope failure occurred on October 20, but no data were available for porewater pressure variation that could describe the initial seepage condition in the slopes prior to the rainfall events. So, the initial seepage conditions were disregarded and the simulation was preceded with the precipitation data of October 19–20. Two days of hourly rainfall data were discretized into 283 time steps of 10 min in length and applied onto the exposed face of the profiles from upward at an equal pressure head and elevation

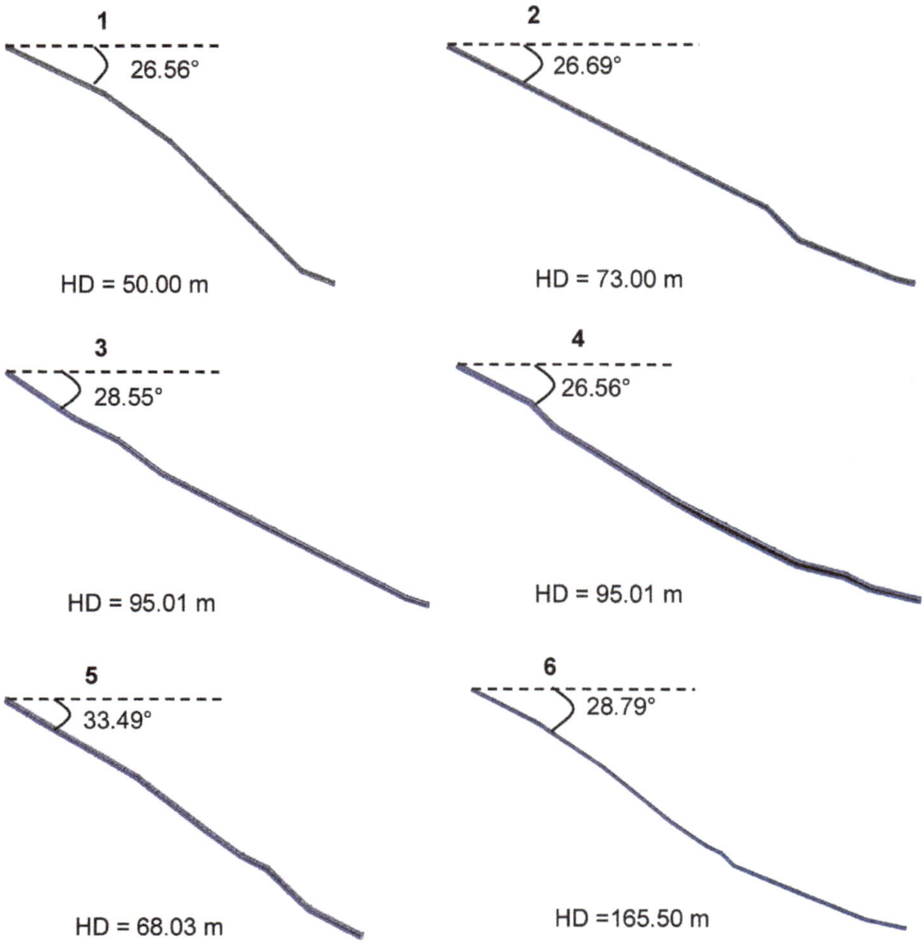

Figure 9. Typical examples of geometrical configurations of slope profiles 1, 2, 3, 4, 5, and 6 (from figure 7).

condition. Transient pore water pressure lines (transient pore water pressure distributions) were developed at each time step. The ground water tables at succeeding time steps had higher depths compared to the preceding ones along each profile length due to the accumulation of rainfall above the bedrock. The peak hourly rainfall intensity was at 12:00 noon on 20 October 2004. There was no eyewitness for the exact time of failure on 20 October, but the peak porewater pressure reached two hours after peak rainfall hour (i.e. at 14:00 rainfall hour) on the same day in all simulations. This implies that the slope failure in the model catchment might have begun at 14:00 hours of rainfall on 20 October (i.e. the 38 hours of continuous rainfall on 19–20 October). Owing to this, the ground water table, which developed at 14:00 hours of rainfall on 20 October 2004, was decided to implement in limit equilibrium analysis. The vertical depth of saturation was recorded at various points along each slope profile in the selected ground water table. With these data, a point map of saturation depths was prepared.

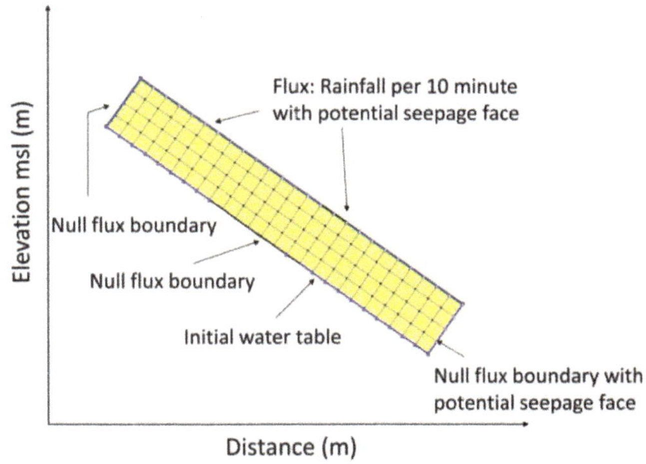

Figure 10. Boundary conditions applied to finite element mesh model.

3.5. Deterministic slope failure hazard modelling

The geo-mechanical and hydrological properties (table 1) of hillslope materials were parameterized in the ILWIS 3.8.1 platform. The point map of soil depth (prepared in field) and saturation depth (prepared from seepage modelling) covering the entire model area were digitized. The spatial correlation of both data was investigated. The spatial correlation provides a semivariogram model about the spatial behaviour of the soil properties (Jorge 2009). The rotational Quadratic semivariogram model was best fitted by both digital soil depth and saturation depth data [figure 11 (a, b, c)], and this model yielded values of nugget, sill, and range for each data type. These three parameters were inputs to the Kriging interpolation along with the respective soil spatial properties. With the ordinary Kriging interpolation method, the randomly distributed point values of soil depth and saturation depth are converted into regularly distributed grid values. To prepare the saturated unit weight and the bulk unit weight map, first the average value of these properties (table 1) were computed for each zone of the model area and then converted into a raster image using ILWIS functions. To prepare the expected and variance maps of the strength parameters (C', C'_r, ϕ'), the uniform distribution of these properties was assumed in the entire model area. For uniformly distributed data, the expected values and variance are computed using the following relationships:

$$E[X] = \frac{(a+b)}{2}, \tag{9}$$

$$V[X] = \frac{(b-a)^2}{12}, \tag{10}$$

where where a and are minimum and maximum values of strength parameters. These values were employed to compute the expected values and variances of the strength parameters for each zone outside ILWIS. The computed expected values and variances were then imported into ILWIS and converted into a generalized continuous

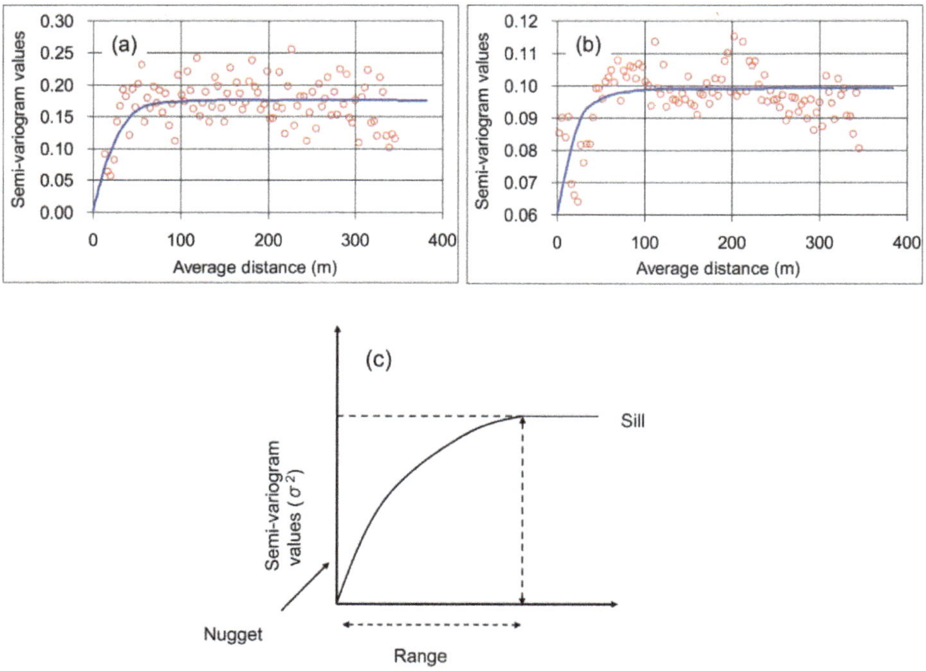

Figure 11. Rotational quadratic semi-variogram models to estimate spatial distribution of (*a*) soil depth and (*b*) saturation depth in model catchment. Nugget, sill, and range were fit parameters for both models, which is illustrated in (*c*).

raster form. The slope gradient map was obtained from the DEM of the model catchment. All these maps were used for deterministic slope failure hazard modelling.

To obtain a robust deterministic slope failure hazard model, a complete parametric analysis was performed by varying shear strength parameters (C', C_r', ϕ'). The expected map and variance map of factor of safety were prepared on a cell-by-cell basis (5 m resolution) using equations 5 and 6. These maps were directly implemented in the z-score computation (equation 7). The z-score herein is the slope failure hazard score. The area under the standardized normal weights of the z-score between $F = 1$ and $Z = -\infty$ gives failure probability. In the case with large voluminous data, as in the present study, the commercially available spreadsheet software can be utilized to calculate probability from the z-score (Dahal et al. 2008b). In this study, the z-score was exported to Microsoft Excel and the NORMSDIST function was employed to compute probability. This process of calculation of failure probability is in accordance with Terlien (1996) and van Westen and Terlien (1996). The probability value computed in excel was imported into ILWIS and rasterized to prepare the slope failure probability map. This map was validated using the detailed slope failure inventory of the model catchment. The hazard classes were determined by visual inspection of the distribution of pixels with higher values of probability from the ROC prediction rate. Generally, high-hazard classes consist of pixels with high probability values (the greatest number of pixels where slope failures occurred) and low hazard classes cover the regions of low probability values (the fewest number of slope failure pixels). Finally, the zonation map was once again crosschecked with the detailed slope failure inventory for spatial agreement.

3.6. *Replication of deterministic model and accuracy check*

After confirming a good predictive success rate from the validation, the results of the deterministic slope failure hazard model were applied to 40 selected test catchments of the study area. For this, first discriminant function modelling was carried out amongst the deterministic model and causal factors of slope failure, and the discriminant function equation was obtained for the model catchment. Using this equation and using the same set of seven DEM-based slope failure causal parameters, the slope failure hazard index maps were obtained in the test catchments. These maps were compared separately with the prevailing slope failure inventories of September and October 2004 to check their prediction accuracy.

4. Results

4.1. *Deterministic modelling*

Figure 12 shows the slope failure probability map obtained after performing full parametric modelling of shear strength parameters. To validate this map against the detailed slope failure inventory, the ROC curve (AUROC) method (Zweig and Campbell 1993) was used. It determines the predictive capability of a model (Yesilnacar and Topal 2005; Van Den Eeckhaut et al. 2006). The ROC makes a diag-

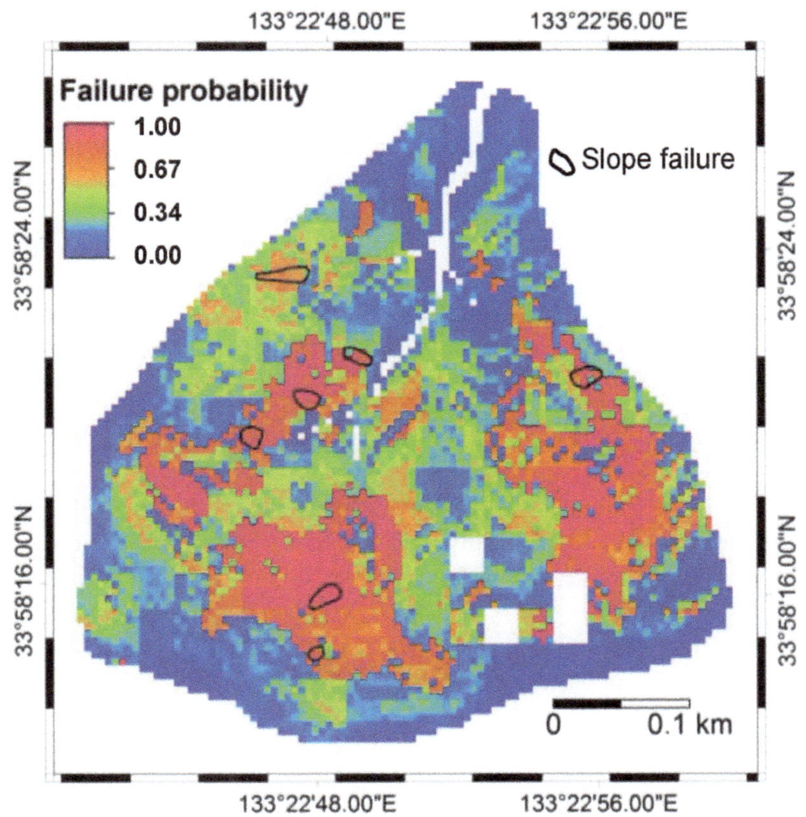

Figure 12. Slope failure probability map.

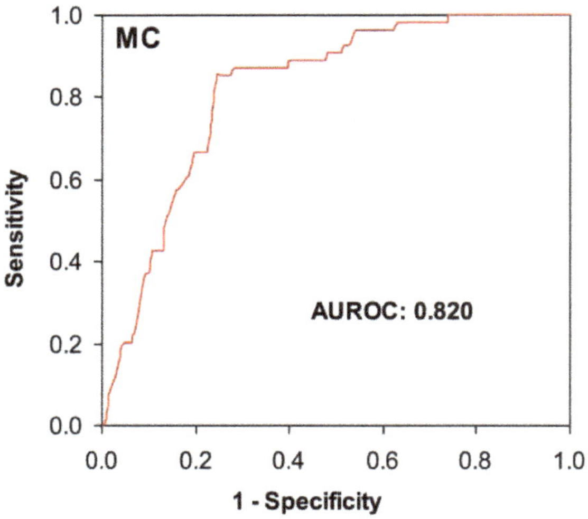

Figure 13. Area under ROC curve estimated from Higashifukubegawa catchment after deterministic slope failure hazard modelling.

nosis for spatial agreement of prevailing slope failures with the failure probability score. The variables used in ROC curve analysis in SPSS are a binary (0-1) map of the slope failure inventory and the failure probability score. Based on the degree of spatial matching between these two variables, the ROC curve method determines the area under the ROC curve (AUROC). This area is the measure of the robustness of the model to predict future slope failures. The AUROC value of the model catchment based on its detailed slope failure inventory was 0.82 (figure 13). The ROC curve also suggests that about 68% of the slope failure numbers were concentrated in areas showing upper 20% of the failure probability index. These reasonably verify the ability of the model to predict future slope failures.

Based on the ROC prediction rate and the coverage of a higher percentage of slope failure at a higher value of probability, the slope failure probability map was classified into four hazard classes: Stable (S, <30% probability value), Metastable (MS, 30%–50% probability value), Quasi-stable (QS, 50%–70% probability value), and Unstable (U, 70%–100% probability value) (figure 14), which is from a low- to a high-hazard level. The classes were assigned the colours green, yellow, blue, and red, respectively. This is the deterministic slope failure hazard model in the Higashifukubegawa catchment. Finally, the zonation map was crossed with the raster map of the detailed slope failure inventory. The result (figure 15) shows that a total of 92.6% of the slope failures were found to occur in the U class, which is very promising. Also, only a few pixels of slope failures were located in the QS and MS classes (which are 1.90% and 5.60% of the total slope failures, respectively). There was no slope failure pixel in S class.

4.2. Discriminant function modelling

As mentioned earlier, the categories of dependent variables and independent variables are required to obtain a discriminant function equation. Here, the four hazard classes of the deterministic model: S, MS, QS, and U were used as categories of

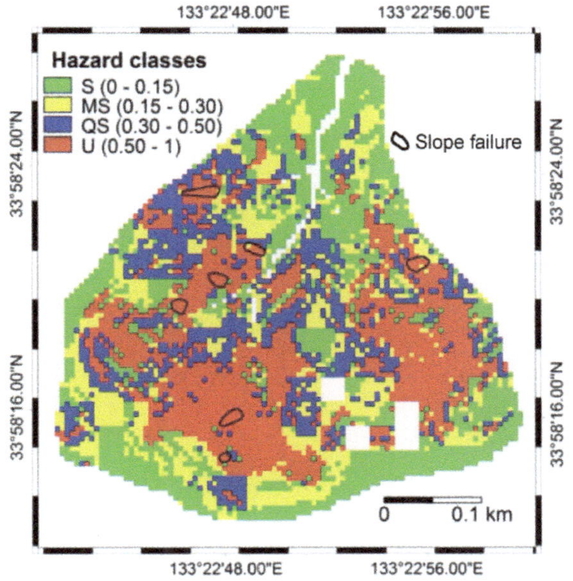

Figure 14. Slope failure hazard zonation map of Higashifukubegawa catchment (deterministic model).

dependent variables. A failure probability number: 1, 2, 3, and 4 was assigned to the classes from a low- to a high-hazard class. Similarly, a set of seven significant and easily available topographical and hydro-geological parameters, namely slope, relief, plan curvature, profile curvature, wetness index, flow accumulation, and drainage density, were selected as the independent variables in model catchment. All of these parameters were derived from DEM (5 m resolution). With inputs of the above-mentioned dependent and independent variables, the discriminant function modelling was performed in the SPSS platform. The results of the discriminant function modelling are presented in tables 2–4.

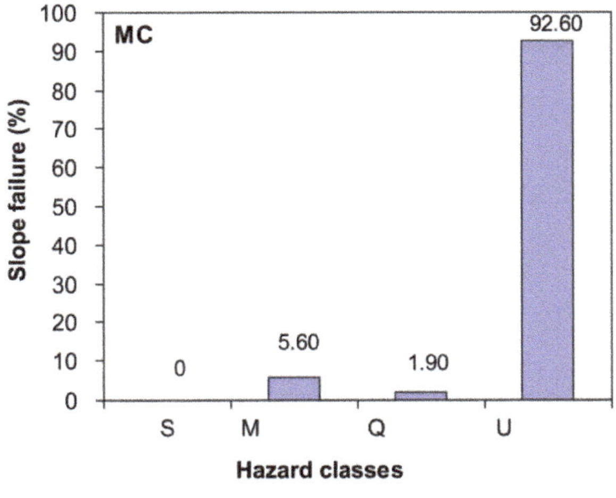

Figure 15. Slope failure distribution in hazard classes of zonation map.

Table 2. Unstandardized canonical discriminant function coefficients.

	Functions		
Parameters	1	2	3
Slope	0.101	0.003	0.012
Relief	−0.0000178	0.017	0.009
Plan curvature	−0.014	−0.047	0.362
Profile curvature	−0.755	0.321	0.328
Wetness index	0.067	−0.118	0.314
Flow accumulation	−0.0003190	−0.001	0.0000708
Draining density	7.565	−8.124	36.413
Constant	−3.711	−1.454	−3.651

The discriminant function model produced three combinations of casual parameters of slope failure or discriminant functions (table 2), which was 1 less than the number of the hazard classes of the deterministic model. All three combinations of the parameters were found significant since the significance of the Chi-square value was less than 0.05 in all combinations (table 3). However, the lowest Wilk's Lambda score (0.734) and significance level (0.00) of the first combination enable us to confirm that the first combination is the most significant among the three (i.e. it most discriminates the hazard classes of the deterministic model or the group means are significantly different for this combination). Based on these results, the first combination of parameters or the discriminant function is included in the discriminant function equation. The discriminating potential of the employed independent slope failure causal parameters to the hazard classes can be understood from the Standardized Canonical Discriminant Function Coefficients listed in table 4

Table 3. Wilk's lambda test.

Function (s)	Wilk's lambda	Chi-square	df	Sig.
1 through	0.705	1932.105	21	0.000
2 through	0.987	74.947	12	3.76E-11
3	0.996	23.394	5	2.84E-04

Table 4. Standardized canonical discriminant function coefficients.

	Functions		
Parameters	1	2	3
Slope	0.988	0.027	0.114
Relief	−0.00100	0.874	0.459
Plan curvature	−0.016	−0.052	0.400
Profile curvature	−0.207	0.088	0.090
Wetness index	0.112	−0.198	0.524
Flow accumulation	−0.10500	0.229	0.023
Draining density	0.174	−0.187	0.837

4.3. *Replication and validation*

To apply the results from discriminant function modelling, the same set of seven DEM-derived slope failure causal parameter maps were prepared in all the test catchments. The discriminant weight values of the parameters of the model catchment were applied to the corresponding parameters of the test catchments and the discriminant function equation was portrayed to compute the discriminant function score or the slope failure hazard index. Figure 16 shows the resulting slope failure hazard index maps of two typical test catchments W20 and W30. The slope failure hazard score was more or less similar in all 40 catchments. To check the prediction accuracy, the resulting hazard maps were crossed with the existing inventories of the September and October slope failures of 2004, and the ROC curves were plotted. The area under the ROC curve gives the prediction rate. Figure 17 exhibits the prediction rate obtained with hazard index maps of catchments W20 and W30. The prediction rate with respect to the September and October slope failures in 40 test catchments is summarized in table 5. From this table, it is understood that the prediction rate was higher with the inventory of the September slope failure in 22 test catchments (W1, W3, W4, W5, W9, W10, W12, W16, W17, W18, W19, W20, W21, W22, W25, W26, W28, W29, W32, W36, W37, and W38) whereas in the remaining test catchments, the prediction rate was higher with the inventory of the October slope failure. Overall, the prediction rates from 0.719 to 0.814 and 0.704 to 0.805 were noted with the inventories of September and October slope failures, respectively, from table 5. This implies that the prediction of the September slope failure was better than the October slope failure by approximately 1%. However, a moderate-to-good prediction accuracy was obtained with both inventories for spatial agreement.

5. Discussion

In this study, deterministic slope failure hazard analysis was carried out in the Higashifukubegawa catchment. But, this kind of analysis can not be conducted in any of the other test catchments since it is very difficult to obtain the detailed geomechanical and hydrological properties of soils required for the analysis. Therefore, it is necessary to explore for replication methodology that incorporates the deterministic model and can predict the slope failure hazard in catchments of similar geological and geomorphological settings. In this study, we have presented and examined the replication process of a deterministic slope failure hazard model in numerous catchments of the selected study area. The deterministic model prepared for the Higashifukubegawa catchment included an area under ROC as 0.82, which is a good success rate. The results of the deterministic model from Higashifukubegawa were then replicated through discriminant function modelling using DEM-based parameters. In this research, it is assumed that the underlying physical processes associated with hillslope failures would be the same in similar geological/geomorphological settings and if the index properties of the soil do not vary significantly. Under such conditions, hillslopes fail in a similar way or under the same combination of parameters.

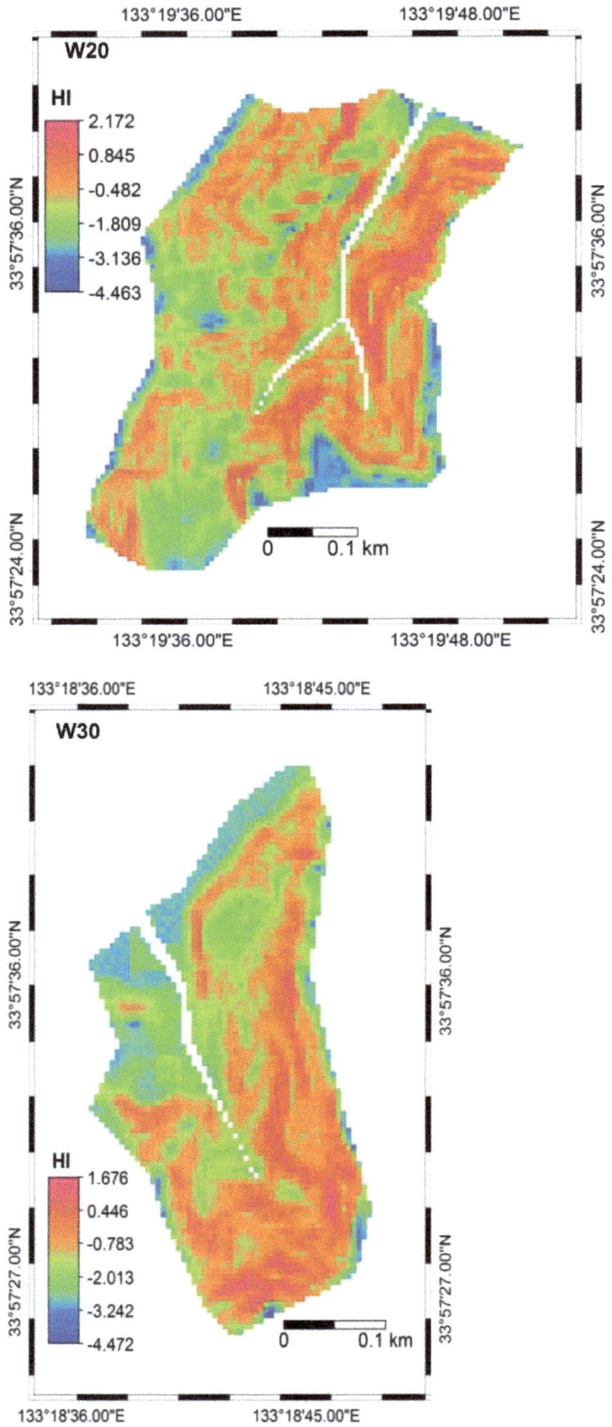

Figure 16. Slope failure hazard index maps of catchments W20 and W30 after application of discriminant function weight values of each parameters from Higashifukubegawa catchment (HI: Hazard Index).

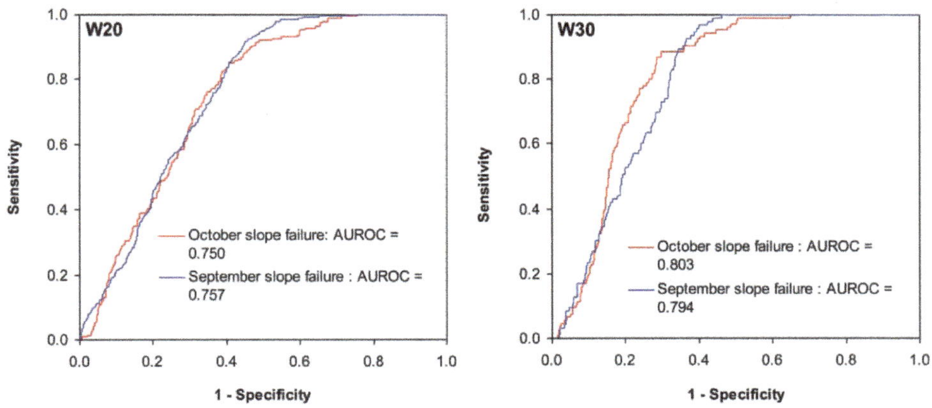

Figure 17. Area under ROC curve estimated from slope failure hazard index score of catchments W20 and W30 after replication of deterministic model.

The discriminant function modelling performed between the deterministic model and a set of seven DEM-based parameters (i.e. slope, relief, plan curvature, profile curvature, wetness index, flow accumulation, and drainage density) in the model catchment resulted in a significant combination of parameters or discriminant function with Wilk's lambda (0.705), the Chi-square value (1932.105), and the significance level of the Chi-square value (0.000). As a result, the model characteristics were transferred into an empirical parametric relationship known as a discriminant function equation. This relationship was applied to 40 test catchments of the study area with the same set of DEM-based parameters and slope failure hazard index maps were obtained. In the results, the prediction accuracies were found to vary over a wide range, from moderate to good with both inventories. In a few catchments, the accuracy was in the same range of model accuracy rate (0.82). From this, it is well understood that if replication of a deterministic model is carried out employing parameters-based multivariate regression modelling, the prediction accuracy can be significantly increased. Replication of slope failure hazard, as in the present study, can be reviewed in a very few catchment-scale studies, such as Ghimire 2011 and Dahal et al. 2012 with a similar reasonable performance. However, this study has first indicated the possibility of the replication of a deterministic slope failure hazard model, although research on the deterministic shallow landslide hazard assessment or physically based modelling of shallow landslides has been practised worldwide for the past few decades (Montgomery and Dietrich 1994; Borga et al. 1998; Moon and Blackstock 2004; Rosso et al. 2006; Meisina and Scarabelli 2007; Dahal et al. 2008b; Godt et al. 2008; Salciarini et al. 2008; Sorbino et al. 2010). The parametric relation obtained in the Higashifukubegawa catchment can be used to predict slope failure hazard conditions in other areas with similar geological/geomorphological characteristics if the DEM-based parameters are available. Further, the replication of a deterministic model could be performed with higher accuracy if enough considerations were given to the quality of the geo-mechanical and hydrological parameters in model catchments as well as the quality of the DEM-based parameters.

Table 5. ROC prediction rates for test catchments.

	Prediction rate	
Catchment ID	October slope failure	September slope failure
W1	0.704	0.743
W2	0.732	0.726
W3	0.731	0.752
W4	0.727	0.740
W5	0.762	0.789
W6	0.751	0.734
W7	0.754	0.736
W8	0.794	0.759
W9	0.790	0.814
W10	0.760	0.814
W11	0.788	0.741
W12	0.761	0.768
W13	0.755	0.719
W14	0.788	0.760
W15	0.773	0.783
W16	0.772	0.788
W17	0.791	0.772
W18	0.741	0.753
W19	0.776	0.780
W20	0.750	0.757
W21	0.805	0.802
W22	0.798	0.784
W23	0.733	0.746
W24	0.763	0.807
W25	0.743	0.730
W26	0.763	0.784
W27	0.734	0.743
W28	0.741	0.789
W29	0.790	0.752
W30	0.803	0.794
W31	0.800	0.796
W32	0.783	0.799
W33	0.727	0.765
W34	0.738	0.773
W35	0.744	0.763
W36	0.735	0.721
W37	0.743	0.735
W38	0.725	0.747
W39	0.783	0.735
W40	0.805	0.770

6. Conclusions

From this work, we conclude with the following statements.

(1) We were successful in connecting hillslope-scale (local-scale) sub-surface hydrology to catchment-scale (landscape-scale) sub-surface hydrology.
(2) We understood that the geo-mechanical and hydrological properties of hillslope materials are the key controlling factors of slope failure hazard in tertiary sedimentary terrain in the Shikoku Region of Japan.

(3) We have obtained 0.719–0.814 and 0.704–0.805 prediction accuracy rates for the spatial prediction of slope failures of the 2004 September and October slope failure inventories, respectively. This reasonably validates the model and replication process.
(4) We have been successful in replicating the model through parameters-based multivariate modelling for the purpose of predicting the slope failure hazard index without using geo-mechanical and hydrological parameters.
(5) Finally, the work in this study indicates that the replication process used here to predict the slope failure hazard condition on a small-catchment scale has a wide applicability in other areas of similar geological and geomorphological characteristics.

Acknowledgements

We are thankful to Mr Masatoshi Anakura and Dr Manita Timilsina (former graduate students of Ehime University) for their help in field/laboratory data preparation.

References

Acharya G, Smedt F De, Long NT. 2006. Assessing landslide hazard in GIS: a case study from Rasuwa, Nepal. Bull Eng Geology Environ. 65:99–107.

Aleotti P, Chowdhury R. 1999. Landslide hazard assessment: summary review and new perspectives. Bull Eng Geology Environ. 58:21–44.

Arnone E, Noto LV, Lepore C, Bras RL. 2011. Physically-based and distributed approach to analyze rainfall-triggered landslides at watershed scale. Geomorphology. 133:121–131.

Baeza C, Corominas J. 2001. Assessment of shallow landslide susceptibility by means of multivariate statistical techniques. Earth Surface Processes Landforms. 26:1251–1263.

Baeza C, Lantada N, Moya J. 2010. Validation and evaluation of two multivariate statistical models for predictive shallow landslide susceptibility mapping of the Eastern Pyrenees (Spain). Environ Earth Sci. 61:507–523.

Barredo J, Benavides A, Hervas J, van Westen CJ. 2000. Comparing heuristic landslide hazard assessment techniques using GIS in the Tirajana basin, Gran Canaria Island, Spain. Int J Appl Earth Observation Geoinformation. 2:9–23.

Bell R. 2007. Lokale und regionale gefahren- und risikoanalyse gravitativer massenbewegungen an der schwäbischen Alb [Local and regional hazard and risk analysis of gravitational mass movements at the Swabian Jura]. Bonn (Germany): University of Bonn German.

Bhandary NP, Dahal RK, Timilsina M, Yatabe R. 2013. Rainfall event-based landslide susceptibility zonation mapping. Nat Hazards. 69:365–388.

Bhandary NP, Yatabe R. 2005. Influence of precipitation pattern on triggering surface-layer failures of slopes during 2004 Niihama disasters. In: Proceedings of the One-day Symposium on Geo-disasters and Geo-environment, Ehime (Japan): Ehime University; p. 63–70.

Bijukchhen SM, Kayastha P, Dhital MR. 2013. A comparative evaluation of heuristic and bivariate statistical modelling for landslide susceptibility mappings in Ghurmi–Dhad Khola, East Nepal. Arab J Geosciences. 6(8):2727–2743.

Borga M, Fontana GD, Ros DD, Marchi L. 1998. Shallow landslide hazard assessment using a physically based model and digital elevation data. J Environ Geol. 35:81–88.

Burrough PA, McDonnell RA. 1998. Principles of geographical information systems. Oxford: Oxford University Press.

Cannon SH, Gartner JE, Rupert MG, Michael JA. 2004. Emergency assessment of debris-flow hazards from basins burned by the Padua fire of 2003, southern, California (US

Geological Survey Open-File Report 2004–1072). Reston (VA): US Geological Survey; p. 14.
Carrara A, Crosta GB, Frattini P. 2003. Geomorphological and historical data in assessing landslide hazard. Earth Surf Process Landf. 28:1125–1142.
Carrara A, Guzzetti F, Cardinali M, Reichenbach P. 1999. Use of GIS technology in the prediction and monitoring of landslide hazard. Nat Hazards. 20:117–135.
Cascini L, Cuomo S, Pastor M, Sorbino G. 2010. Modeling of rainfall-induced shallow landslides of the flow-type. J Geotech Geoenviron Eng. 136(1):85–98.
Ching J, Liao HJ, Lee JY. 2011. Predicting rainfall-induced landslide potential along a mountain road in Taiwan. Geotechnique. 61:153–166.
Crosta G. 1998. Regionalization of rainfall thresholds: an aid to landslide hazard evaluation. Environ Geol. 35:131–145.
Crozier MJ. 1999. Prediction of rainfall-triggered landslides: a test of the Antecedent Water Status Model. Earth Surf Process Landf. 24(9):825–833.
Dahal RK, Hasegawa S, Bhandary NP, Poudel PP, Nonomura A, Yatabe R. 2012. A replication of landslide hazard mapping at catchment scale. Geomatics, Nat Hazards Risk. 3: 161–192.
Dahal RK, Hasegawa S, Nonomura A, Yamanaka M, Dhakal S. 2008b. DEM-based deterministic landslide hazard analysis in the Lesser Himalaya of Nepal. Georisk: Assess Manag Risk Eng Syst Geohazards. 2:161–178.
Dahal RK, Hasegawa S, Nonomura A, Yamanaka Y, Dhakal S, Paudyal P. 2008a. Predictive modelling of rainfall-induced landslide hazard in the Lesser Himalaya of Nepal based on weights-of-evidence. Geomorphology. 102:496–510.
Dahal RK, Hasegawa S, Yamanaka M, Bhandary NP, Yatabe R. 2011. Rainfall-induced landslides in the residual soil of andesitic terrain, western Japan. J Nepal Geol Soc. 42: 127–142.
Dahal RK, Hasegawa S, Yamanaka M, Nishino K. 2006. Rainfall triggered flow-like landslides: understanding from southern hills of Kathmandu, Nepal and northern Shikoku, Japan. In: Proceedings of the 10th Int Congr of IAEG; 2006 Sep 6–10; Nottingham, UK. London: The Geological Society of London. IAEG2006 Paper no. 819, p. 1–14. (CD-ROM).
Dai FC, Lee CF, Ngai YY. 2002. Landslide risk assessment and management: an overview. Eng Geol. 64(1):65–87.
Daitou Techno Green. 2009. Hasegawa in situ permeability tester, operating manual. v. 3.1. Available from: http://www.daitoutg.co.jp/prd/pdf/tousui0703.pdf
D'Odorico P, Fagherazzi S. 2003. A probabilistic model of rainfall-triggered shallow landslides in hollows: a long-term analysis. Water Resour Res. 39:1262. doi:10.1029/2002WR001595.
Dong JJ, Tung YH, Chen CC, Liao JJ, Pan YW. 2009. Discriminant analysis of the geomorphic characteristics and stability of landslide dams. Geomorphology. 110:162–171.
Eeckhaut MVD, Reichenbach P, Guzzetti F, Rossi M, Poesen J. 2009. Combined landslide inventory and susceptibility assessment based on different mapping units: an example from the Flemish Ardennes, Belgium. Nat Hazards Earth Syst Sci. 9:507–521.
Gasmo JM, Rahardjo H, Leong EC. 2000. Infiltration effects on stability of a residual soil slope. Comput Geotechnics. 26:145–165
GeoStudio. 2005. GeoStudio Tutorials includes student edition lessons. 1st ed. Calgary (Canada): Geo-Slope International.
Ghimire M. 2011. Landslide occurrence and its relation with terrain factors in the Siwalik Hills, Nepal: case study of susceptibility assessment in three basins. Nat Hazards. 56:299–320.

Ghosh S, van Westen CJ, Carranza EJM, Jetten VG, Cardinali M, Rossi M, Guzzetti F. 2012. Generating event-based landslide maps in a data-scarce Himalayan environment for estimating temporal and magnitude probabilities. Eng Geol. 128:49–62.

Godt JW, Baum RL, Savage WZ, Salciarini D, Schulz WH, Harp EL. 2008. Transient deterministic shallow landslide modeling: requirements for susceptibility and hazard assessments in a GIS framework. Eng Geol. 102:214–226.

Guzzetti F, Reichenbach P, Ardizzone F, Cardinali M, Galli M. (2006). Estimating the quality of landslide susceptibility models. Geomorphology. 81:166–184.

Guzzetti F, Reichenbach P, Cardinali M, Galli M, Ardizzone F. 2005. Probabilistic landslide hazard assessment at the basin scale. Geomorphology. 72:272–299.

Hadmoko DS, Lavigne F, Sartohadi J, Hadi P, Winaryo. 2010. Landslide hazard and risk assessment and their application in risk management and landuse planning in eastern flank of Menoreh Mountains, Yogyakarta Province, Indonesia. Nat Hazards. 54:623–642.

Hammond C, Hall D, Miller S, Swetik P. 1992. Level I Stability Analysis (LISA), documentation for Versito 2.0, Gen. Tech. Rep. INT-285. Ogden (UT): Department of Agriculture, Forest Service, Intermountain Research Station; p. 190.

Harp EL, Reid ME, McKenna JP, Michael JA. 2009. Mapping of hazard from rainfall-triggered landslides in developing countries: examples from Honduras and Micronesia. Eng Geol. 104(3–4):295–311.

Harris SJ, Orense RP, Itoh K. 2012. Back analyses of rainfall-induced slope failure in Northland Allochthon formation. Landslides. 9:349–356.

Hashimoto M, editor. 1991. Geology of Japan. Tokyo: Terra Sci. Pub.; p. 249.

He S, Pan P, Dai L, Wang H, Liu J. 2012. Application of kernel-based Fisher discriminant analysis to map landslide susceptibility in the Qinggan River delta, Three Gorges, China. Geomorphology. 171–172:30–41.

Iverson RM, Reid ME, LaHusen RG. 1997. Debris-flow mobilization from landslides. Annu Rev Earth Planetary Sci. 25:85–138.

Jagielko L, Martin YE, Sjogren DB. 2012. Scaling and multivariate analysis of medium to large landslide events: Haida Gwaii, British Columbia. Nat Hazards. 60:321–344.

Jamaludin S, Huat BBK, Omar H. 2006. Evaluation of slope assessment systems for predicting landslides of cut slopes in granitic and meta-sediment formations. American J Environ Sci. 2(4):135–141.

Jia N, Mitani Y, Xie M, Djamaluddin I. 2012. Shallow landslide hazard assessment using a three-dimensional deterministic model in a mountainous area. Comput Geotechnics. 45:1–10.

Jorge LAB. 2009. Soil erosion fragility assessment using an impact model and geographic information system. Sci Agric. (Piracicaba, Braz.). 66:658–666.

Lee CT, Huang CC, Lee JF, Pan KL, Kin ML, Dong JJ. 2008. Statistical approach to earthquake induced landslide susceptibility. Eng Geol. 100:43–58.

Meisina C, Scarabelli S. 2007. A comparative analysis of terrain stability models for predicting shallow landslides in colluvial soils. Geomorphology. 87:207–223.

Montgomery DR, Dietrich WE. 1994. A physically based model for the topographic control on shallow landsliding. Water Resour Res. 30:1153–1171.

Moon V, Blackstock H. 2004. A Methodology for assessing landslide hazard using deterministic stability models. Nat Hazards. 32(1):111–134.

Nagarajan R, Roy A, Vinod Kumar R, Mukherjee A, Khire MV. 2000. Landslide hazard susceptibility mapping based on terrain and climatic factors for tropical monsoon regions. Bull Eng Geol Environ. 58:275–287.

Neupane RR. 2005. Experimental analysis of strength of local grass and shrub roots used for slope stabilization, Thesis No. 059/MSW/418. Nepal: Master of Science in Water Resources Engineering, Institute of Engineering, Tribhuvan University; p. 85.

Ng CWW, Shi Q. 1998. A numerical investigation of the stability of unsaturated soil slopes subjected to transient seepage. Comput Geotechnics. 22:1–28.

Pavel M, Fannin RJ, Nelson JD. 2008. Replication of a terrain stability mapping using an artificial neural network. Geomorphology. 97:356–373.

Pradhan B, Oh H-J, Buchroithner M. 2010, Weights-of-evidence model applied to landslide susceptibility mapping in a tropical hilly area. Geomatics Nat Hazards Risk. 1:199–223.

Reid ME, Iverson RM. 1992. Gravity-driven groundwater flow and slope failure potential: 2. Effects of slope morphology, material properties, and hydraulic heterogeneity. Water Resour Res. 28:939–950.

Ross SM. 2004. Introduction to probability and statistics for engineers and scientists. 3rd ed. Delhi (India): Academic Press (an imprint of Elsevier).

Rosso R, Rulli MC, Vannucchi G. 2006. A physically based model for the hydrologic control on shallow landsliding. Water Resour Res. 42:W06410. doi: 10.1029/2005WR004369.

Saha AK, Gupta RP, Arora MK. 2002. GIS-based landslide hazard zonation in the Bhagirathi (Ganga) valley, Himalayas. I J Remote Sens. 23(2):357–369.

Salciarini D, Godt JW, Savage WZ, Baum RL, Conversini P. 2008. Modeling landslide recurrence in Seattle, Washington, USA. Eng Geol. 102:227–237.

Segoni S, Rossi G, Catan F. 2012. Improving basin scale shallow landslide modelling using reliable soil thickness maps. Nat Hazards. 61:85–101.

Sidle RC. 1991. A conceptual model of changes in root cohesion in response to vegetation management. J Environ Qual. 20:43–52.

Soeters R, van Westen CJ. 1996. Slope instability recognition, analysis, and zonation. In: Turner AK, Schuster RL, editors, Landslides — investigation and mitigation. Transp. Res. Board Special Rep. Vol. 247. Washington DC: National Academy Press; p. 129–177.

Sorbino G, Sica C, Cascini L. 2010. Susceptibility analysis of shallow landslides source areas using physically based models. Nat Hazards. 53:313–332.

Talebi A, Uijlenhoet R, Troch PA. 2008. Application of a probabilistic model of rainfall-induced shallow landslides to complex hollows. Nat Hazards Earth Syst Sci. 8:733–744.

Tarolli P, Borga M, Chang KT, Chiang SH. 2011. Modeling shallow landsliding susceptibility by incorporating heavy rainfall statistical properties. Geomorphology. 133:199–211.

Taylor DW. 1948. Fundamentals of soil mechanics. New York (NY): Wiley.

Terlien MTJ. 1996. Modelling spatial and temporal variations in rainfall-triggered landslides. [dissertation]. Enschede (The Netherlands): ITC Publication.

Terlien MTJ, Van Westen CJ, Van Asch TW. 1995. Deterministic modelling in GIS-based landslide hazard assessment. In: Carrara A, Guzzetti F, editors. Advances in natural and technological hazard research. Dordrecht: Kluwer Academic Publishers; p. 51–77.

Terzaghi K, Peck RB. 1967. Soil mechanics in engineering practice. Hoboken (NJ): Wiley Intersci.

Trandafir AC, Sidle RC, Gomi T, Kamai T. 2008. Monitored and simulated variations in matric suction during rainfall in a residual soil slope. Environ Geol. 55:951–961.

Van Asch Th.WJ, Buma J, Van Beek LPH. 1999. A view on some hydrological triggering systems in landslides. Geomorphology. 30:25–32.

Van Den Eeckhaut M, Vanwalleghem T, Poesen J, Govers G, Verstraeten G, Vanderkerckhove L. 2006. Prediction of landslide susceptibility using rare events logistic regression: a case-study in Flemish Ardennes (Belgium). Geomorphology. 76:392–410.

van Westen CJ, Terlien TJ. 1996. An approach towards deterministic landslide hazard analysis in GIS. A case study from Manizales (Colombia). Earth Surf Process Landf. 21:853–868.

Wang X, Zhang L, Wang S, Lari S. 2013. Regional landslide susceptibility zoning with considering the aggregation of landslide points and the weights of factors. Landslides. doi: 10.1007/s10346-013-0392-6.

Ward TJ, Li RM, Simons DB. 1981. Use of a mathematical model for estimating potential landslide sites in steep forested basin. In: Davis TRH, Pearce AJ, editors. Erosion and sediment transport in Pacific Rim steep lands. International Hydrological Science Publ No. 132. Wallingford, (Oxfordshire): Institute of Hydrology; p. 21–41.

Wieczorek GF, Glade T. 2005. Climatic factors influencing occurrence of debris flows. In: Jakob M, Hungr O, editors. Debris-flow hazards and related phenomena. Berlin: Springer; p. 325–362.

Wu W, Sidle RC. 1995. A distributed slope stability model for steep forested basins. Water Resour Res. 31:2097–2110.

Yesilnacar E, Topal T. 2005. Landslide susceptibility mapping: a comparison of logistic regression and neural networks methods in a medium scale study, Hendek region (Turkey). Eng Geol. 79:251–266.

Zolfaghari A, Heath AC. 2008. A GIS application for assessing landslide hazard over a large area. Comput Geotechnics. 35:278–285.

Zweig MH, Campbell G. 1993. Receiver-operating characteristics (ROC) plots. Clin Chem. 39:561–577.

5

Experimental assessment of post-processed kinematic Precise Point Positioning method for structural health monitoring

CEMAL OZER YIGIT*

Gebze Institute of Technology, Department of Geodetic and Photogrammetric Engineering, Gebze 41400, Kocaeli, Turkey

Monitoring the response of engineering structures, such as tall buildings, tower and large-scale bridges, under severe loading conditions, such as strong earthquake or wind storm, is an important requirement to verify their design and construction and to evaluate structural condition and reliability. In the last two decades, high-rate real-time or post-processed kinematic differential Global Positioning System (DGPS) has been widely used in dynamic displacement measurements of civil engineering structures. In recent years, interest has increased for Precise Point Positioning (PPP) due to its capability to generate positioning solutions as accurate as DGPS. In this study, the potential of post-processed kinematic PPP in terms of monitoring dynamic displacement response of a structure has been explored based on free damped oscillation events obtained from a model structure, which is able to vibrate in the fundamental and higher modes of vibration. A number of experiments have been carried out and five events, each of which is different character, have been selected to compare PPP results with DPGS results in the time and frequency domain. The results clearly demonstrate that the PPP method, like the DGPS method, offers great potential for the measurement of horizontal and vertical dynamic movement of structures. The impact of a short period (one minute) of observation length on the result of the kinematic PPP method was also investigated in terms of sensing the dynamic movement of a structure. Twenty selected one-minute data-sets extracted from a one-hour original data-set were processed by Canadian spatial reference system PPP and each one-minute PPP solution was compared with the corresponding segment obtained from the one-hour PPP solution. The results show that the one-minute PPP solution is able to extract the fundamental natural frequency of the oscillation in the horizontal and vertical component just like the one-hour PPP solution after the offset is removed and the lower frequency trend component is filtered out.

1. Introduction

High-rate Global Positioning System (GPS) positioning has been recognized as a powerful tool in measuring dynamic displacements of engineering structures under wind loading or earthquake loading. Therefore, the GPS observations have been widely used in structural health monitoring (SHM) projects over the past several decades to monitor structural dynamic deformations with a precision of the order of millimetres. Nowadays, relative displacements can be

*Email:cyigit@gyte.edu.tr

measured at rates of 10–50 Hz and even higher up to 100 Hz by using ultra-high rate GPS receivers. The accuracy of the dynamic displacement measurement using the DGPS is at sub-centimetre level. The differential real-time or post-processed kinematic positioning method that requires the combination of observations from a minimum of two Global Navigation Satellite Systems (GNSS) receivers, with at least one occupying a station with known coordinates, is widely used to measure relative displacements.

Until recently, numerous studies have been conducted concerning the technical feasibility of using DGPS to measure dynamic displacements of high-rise buildings and tall slender structures (Çelebi 2000; Tamura et al. 2002; Kijewski-Correa et al. 2006; Li et al. 2006; Breuer et al. 2008; Park et al. 2008; Yigit et al. 2010; Pehlivan et al. 2013; Yi et al. 2013), long-span suspension and cable-stayed bridges (Nakamura 2000; Xu et al. 2002; Roberts et al. 2004; Erdoğan et al. 2007; Meng et al. 2007; Yi et al. 2013a), and short-span bridges (Moschas & Stiros 2011, 2014; Psimoulis & Stiros 2013). Literature reviews on the application of GPS-based monitoring for high-rise structures and for SHM are available (Im et al. 2013; Yi et al. 2013b). In addition to the full-scale monitoring studies, there is a number of experimental studies for assessing the accuracy and dynamic performance of DGPS in terms of SHM that include but are not limited to those by Chan et al. (2006), Nickitopoulou et al. (2006), Psimoulis et al. (2008), Psimoulis and Stiros (2008), and Wang et al. (2011).

In recent years, interest has increased for PPP since it is quite attractive both in terms of performance and ease of use. PPP (Zumberge et al. 1997; Kouba & Heroux 2001) is a powerful positioning method aimed at the processing of measurements from a single GPS/GNSS receiver to compute high-accurate position in a global reference frame. PPP is becoming a cost-effective technique to achieve centimetre-level accuracy (Li et al. 2013a).

PPP has been demonstrated to be a powerful and efficient method for many scientific research and for civilian applications (Li & Zhang 2013), such as crustal deformation monitoring (Savage et al. 2004; Hammond & Thatcher 2005; Calais et al. 2006; Ohta et al. 2008), GPS seismology (Kouba 2003; Avallone et al. 2011), earthquake early warning system (Li et al. 2013b), GPS buoy (Kuo et al. 2012; Fund et al. 2013), precise orbit determination of low Earth orbiting satellites (Bock et al. 2003) and the precise positioning of mobile objects in land, air and marine (Gao et al. 2005; Zhang & Andersen 2006; Geng et al. 2010; Alkan & Öcalan 2013).

Kouba (2003) first applied PPP to process the high-rate GPS data for the 2002 Denali Fault earthquake and demonstrated that PPP method using the clock and orbit product of the International Global Navigation Satellite Systems Service (IGS) can be used for detecting seismic waves and satisfy the requirement of the GPS seismology. Recently, Shi et al. (2010) evaluated the seismic deformation of the Mw 8.0 Wenchuan earthquake based on station displacement derived from 1 Hz GPS data collected at a set of station in China by using PPP approach. They investigated the impact of integer ambiguity resolution and station environment-dependent effect to yield more accurate positioning results. Their results demonstrated that GPS can sense the rapid position oscillation of about 2 cm in amplitude in the case of the position accuracy of horizontal components of better than 1 cm.

Most recently, Xu et al. (2013) evaluated the accuracy of high-rate PPP for measuring seismic wave motion generated by the shake table allowing motion in any of six degrees of freedom. They compared PPP results with inertial measurement unit and reported that high-rate PPP can produce absolute horizontal displacement waveforms at the accuracy of 2–4 mm and absolute vertical displacement waveforms at the sub-centimetre level of accuracy within a short period of time. In addition, Moschas et al. (2013) tested the potential of PPP based on two events for monitoring horizontal semi-static and harmonic axial oscillations with frequency of 3.3 Hz. They also compared the PPP results, obtained from two softwares, with accelerometer, robotic total station and DGPS results. According to their findings, PPP can well constrain the time history of dynamic and semi-static displacements of the order of 1 or 2 cm in the horizontal component.

This paper assesses the potential of using post-processed kinematic PPP (hereafter referred to simply as kinematic PPP) for SHM on the basis of the five different events selected from a number of experiments with respect to their amplitude and frequency characteristic. PPP results from each event compared to the corresponding post-processed kinematic DGPS (hereafter referred to as kinematic DGPS) results in the time and frequency domain. The effect of short-duration (one minute) GPS records on the results of kinematic PPP for SHM applications is also investigated. Twenty one-minute PPP results were compared with corresponding one-hour PPP results.

The experimental model structure used in this study is similar to that of Çelebi and Şanli (2002), which investigated the feasibility of using DGPS technology to monitor tall buildings (and other long-period structures) in horizontal component. However, the experimental procedure used in the present study differs from previous ones in two aspects.

First, the model structure can oscillate at not only first mode natural frequency but also second mode natural frequency, meaning that it is able to generate two sinusoidal free damped oscillations with different frequencies and amplitudes. The first and second mode natural frequencies of the model structure are around 0.39 and 4.33 Hz, respectively, which covers the main modal frequencies of the majority of civil engineering structures (Psimoulis & Stiros 2008).

Second, not only the horizontal component but also the vertical component are taken into account to evaluate the dynamic measurement quality and reliability of PPP method for long (or short)-span and cable-stayed bridges monitoring applications.

The rest of the paper is organized as follows: the principle of the post-processed PPP technique is introduced in Section 2. In section 2, Web-based PPP engines and online Canadian spatial reference system Precise Point Positioning (CSRS-PPP) software, which is used to process GPS data of the current study, are briefly presented. The experimental implementation is explained in detail in Section 3. Section 4 deals with preparatory data processing regarding coordinate transformations. The experimental results are presented and discussed in Section 5. Finally, the conclusions are drawn in Section 6.

2. Post-processed PPP technique

2.1. PPP principles

Compared to well-established DGPS techniques, PPP has certain advantages that make it a viable technique for a broad range of positioning applications. Its most important feature is the high positional accuracy obtained using only a single receiver at the user location. Also, PPP provides a positioning solution in a global reference system, the same one as the satellite orbits. Such a reference system is free of local datum inconsistencies associated with differential positioning that relies on local datum coordinates.

However, despite these significant assets of the method, PPP still suffers a number of deficiencies. First, because it relies on undifferenced code and carrier phase measurements, major error sources associated with satellite clock and orbital parameters cannot be cancelled out (not even mitigated) as is the case for differential positioning. Instead, in order to overcome this limitation, PPP uses precise satellite clock corrections and orbits (Gao & Kongzhe 2004; Hofmann-Wellenhof et al. 2008; Pengfei et al. 2011). Such information is currently available from organizations, such as the IGS that provides freely ultra-rapid precise ephemerides tied directly to the International Terrestrial Reference System (ITRF). Second, conventional PPP has involved the use of ambiguity float solutions which require long observation time to converge to centimetre accuracy. This has a minimal effect for static applications; howbeit, it is a major limitation for real-time applications (Retscher 2002). Although the fixing of integer ambiguities is a particularly difficult challenge for the undifferenced measurement processing used by PPP (Rizos et al. 2012), three PPP integer ambiguity resolution methods, which is single difference between satellites method, the integer phase clock model and the decoupled clock model, so far have been developed by GFZ (GeoForschungsZentrum Postdam), CNES (Centre National d'Etudes Spatiales) and NRCan, respectively (Shi & Gao 2014). Another major concern in PPP relates to ionospheric delays that, as opposed to DGNSS, cannot be assumed to cancel out. Using precise orbits and clock corrections, the first-order ionosphere-free code and phase combinations for dual frequency receivers are as follows (Hofmann-Wellenhof et al. 2008):

$$\frac{R_1 f_1^2}{f_1^2 - f_2^2} - \frac{R_2 f_2^2}{f_1^2 - f_2^2} = \rho + c\, dt_r + d_{\text{trop}} + \varepsilon_R \quad (1)$$

$$\frac{\lambda_1\, \Phi_1 f_1^2}{f_1^2 - f_2^2} - \frac{\lambda_2\, \Phi_2\, f_2^2}{f_1^2 - f_2^2} = \rho + c\, dt_r + d_{\text{trop}} + \frac{\lambda_1 N_1 f_1^2}{f_1^2 - f_2^2} - \frac{\lambda_2 N_2 f_2^2}{f_1^2 - f_2^2} + \varepsilon_\Phi \quad (2)$$

where R_1, R_2 and Φ_1, Φ_2 are the measured pseudoranges and carrier phases, respectively; ρ is the true geometric range; f_1, f_2 and λ_1 N_1, λ_2 N_2 stand for the frequencies and phase ambiguity terms of the carrier waves; c stands for the speed of light in vacuum; dt_r and d_{trop} denote the receiver clock error and tropospheric delay; and finally, ε_R , ε_Φ are measurement noise terms. Notwithstanding these formulations increase redundancy and aid to the estimation process, they only represent a first-order ionosphere-free linear combination that cannot fully mitigate ionospheric delays (El-Mowafy 2011).

Finally, but not least, for high accuracy static applications, advanced modelling is required to accommodate for global effects, such as tectonic plate motion, solid earth tide displacements, polar tides and ocean loading. Additional corrections should also be applied to account for satellite and receiver antenna phase wind-up effects. A comparison between the measurement biases and errors that need to be applied or accounted for in typical PPP and DGPS positioning techniques can be found in Rizos et al. (2012).

2.2. *Web-based PPP processing engines*

Today, a number of online and free PPP post-processed services are available over the Internet. Most of them are developed and operated by top-class University/institutions and leading research centres worldwide, such as the Jet Propulsion Laboratory (JPL) at the National Aeronautics and Space Administration (NASA), the Geodetic Survey Division (GSD) of Natural Resources Canada (NRCan), the University of New Brunswick (UNB) and the GMV Aerospace and Defense business group. The use of these services presupposes the end user to upload GNSS (RINEX) observation files via the Internet. Depending on the quality and length of the data provided, the service generates undifferenced, float solutions made available to the end user free of charge (El-Mowafy 2011; Ghoddousi-Fard & Dare 2006). A brief overview of the main online PPP services and associated software is as follows:

APPS (automatic precise positioning service) is a Web-based PPP service developed by JPL-NASA (http://www.apps.gdgps.net). It relies on the GIPSY-OASIS software for processing static and kinematic GPS measurements and supports inputs in RINEX format ver. 2.10 and 2.11 as well as GIPSY TDP type files.

GAPS (GPS analysis and positioning software) is a GPS positioning and analysis software developed at the University of New Brunswick (UNB), Canada. This service provides online PPP services and is available through the UNB website at http://www.gaps.gge.unb.ca/gaps_summary.txt. Detailed information about this system can be found in Leandro et al. (2008).

magicGNSS, introduced by GMV Aerospace and Defence, is a Web-based application (http://www.magicgnss.gmv.com/) that provides a suite of tools for GNSS data processing. Unlike APPS and GAPS services, it also processes GLONASS observables after 1 January 2010 (Píriz et al. 2009).

CSRS-PPP service is developed and operated by the Geodetic Survey Division of the Natural Resources Canada (NRCan-GSD). It is available via the GSD web page (http://webapp.geod.nrcan.gc.ca/geod/tools-outils/ppp.php). In this study, data processing relies on CSRS-PPP software. Therefore, further details on this service are given in following section.

2.3. *Online global GPS processing service CSRS-PPP*

NRCan's PPP is a free online post-processing service that allows GPS users all over the world to compute better-accuracy positions from their GPS raw observation data. PPP can process raw GPS data from single or dual-frequency receivers in either static mode or kinematic mode (epoch-by-epoch). It can be also used to process both static and kinematic GNSS data (GLONASS starting on week 1650) in order to obtain receiver coordinates. Although CSRS-PPP was based entirely on IGS products when first launched in 2003, it now makes use of a mixture of IGS and NRCan products. NRCan products, computed using data from the IGS global tracking network, are needed to provide a rapid access to 30-second GPS and GLONASS clock products (Donahue et al. 2012). After submitting raw data online, users receive via email corrected coordinates in either the NAD83 or the ITRF reference system. There is no minimum length for a GPS observation session used in CSRS-PPP (El-Mowafy 2011)

Parameters estimated in PPP are station positions (in static or kinematic mode), station clock states, local troposphere zenith delays and carrier phase ambiguities.

The best position solution accuracies, reaching the few centimetres in horizontal coordinates and less than 10 cm in vertical coordinates (RMS), are obtained by processing GPS dual-frequency pseudorange and carrier phase observations with IGS precise satellite orbit and clock products. NRCan PPP can achieve these using accurate models for all the physical phenomena involved. Further details on the PPP algorithms, models and specifications can be found in Kouba and Héroux (2001).

PPP generates two forms of output; both short and detailed summary reports and graphical time series plots. The CSRS-PPP summary report contains a header record followed by three sections. Section 1 is a file summary that specifies the input, output and internal filenames used during processing. Section 2 gives the processing parameters extracted from the internal files (maintained centrally at the CSRS-PPP site). It reports (1) cycle slip filter thresholds, (2) satellite antenna phase centre offsets, (3) user antenna phase centre offsets, (4) reference frame transformation parameters between ITRF and NAD83-CSRS, (5) ocean loading coefficients and (6) surface meteorological data. Section 3 is a session processing summary that reports (1) processing options, (2) observation session, (3) coordinate estimates, (4) coordinate differences, (5) receiver clock estimates and (6) observation residuals.

3. Description of high-rate GPS experiment

To evaluate the accuracy of high-rate PPP and to compare its results with DGPS, we performed a number of tests using a model structure and two dual-frequency geodetic-type GPS antenna/receivers. Figure 1 depicts the overall set-up. A slender steel flat bar is selected to simulate a tall, flexible building. The length, width and thickness of the bar are 205, 3.95 and 0.55 cm, respectively. The bar is able to oscillate in one

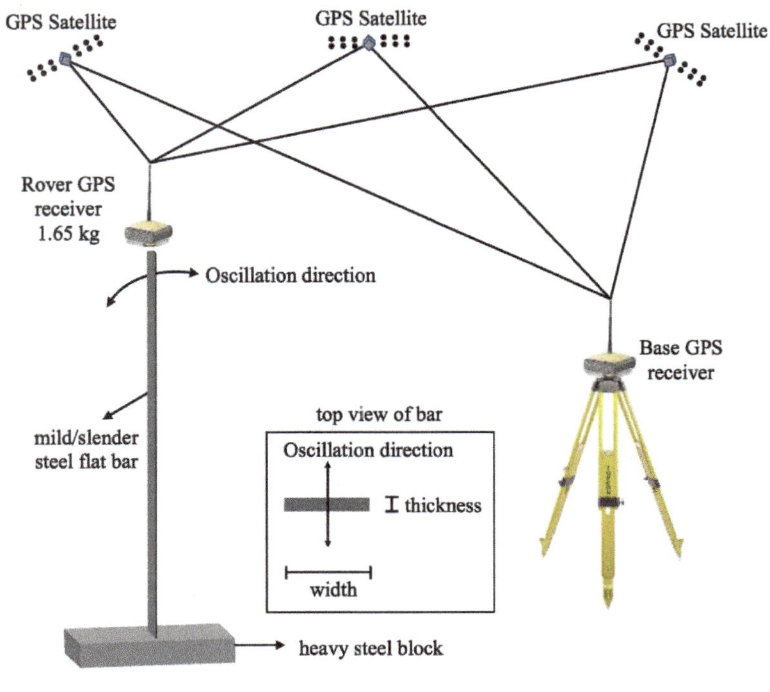

Figure 1. Schematic view of experimental set-up.

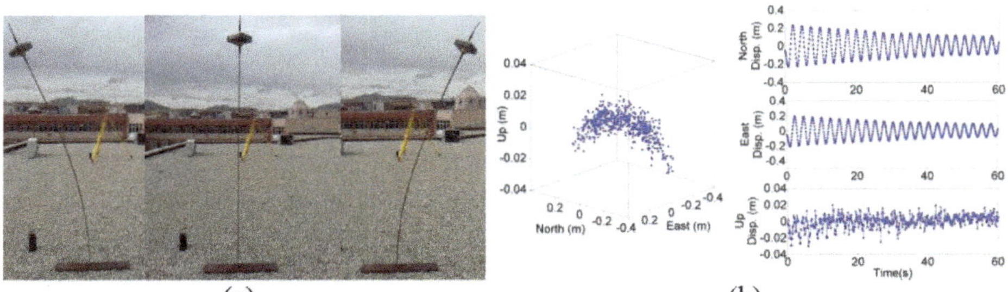

(a) (b)

Figure 2. (a) Photographs taken during one of the oscillation events performed: the photo in the middle shows the inactive (motionless) position. The photos on the left and the right show the maximum displacement positions while the bar is going back and forth. (b) 3D plot obtained from PPP processing based on the data of one of the oscillation events and their time series in the north, east and up component. Note that the vertical axis limit of the plots is 10 times smaller than the horizontal axis limit in order to realize vertical oscillation, which depends on magnitude of horizontal displacement due to bending nature of bar.

direction only (see figure 1), meaning that the perpendicular direction to oscillation direction is motionless. The bar was fixed at a heavy steel block to prevent it from slipping during the experiment and the GPS antenna/receiver was attached at its top. The reference GPS receiver was set up 10 m away from the rover GPS receiver since the accuracy of coordinate estimate based on kinematic DGPS method depends on inter-station baseline lengths. Therefore, most of the GPS errors are eliminated and the coordinates of rover can be determined very precisely. Figure 2 shows three photographs taken during the experiment and three-dimensional (3D) plot and time series plot in the east, north and up component. The first three vibration mode shapes of a slender bar clamped at one end are depicted in figure 3.

DGPS positioning methods require at least one or more reference station with precisely known coordinates. PPP provides a positioning solution in a global reference frame such as the International Terrestrial Reference Frame (ITRF) while differential positioning techniques is based on nearby reference stations with coordinates expressed in a national or global datum. In this study, in order to obtain the coordinates in the same datum from DGPS and PPP techniques, static GPS observations at the reference station were collected 24-hour at 30-second interval using an

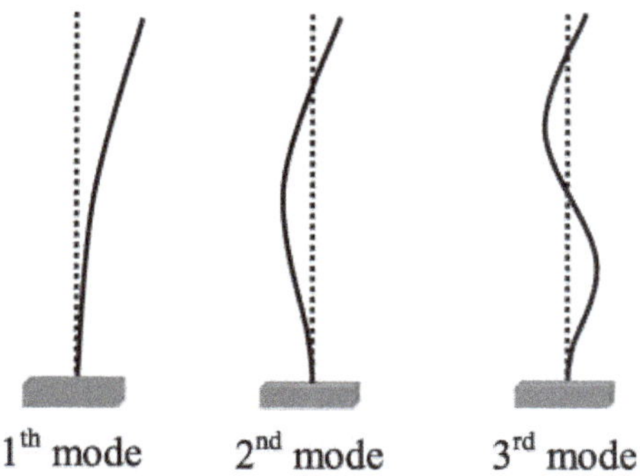

Figure 3. The first three vibration mode shapes of a slender bar clamped at one end.

elevation mask of 10° the day before the experiment was started. And then, the coordinates of reference station being used in differential kinematic processing were estimated based on the daily GPS data by using the CSRS-PPP in static mode. Therefore, the coordinates estimated from the two techniques will be based on the same coordinate system (e.g., ITRF datum), enabling us to compare their results.

The bar experiment were conducted on the roof of engineering faculty of the Selcuk University in March 2010 and lasted 26 minutes. The GPS data were collected using Topcon™ Hiper Pro in kinematic survey mode and recorded at 10 Hz sampling rate. The advertised positional accuracy according to the manufacturer was 10 mm + 1 ppm (x baseline length) horizontally and 15 mm + 1 ppm (x baseline length) vertically in kinematic survey mode. The satellite elevation cut-off angle was 10°. The experiment was conducted under open sky conditions and minimum multipath environment during calm weather in order to prevent the experimental data from external disturbance. Eight satellites were visible during the experiment.

The GPS data from the reference and rover antenna/receiver on the bar were processed in the differential kinematic mode by using the Leica Geo Office 3.0 software. In other words, the coordinates of the rover antenna/receiver were estimated in a post-processing differential kinematic mode, based on the reference station. The GPS data from the rover antenna/receiver on the bar were processed in the kinematic PPP mode by using the online software CSRS-PPP.

4. Preparatory data processing

4.1. Geocentric to topocentric coordinate transformation

The coordinates of the GPS receiver are computed in the ITRF, the earth-centred earth-fixed Cartesian coordinate system. However, in SHM, direct use of the geocentric Cartesian coordinate is not suitable for evaluating both the horizontal and the vertical displacement response of the structure due to its nature. To overcome this problem, geocentric coordinates XYZ must be transformed to a topocentric coordinate system ENU (east, north, up), also referred to as local 3D Cartesian system.

Given the geocentric coordinates of the topocentric origin (X_0, Y_0, Z_0), geocentric coordinates (X_t, Y_t, Z_t) at the time t can be converted into topocentric coordinates (E_t, N_t, U_t) by using the following formula:

$$\begin{bmatrix} E_t \\ N_t \\ U_t \end{bmatrix} = \begin{bmatrix} -\sin(\lambda_0) & \cos(\lambda_0) & 0 \\ -\sin(\phi_0)\cos(\lambda_0) & -\sin(\phi_0)\sin(\lambda_0) & \cos(\phi_0) \\ \cos(\phi_0)\cos(\lambda_0) & \cos(\phi_0)\sin(\lambda_0) & \sin(\phi_0) \end{bmatrix} \begin{bmatrix} X_t - X_0 \\ Y_t - Y_0 \\ Z_t - Z_0 \end{bmatrix} \tag{3}$$

where ϕ_0 and λ_0 is geodetic latitude and longitude of the topocentric origin, respectively, which is derived from their geocentric coordinates (X_0, Y_0, Z_0).

In this study, the origin of the topocentric coordinate system is selected by the geometric centre of the bar with motionless position, whose geocentric position was estimated based on the daily GPS data by using the CSRS-PPP in static mode. The GPS measurements were carried out simultaneously with the reference station the day before the experiment was started.

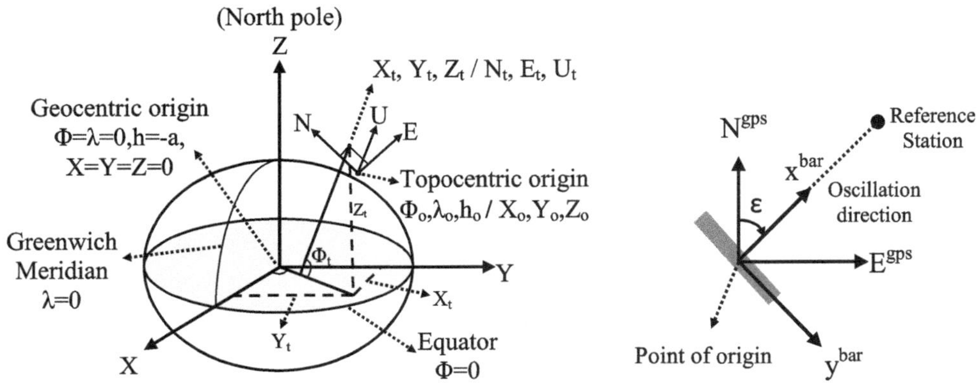

Figure 4. Geocentric and topocentric systems (left) and the GPS and Bar coordinate axes orientation with respect to the bar cross-section (right).

4.2. 2D similarity transformation

Since *N* and *E* axes of the topocentric system pointed to the north and east directions, respectively, the GPS-derived displacement of the bar in horizontal plane will be expressed in the *N* and *E* axes, depending on the orientation of oscillation direction with respect to the topocentric system. As can be seen from figure 4 (right), the GPS coordinate and the bar coordinate axis systems are different from each other. The topocentric GPS coordinates (N^{gps}, E^{gps}) are projected onto the bar coordinate system (x^{bar}, y^{bar}) using the following formulas:

$$\begin{aligned} x^{bar} &= N^{gps}\cos(\varepsilon) + E^{gps}\sin(\varepsilon) \\ y^{bar} &= -N^{gps}\sin(\varepsilon) + E^{gps}\cos(\varepsilon) \end{aligned} \quad (4)$$

where ε is the rotation angle between the topocentric GPS and the bar coordinate systems. Note that the angle ε is the grid bearing of the bar-reference line since the x^{bar} axis is oriented to the reference station (see figure 4 [right]). Thus, the angle ε was calculated using the coordinates of the reference station and the bar with motionless position estimated from the daily GPS measurement as mentioned in the previous sub-section.

The x^{bar} axis coincided with the oscillation direction expressing the measurements' signal. The y^{bar} axis, perpendicular to the x^{bar} axis, expresses the GPS measurements' noise (Psimoulis & Stiros 2007) since the movement is in one direction. Figure 5 shows the GPS-derived displacement and the corresponding fast Fourier transform (FFT) spectrum before and after the two-dimensional (2D) similarity transformation. It is interesting to note that the average amplitude of the peak frequency (0.39 Hz) from the FFT spectrum of time series expressed in the topocentric system is 14 cm for N^{gps}, whereas the corresponding average amplitude after transformation is 18 cm for X^{bar}, which reflects the actual movement along the oscillation axis. When compared to the perpendicular axis, the peak frequency (0.39 Hz with amplitude of 12 cm) in the E^{gps} direction disappeared and left only the white noise of the GPS measurements after the transformation of the coordinate system. Therefore, the importance and necessity of 2D transformation is evident in order to evaluate the dynamic behaviour of the structures based on actual displacement along the oscillation axis. It is also important to

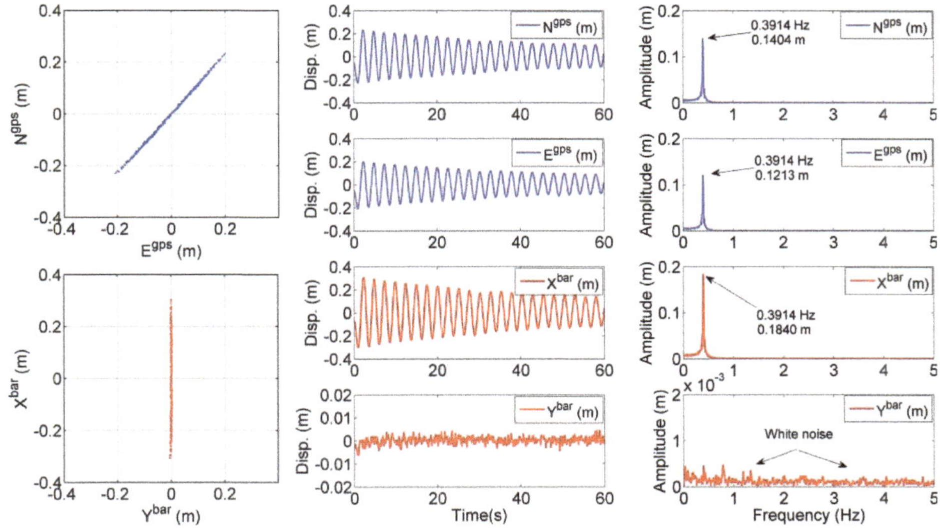

Figure 5. Displacement time series and corresponding FFT spectrums in both local topocentric coordinate systems (N^{gps}, E^{gps}) and the bar coordinate system (x^{bar}, y^{bar}).

decompose the observed displacements of structure into the two perpendicular directions in order to understand the response to the dynamic load if the structure oscillates along two main directions. Hereafter, for simplicity, the displacement of the X^{bar} and U^{bar} (or U^{gps}) axes in the present study is presented and referred to as 'horizontal displacement' and 'vertical displacement', respectively.

5. Experimental results and discussion

This section contains the results of an extensive experimental study that examines the ability of the PPP method to monitor structural dynamic movement and compares the dynamic performance of PPP with that of DGPS. In this study, kinematic DGPS results are regarded as reference because its accuracy and monitoring ability/performance of static, quasi-static and dynamic movements of structures are well documented in the literature (Çelebi et al. 2000; Tamura et al. 2002; Li et al. 2006; Nickitopoulou et al. 2006). A number of experiments were carried out and repeated several times to obtain various combinations of the oscillation characteristics.

The experiments were produced in the following ways. (1) The upper (free) end of the bar was pushed by hand from the equilibrium position by about 30–600 mm and then was left to oscillate freely. This induces oscillation to the bar with a fundamental (first) natural frequency of the order of 0.39 Hz. In order to produce oscillation of the bar with a second natural frequency of the order of about 4.30 Hz, the following two techniques were used: (2) it was quickly hit through the centre of the bar while oscillating with fundamental natural frequency and (3) it was very quickly hit by hand through the centre of the bar at the equilibrium position.

Both horizontal and vertical coordinate changes with respect to their mean values obtained from the DGPS and PPP processing are shown in figure 6, consisting of about 12 free decayed oscillation events. It can be clearly seen that DGPS-derived displacement in the horizontal and vertical components is very much consistent, whereas the PPP-derived displacement fluctuates around their mean value.

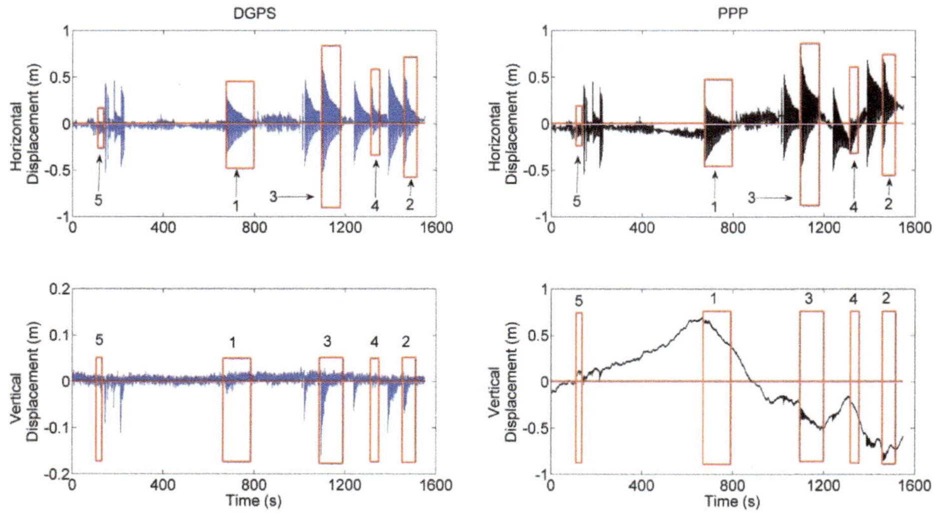

Figure 6. Horizontal and vertical displacements from DGPS processing (left) and PPP processing (right) based. The experiment lasted 26 minutes and contained around 12 free decayed oscillation events. Red rectangles represent the events discussed in the present study. Note that the vertical axis limit for right plot (PPP) is five times larger than that of the left plot (DGPS).

Long-period fluctuation from the mean in the horizontal and vertical component range around ±20 and ±80 cm, respectively. The larger variation is due to the position convergence process (Cai 2009). Therefore, longer time is needed for reaching converged solution. The convergence time, which represents the time to reach the desired accuracy level, is quite dependent on many factors such as the number and geometry of visible satellites, user environment and dynamics, observation quality, sampling rate and algorithm (Bisnath & Gao 2008). One of the possible solutions to reduce such a large amount fluctuation is to extend the period of data-taking depending on conditions. Such fluctuation is a very serious problem if the static and semi-static movements are fundamental for the assessment of structures under examination, which completely masks static and semi-static motion to be measured. However, since this study is focused on the dynamic movement over a short period of time, e.g., within 90–16 seconds, long-term fluctuation over a long period of time, e.g., around 26 minutes, is not important. Therefore, offsets and trends for each event were independently removed from DPGS and PPP-derived displacement time series in order to extract the useful signal describing the dynamic movements of the bar.

In the present study, five events indicated with rectangle in figure 6 were selected based on their oscillation characteristics, such as initial amplitude and mode shape (fundamental mode and higher mode). Table 1 summarizes these events' features (duration, mode shape and initial displacement).

5.1. Free decayed oscillation with fundamental mode (event 1)

In event 1, oscillation was produced as follows: the upper end of the bar (just below the GPS receiver) was pushed by hand from the equilibrium position and was then left to oscillate freely. The initial amplitude of the oscillation in the horizontal component is about 30 cm, and the corresponding initial amplitude of the oscillation in

Table 1. Properties of five events discussed in the present study.

Event No	Duration	Mode	Initial disp. (horizontal)	Initial disp. (vertical)
Event 1	90 s	1st	30 cm	2.2 cm
Event 2	60 s	$(1\text{st})^{10\text{ s}} + (1^{\text{st}} + 2\text{nd})^{50\text{ s}}$	23 cm + 3 cm	1.3 cm
Event 3	60 s	1st + 2nd	64 cm + 4 cm	10.3 cm
Event 4	25 s	1st + 2nd	15 cm + 3 cm	0.6 cm
Event 5	16 s	1st + 2nd	4 cm + 4 cm	–

the vertical component is about 2.2 cm, the magnitude of which is limit to vertical accuracy of kinematic DGPS (2 cm). Figure 7 shows the time series of the DGPS and PPP-derived displacements and their FFT spectrums in the range between 0 and 5 Hz. As can be seen from figure 7, the PPP-derived displacement in the horizontal component shows good agreement with that of the DGPS-derived displacement, whose standard deviation, maximum and minimum of the difference between them are ±4.6, 12 and −16 mm, respectively (see table 2). In addition to this, the corresponding FFT spectrum results from the two methods agree well with each other, the peak frequency of which is 0.391 Hz (first natural frequency) with 135 mm of amplitude. In contrast to the result in the horizontal component, the PPP-derived displacement in the vertical component fluctuates around zero in addition to dynamic movement with white noise, while the DGPS-derived displacement shows only dynamic movement with white noise along zero. Standard deviation, maximum and minimum of the difference between them are ±10.0, 45 and −29 mm, respectively (see table 2). However, their FFT results show a similar form in terms of the ability to capture dynamic movement if the dissimilarity of their lower frequency components between 0 and 0.05 Hz is ignored. The peak at 0.792 Hz with 3.7 and

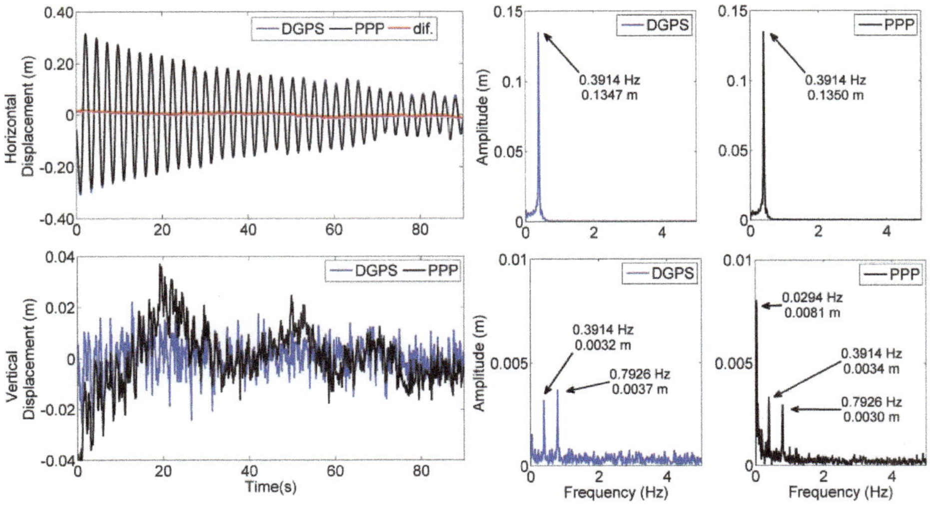

Figure 7. Horizontal and vertical displacement time series from DGPS and PPP processing and their FFT spectrum for event 1.

Table 2. Standard deviation, maximum and minimum values of the differences between DPGS and PPP.

	Horizontal (mm)			Vertical (mm)		
Event No	Std	Max	Min	Std	Max	Min
Event 1	4.6	12	−16	10.0	45	−29
Event 2	2.9	7	−8	24.4	39	−51
Event 3	5.9	14	−15	10.3	25	−34
Event 4	3.8	10	−10	8.2	19	−17
Event 5	3.7	8	−8	8.6	20	−18

3.0 mm of amplitude for DGPS and PPP, respectively, is the first natural frequency in the vertical component, which is twice the first natural frequency in the horizontal component. This relationship between them is because the oscillation in the vertical component completes two cycles while the corresponding oscillation in the horizontal component completes one cycle provided that the bar is along the gravity direction. In addition to the peak at 0.792 Hz, there is a peak at 0.391 Hz with 3.2 and 3.4 mm of amplitude for DGPS and PPP, respectively, which is corresponding to the horizontal oscillation frequency. This is most likely because the vertical oscillation is formed as a part of the horizontal oscillation due to the bending of the bar. For whatever reason, the two methods give similar results.

5.2. Transition from fundamental mode to higher mode (event 2)

In order to further examine the ability and performance of PPP methods in case of higher mode of structures, the second mode frequency of the bar was triggered by hitting quickly through its centre while oscillating with fundamental mode. Figure 8

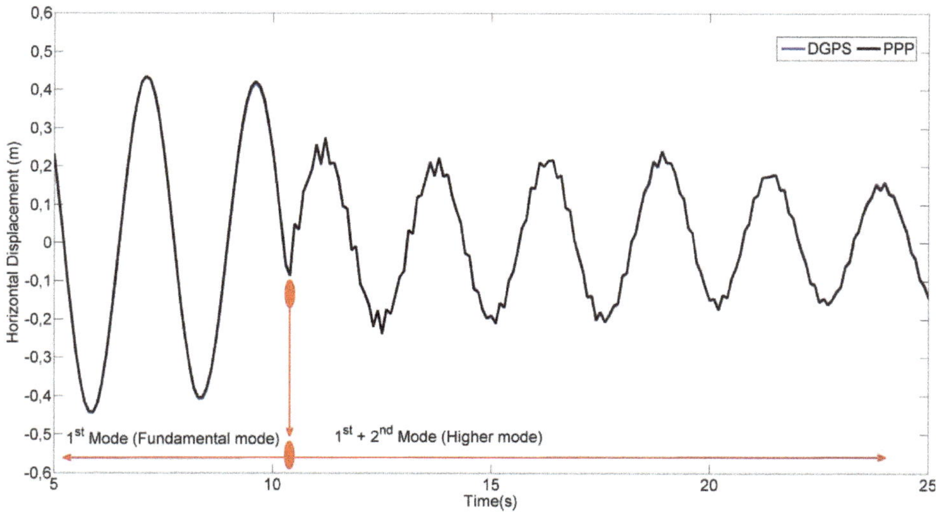

Figure 8. The time history of horizontal displacement transition from fundamental mode to higher mode.

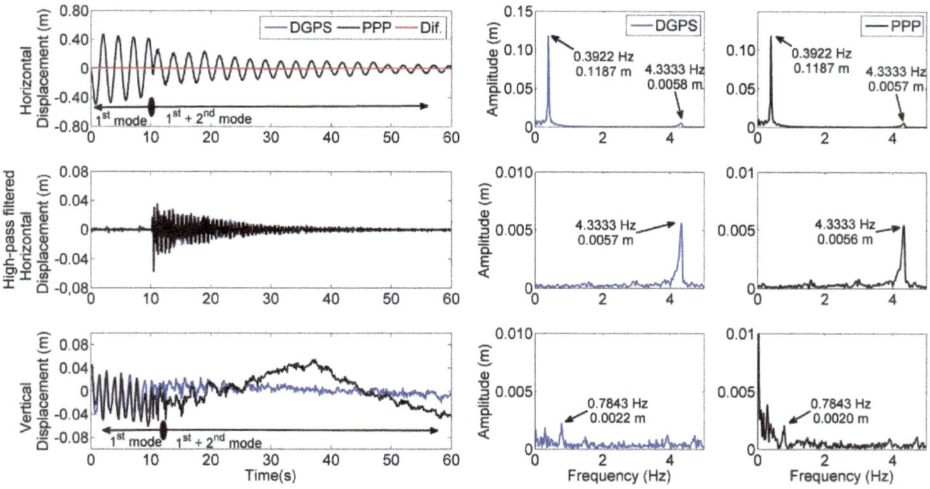

Figure 9. Horizontal and vertical displacement time series from DGPS and PPP processing and their FFT spectrum for event 2. Middle plots represents high-pass filtered time series and their FFT spectrum.

shows the translation from the fundamental mode to the higher mode, also called superposition principle of wave, meaning that a time series with two or more sine waves of different frequency and amplitude. From inspection of the time series, the fundamental period of oscillation after stimulating the higher frequency remains same with different amplitudes and phases, though the wave profiles are not smooth, indicating the presence of other frequencies or non-linearities in the data.

The FFT is a general method for examining the overall energy distributions, but it can only be used for the analysis of stationary signals. In other words, the FFT analysis of non-stationary signals cannot describe the local transient features due to averaging over duration of the signal. Therefore, the FFT results presented in figure 9 were based on the superimposed segment of the time series (from 10.4 to 60 seconds).

The initial amplitude of the oscillation in the horizontal component is about 23 cm for the first natural frequency and about 3 cm for the second natural frequency (see figure 9). The standard deviation, maximum and minimum of the difference between PPP- and GPS-derived displacements in the horizontal component are ±2.9, 7 and −8 mm, respectively (see table 2). Their FFT spectrums give two peaks at 0.392 Hz with 119 mm of amplitude for the first mode and at 4.333 Hz with 6 mm of amplitude for the second mode; the higher frequency peak is much smaller than the lower frequency peak depending on the size of the amplitude. The middle panel in figure 9 shows the tracking of the second natural frequency with white noise isolated from the first mode frequency by using a fourth-order Chebyshev high-pass filter with cut-off frequency of 1 Hz. It can be observed that the amplitude decay of the shorter period oscillation is faster than that of the fundamental (longer) period oscillation. According to these results from both the time and frequency domain, like DGPS, PPP is good enough to detect not only lower mode frequency but also higher mode frequency with small amplitude. Unlike the result in the horizontal component, the PPP-derived displacement in the vertical component fluctuates, while the DGPS-derived displacement is consistent. The standard deviation, maximum and minimum of the difference between them are ±24.4, 39 and −51 mm, respectively (see table 2). However, their FFT results give a peak at 0.784 Hz with 2 mm of amplitude regardless its statistical significance, although the initial amplitude of the vertical oscillation is about 1.3 cm, which is really small enough to detect. This indicates that PPP in the vertical component is as precise as DGPS.

5.3. Free decayed oscillation with higher mode (two sinusoidal waves of different frequency and amplitude (events 3, 4 and 5)

In these events, in order to generate higher mode frequencies of oscillation, namely superposition wave form, it was very quickly hit by hand through the centre of the bar at the equilibrium position.

In event 3, the initial amplitude of the oscillation in the horizontal component is about 64 and 4 cm for the first and second natural frequencies, respectively (figure 10). The corresponding initial amplitude in the vertical component is about 10 cm for the first natural frequency. The standard deviation of the differences between the PPP- and DGPS-derived displacements is ±6 mm for horizontal component and ±10 mm for vertical component, the amount of which is quite small. The FFT spectrums of the PPP and DGPS time series in the horizontal component give two spectral peaks at 0.391 Hz with an amplitude of 320 mm (first mode) and at 4.247 Hz with amplitude of 6 mm (second mode).

For the vertical component, the FFT spectrums from the two methods show a similar pattern in terms of spectral peaks and their amplitudes. As can be seen in figure 10, the most dominant peak is at 0.783 Hz with amplitude of 17 mm,

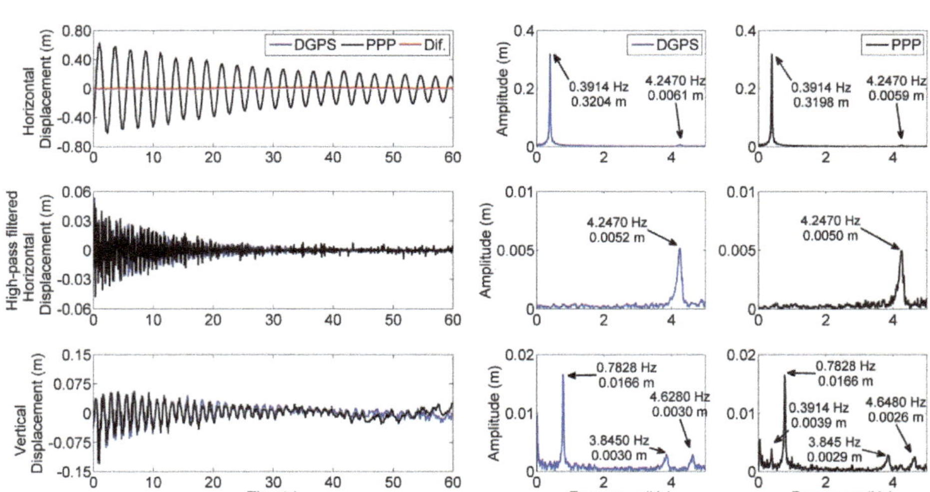

Figure 10. Horizontal and vertical displacement time series from DGPS and PPP processing and their FFT spectrum for event 3. Middle plots represents high-pass filtered time series and their FFT spectrum.

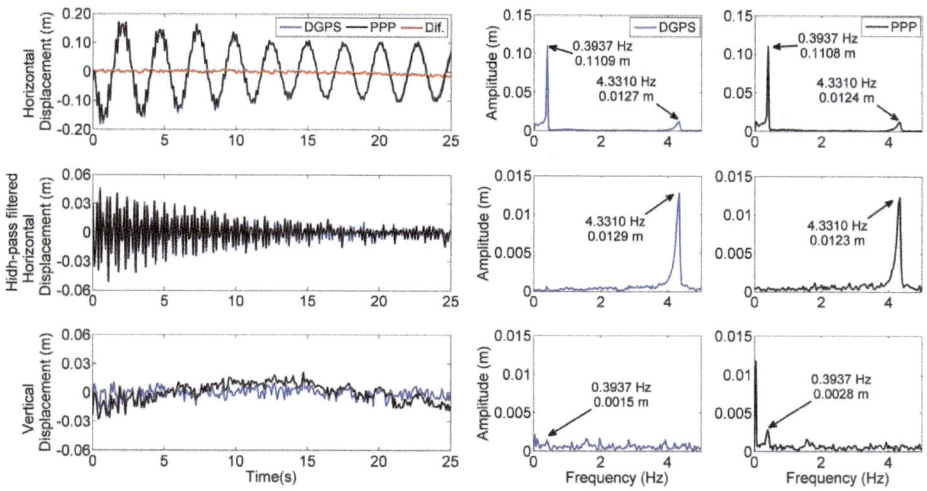

Figure 11. Horizontal and vertical displacement time series from DGPS and PPP processing and their FFT spectrum for event 4. Middle plots represents high-pass filtered time series and their FFT spectrum.

which is the first mode frequency for the vertical component. In addition, there are less significant peaks at around 0.391 (invisible for DGPS), 3.845 and 4.648 Hz with amplitudes of 4, 3 and 3 mm, respectively. 3.845 and 4.648 Hz are likely because of the second natural oscillation frequency in the horizontal component or spectral leakage effects due to the finite observation time, while 0.39 Hz is due to the first natural oscillation frequency in the horizontal component as explained before. Please note that the average of these two frequencies is 4.246 Hz, corresponding to the second natural frequency in the horizontal component.

In event 4, the initial amplitude of the oscillation in the horizontal component is about 15 and 3 cm for the first and second natural frequencies, respectively (figure 11). The standard deviation of the differences between the PPP- and DGPS-derived displacements is ±4 mm for horizontal component and ±8 mm for the vertical component. The FFT spectrums of the PPP and DGPS time series in the horizontal component give two spectral peaks at 0.394 Hz with amplitude of 111 mm (first mode) and at 4.331 Hz with amplitude of 13 mm (second mode). Once again, the PPP result is at good agreement with that of DGPS. Regarding the vertical component, there is no visible oscillation, since the corresponding initial amplitude in the vertical component is about 0.6 cm for the first natural frequency, which is below the measurement accuracy of approximately 1–2 cm for the vertical component. As it can be seen from their FFT spectrums, there is also no significant peak regarding the fundamental frequency due to the signal buried in the inherent GPS noise.

In event 5, the initial amplitude of the oscillation in the horizontal component is about 3 and 4 cm for the first and second natural frequencies, respectively (figure 12). The standard deviation of the differences between the PPP and DGPS-derived displacements is ±4 mm for the horizontal and ±9 mm for the vertical component. The FFT spectrums of the PPP and DGPS time series in the horizontal

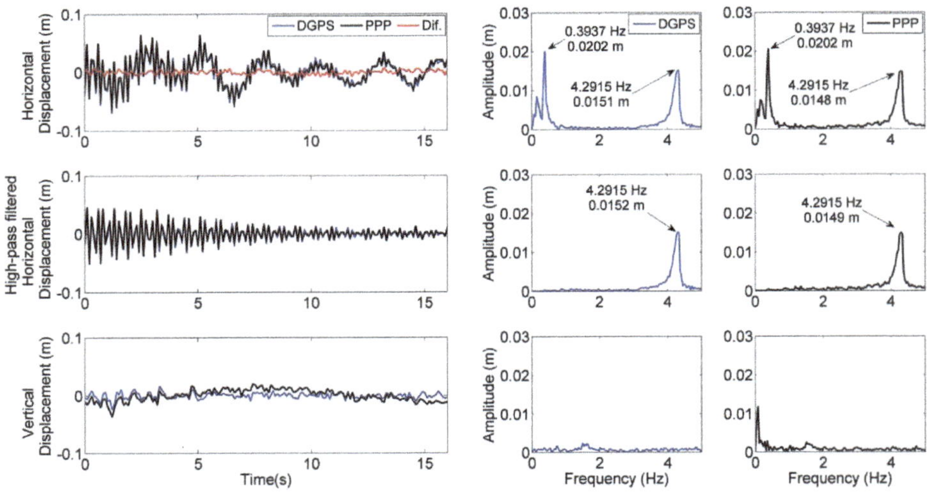

Figure 12. Horizontal and vertical displacement time series from DGPS and PPP processing and their FFT spectrum for event 5. Middle plots represents high-pass filtered time series and their FFT spectrum.

component give two spectral peaks at 0.394 Hz with amplitude of 20 mm (first mode) and at 4.292 Hz with amplitude of 15 mm (second mode). Once again, the PPP result is at good agreement with that of DGPS. Like event 4, there is no visible oscillation in the vertical component as expected.

5.4. *The impact of the length of a data span on the result of kinematic PPP*

It is worth mentioning that the author of the present study performed more additional experiments in a similar way as in the first experiment in order to better understand and see the impact of the length of a data span on the results of kinematic PPP. For this reason, this section is devoted to present the results of one selected experiment.

The length of the selected experiment, consisting of around 25 free decayed oscillation events, is one hour. GPS data were collected with the same equipment and settings as used in previous experiment. The data were processed by CSRS-PPP using precise IGS products. Then, the original RINEX file was split into 60 subsets, each of which is one-minute long. Each subset was separately processed by CSRS-PPP. Due to the limited space, only 20 consecutive subset results were provided here. Figure 13 shows the time series of each subset (in red) and entire data-set (in blue) for all components (north, east and up).

As can be seen in figure 13, time series from each one-minute data solution with respect to time series from entire data solution has a mean offset value, which is most likely due to the float ambiguity solution, among others. The largest apparent position offset is experienced in the up component. They reach up to around 5 metres. In the north and east components, they usually do not exceed 1 metre. As explained and stressed before, these offset values are not an issue if the focus is on dynamic movement. In order to compare one-hour and one-minute solutions more detail in terms of their dynamic monitoring performance, calculated mean position offsets

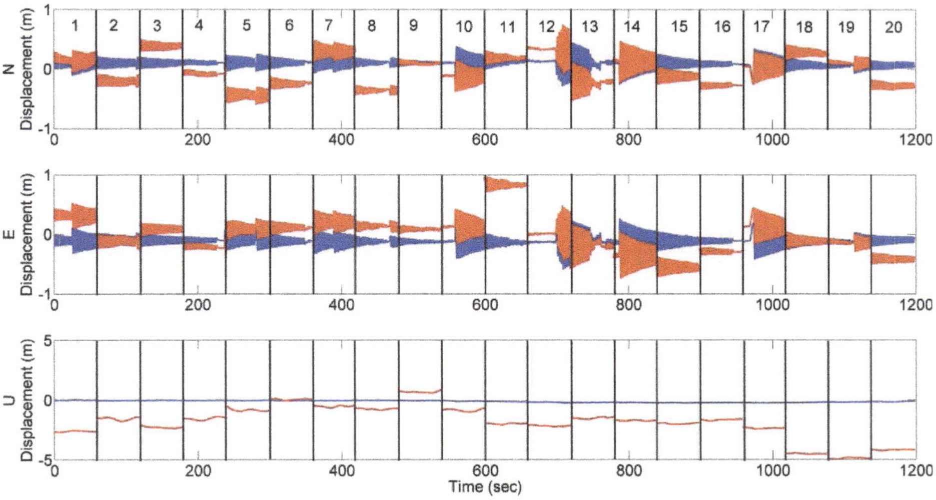

Figure 13. North, east and up displacements time series from PPP processing based on the short period (one minute) of observation (red) and long period (one hour) of observation (blue). The graphs present 20 minutes of the original data. Each one minute discussed in the present study is separated by a vertical line.

were removed from each one-minute solution. Results have demonstrated that each one-minute time series followed the same dynamic movement as good as obtained from the entire solution. The only apparent difference between them is a generally linear trend with a very small slope in the north and east components while low-frequency (0–0.05 Hz) trend in the up component. These trends are probably because of position filter used in the software, among others. The standard deviation of the difference between them after eliminating the mean offset does not generally exceed 1 cm in the north and east components, but 11 cm in the up component. Note that these standard deviations were calculated in the presence of the trend component. Table 3 also provides the standard deviation of the differences after removing both offset and linear trend component. It can be seen in table 3, they were slightly decreased, indicating that the time series obtained from very short period data is consistent with the time series obtained from long-period data in terms of tracing dynamic movement.

In addition to the comparison performed in the time domain, the FFT spectrum of each subset was compared to the FFT spectrum obtained from the corresponding segment of the entire data solution. Their FFT spectrums presented similar structures in terms of sensing the oscillation frequency of the bar in all components. However, in addition to the linear trend removal, lower frequencies (0–0.05 Hz) of each one-minute PPP solution were filtered out by applying Chebyshev high-pass filter with cut-off frequency of 0.05 Hz in order to extract the fundamental frequency of oscillation in the vertical component. With respect to both the time and frequency domain analysis, PPP results based on short- and long-period data showed similar dynamic performance after removing existing offset and trend components between them. Therefore, it is possible to sense dynamic movement using PPP algorithm regardless of the GPS data length.

Table 3. Mean offset between one-hour and one-minute PPP processing, standard deviation of the differences after removing offset and linear trend.

	Mean offset between l-hour and 1-minute PPP			Standard deviation of the difference after removing offset			Standard deviation of the difference after removing offset and linear trend		
Event no.	*N* (meter)	*E* (meter)	*U* (meter)	*N* (meter)	*E* (meter)	*U* (meter)	*N* (meter)	*E* h(meter)	*U* (meter)
1	0.063	0.423	−2.616	0.016	0.005	0.033	0.015	0.005	0.033
2	−0.301	−0.015	−1.529	0.010	0.008	0.114	0.008	0.008	0.112
3	0.287	0.186	−2.270	0.005	0.005	0.068	0.004	0.004	0.043
4	−0.182	−0.112	−1.520	0.008	0.009	0.111	0.006	0.007	0.088
5	−0.543	0.190	−0.807	0.007	0.007	0.093	0.002	0.007	0.077
6	−0.351	0.226	0.089	0.005	0.004	0.049	0.005	0.003	0.048
7	0.187	0.326	−0.488	0.004	0.005	0.062	0.003	0.005	0.062
8	−0.462	0.258	−0.723	0.005	0.005	0.057	0.004	0.005	0.057
9	−0.004	0.210	0.722	0.012	0.003	0.056	0.011	0.003	0.055
10	−0.228	0.246	−0.766	0.012	0.005	0.090	0.009	0.005	0.090
11	0.059	0.964	−1.860	0.008	0.003	0.049	0.005	0.002	0.044
12	0.208	0.138	−1.992	0.006	0.004	0.045	0.005	0.004	0.044
13	−0.327	−0.062	−1.306	0.006	0.003	0.056	0.002	0.003	0.050
14	0.029	−0.221	−1.505	0.004	0.007	0.026	0.003	0.003	0.025
15	−0.216	−0.433	−1.738	0.003	0.004	0.049	0.002	0.002	0.049
16	−0.358	−0.177	−1.426	0.005	0.005	0.046	0.002	0.003	0.036
17	−0.037	0.246	−2.124	0.005	0.003	0.030	0.004	0.003	0.029
18	0.210	0.005	−4.309	0.004	0.004	0.045	0.004	0.002	0.023
19	0.045	−0.018	−4.723	0.010	0.012	0.046	0.007	0.004	0.045
20	−0.342	−0.312	−4.061	0.011	0.012	0.030	0.007	0.004	0.027
Mean	−0.113	0.103	−1.748	0.007	0.006	0.058	0.005	0.004	0.052

6. Summary and conclusion

Kinematic DGPS, which requires simultaneously observed pseudo-range and carrier-phase data from one or more reference stations with known coordinates, has been widely used in precise GPS applications such as structural health (and vibration) monitoring over the last two decades. In recent years, interest has increased for the PPP method, which is based on the processing of undifferenced observations from a single GPS receiver employing readily available satellite orbit and clock corrections because it eliminates the need to acquire simultaneous tracking data from a reference (base) station or a network of stations.

This study is one of the first attempts to evaluate the performance of kinematic PPP for both horizontal and vertical components in terms of SHM applications. The main goal of this paper was to investigate the kinematic PPP method in terms of monitoring high-frequency dynamic movement of an engineering structure based on 10 Hz data from one dual frequency GPS receiver. To meet this goal, a large number of free decayed oscillation experiments using simple cost-effective model structure were performed. The performance of the kinematic PPP method based on the five events selected for this study has been assessed through a comparison with that of the kinematic DGPS methods in the time and frequency domain. The results of the experiments show that the PPP method is very efficient and can satisfy SHM applications as well as DGPS method in terms of extracting dynamic oscillation frequencies in the horizontal (0.39 and 4.33 Hz) and vertical (0.78 Hz) components after removing offset and trend from PPP-derived time series.

The major outcome of this study is that stand-alone GPS data processed by the PPP method can monitor dynamic displacements with initial amplitude about 13 mm in the vertical component (see event 2) although the accuracy of the vertical component is worse than the horizontal ones. This proves that the kinematic PPP can track high-frequency dynamic movement as precisely as the kinematic DGPS in both horizontal and vertical component.

In addition, another experiment was performed to investigate whether very short duration GPS records (one minute) submitted to the PPP software is capable of capturing dynamic movement just as long duration GPS records. Therefore, selected 20 one-minute PPP solutions were compared to the corresponding segments extracted from the one-hour PPP solution. Results clearly show that differences between one-minute and one-hour PPP solutions are offset in both components and lower frequency (0–0.05 Hz) trend in mostly the vertical component. However, their spectral peaks in the FFT spectrum regarding oscillation frequency show good agreement to each other after removing offset in the time domain and filtering out the lower frequency trend in the frequency domain by using the Chebyshev high-pass filter with cut-off frequency of 0.05 Hz.

In conclusion, the post-processed kinematic PPP method even in the case of short duration GPS records can provide useful information regarding dynamic displacements and oscillation frequencies of a wide range of engineering structures. Hence, it can be used as an efficient alternative to the kinematic DGPS method and complementary to other sensors, such as accelerometer, in studies of assessment of structural health and integrity of engineering structures and of identification of their dynamic characteristics (fundamental mode frequency, damping ratio). Depending on the availability to real-time product for PPP users, development of algorithm and increasing of its accuracy, it is expected to use the real-time PPP method for real-time monitoring applications in the near future.

Acknowledgements
The author is very grateful to the chief editor and three anonymous reviewers for their constructive comments, which helped to improve the manuscript. The author also thanks production editor for editing and the Natural Resources Canada for providing CSRS-PPP service.

References

Alkan RM, Öcalan T. 2013. Usability of the GPS Precise Point Positioning technique in marine applications. J Navig. 66:579–588.

Avallone A, Marzario M, Cirella A, Piatanesi A, Rovelli A, Di Alessandro C, D'Anastasio E, D'Agostino N, Giuliani R, Mattone M. 2011. Very high rate (10 Hz) GPS seismology for moderate-magnitude earthquakes: the case of the Mw 6.3 L'Aquila (central Italy) event. J Geophys Res. 116:1–14.

Bisnath S, Gao Y. 2008. Current state of Precise Point Positioning and future prospects and limitations. Int Assoc Geod Symp. 133:615–623.

Bock H, Hugentobler U, Beutler G. 2003. Kinematic and dynamic determination of trajectories for low Earth satellites using GPS. In: Reigber C, Luhr H, Schwintzer P. editors. First CHAMP mission results for gravity magnetic and atmospheric studies. Heidelberg: Springer; p. 65–69.

Breuer P, Chmielewski T, Górski P, Konopka E, Tarczyński L. 2008. The Stuttgart TV Tower – displacement of the top caused by the effects of sun and wind. Eng Struct. 30:2771–2781.

Cai C. 2009. Precise Point Positioning using dual-frequency GPS and GLONASS measurement [master thesis]. Calgary: University of Calgary.

Calais E, Han JY, Demets C, Nocquet JM. 2006. Deformation of the North American plate interior from a decade of continuous GPS measurements. J Geophys Res. 111:1–23.

Çelebi M. 2000. GPS in dynamic monitoring of long-period structures. Soil Dyn Earthquake Eng. 20:477–483.

Çelebi M, Sanli A. 2002. GPS in pioneering dynamic monitoring of long-period structures. Earthquake Spectra. 18:47–61.

Chan WS, Xu YL Ding XL, Xiong YL, Dai WJ. 2006. Assessment of dynamic measurement accuracy of GPS in three directions. J Surv Eng-ASCE. 132:108–117.

Donahue B, Mireault Y, Ghoddousi-Fard R, Lahaye F, Tétreault P, Caissy M. 2012. Current status and future plans at the Natural Resources Canada (NRCan) analysis centre. Paper presented at: IGS Workshop 2012; Poland.

El-Mowafy A. 2011. Analysis of web-based GNSS post-processing services for static and kinematic positioning using short data spans. Surv Rev. 43:535–549.

Erdoğan H, Akpınar B, Gülal E, Ata E. 2007. Monitoring the dynamic behaviors of the Bosporus Bridge by GPS during Eurasia Marathon. Nonlinear Proc Geophys. 14:513–523.

Fund F, Perosanz F, Testut L, Loyer S. 2013. An integer Precise Point Positioning technique for sea surface observations using a GPS buoy. Adv Space Res. 51:1311–1322.

Gao Y, Kongzhe C. 2004. Performance analysis of Precise Point Positioning using real-time orbit and clock products. J Global Positioning Syst. 3:95–100.

Gao Y, Wojciechowski A, Chen K. 2005. Airborne kinematic positioning using Precise Point Positioning methodology. Geomatica. 59:275–282.

Geng J, Teferle FN, Meng X, Dodson AH. 2010. Kinematic Precise Point Positioning at remote marine platforms. GPS Solutions. 14:343–350.

Ghoddousi-Fard R, Dare P. 2006. Online GPS processing services: an initial study. GPS Solutions. 10:12–20.

Hammond WC, Thatcher W. 2005. Northwest basin and range tectonic deformation observed with the global positioning system: 1999–2003. J Geophys Res Lett. 110: 1–12.

Hofmann-Wellenhof B, Lichtenegger H, Walse E. 2008. GNSS: Global Navigation Satellite Systems – GPS, GLONASS, GALLILEO and more. New York: Springer.

Im SB, Hurlebaus S, Knag YJ. 2013. Summary review of GPS technology for structural health monitoring. J Struct Eng-ASCE. 139:1654–1664.

Kijewski-Correa T, Kilpatrick J, Kareem A, Kwon D-K, Bashor R, Kochly M, Young BS, Abdelrazaq A, Galsworthy J, Isyumov N, et al. 2006. Validating wind-induced response of tall buildings: synopsis of the Chicago full-scale monitoring program. J Struct Eng-ASCE. 132:1509–1523.

Kouba J. 2003. Measuring seismic waves induced by large earthquakes with GPS. Stud Geophys Geod. 47:741–755.

Kouba J, Hérou P. 2001. Precise Point Positioning using IGS orbit and clock products. GPS Solutions. 5:12–28.

Kuo C, Chiu K, Chiang K, Cheng K, Lin L, Tseng H, Chu F, Lan W, Lin H. 2012. High-frequency sea level variations observed by GPS buoys using Precise Point Positioning technique. Terr Atmos Ocean Sci. 23:209–218.

Leandro RF, Langely RB, Santos MC. 2008. GNSS data analysis in GAPS, the GPS analysis and positioning software using IGS products. Proceedings of IGS Analysis Center Workshop; Miami Beach, FL.

Li P, Zhan X. 2013. Integrating GPS and GLONASS to accelerate convergence and initialization times of Precise Point Positioning. GPS Solutions. doi:10.1007/s10291-013-0345-5

Li T, Wang J, Laurichesse D. 2013a. Modeling and quality control for reliable Precise Point Positioning integer ambiguity resolution with GNSS modernization. GPS Solutions. doi:10.1007/s10291-013-0342-8

Li X, Ge L, Ambikairajah E, Rizos C, Tamura Y, Yoshida A. 2006. Full-scale structural monitoring using an integrated GPS and accelerometer system. GPS Solutions. 10:233–247.

Li X, Ge M, Zhang X, Zhang Y, Guo B, Wang R, Klotz J, Wickert J. 2013b. Real-time high-rate co-seismic displacement from ambiguity-fixed Precise Point Positioning: application to earthquake early warning. Geophys Res Lett. 40:295–300.

Meng X, Dodson AH, Roberts GW. 2007. Detecting bridge dynamics with GPS and triaxial accelerometers. Eng Struct. 29:3178–3184.

Moschas F, Avallone A, Saltogianni V, Stiros S. 2013. Application of the Precise Point Positioning (PPP) algorithm for the monitoring of dynamic deformations with a single GPS receiver. 2nd Joint International Symposium on Deformation Monitoring (JISDM); Nottingham.

Moschas F, Stiros S. 2011. Measurement of the dynamic displacements and of the modal frequencies of a short-span pedestrian bridge using GPS and an accelerometer. Eng Struct. 33:10–17.

Moschas F, Stiros SC. 2014. Three-dimensional dynamic deflections and natural frequencies of a stiff footbridge based on measurements of collocated sensors. Struct Control Health Monit. 21:23–42.

Nakamura S. 2000. GPS measurement of wind-induced suspension bridge girder displacements. J Struct Eng-ASCE. 126:1413–1419.

Nickitopoulou A, Protopsalti K, Stiros S. 2006. Monitoring dynamic and quasi-static deformations of large flexible engineering structures with GPS: accuracy, limitations, and promises. Eng Struct. 28:1471–1482.

Ohta Y, Ohzono M, Miura S, Iinuma T, Tachibana K, Takatsuka K, Miyao K, Sato T, Umino N. 2008. Coseismic fault model of the 2008 Iwate-Miyagi Nairiku earthquake deduced by a dense GPS network. Earth Planets Space. 60:1197–1201.

Park HS, Sohn HG, Kim IS, Park JH. 2008. Application of GPS to monitoring of wind-induced responses of high-rise buildings. Struct Des Tall Spec Build. 17:117–132.
Pehlivan H, Aydin Ö, Gülal E, Bilgili E. 2013. Determining the behaviour of high-rise structures with geodetic hybrid sensors. Geomatics Nat Hazard Risk. doi:10.1080/19475705.2013.854280
Pengfei C, Wei L, Jinzhong B, Hanjiang W, Yanhui C, Hua W. 2011. Performance of Precise Point Positioning (PPP) based on uncombined dual-frequency GPS observables. Surv Rev. 43:343–350.
Píriz R, Calle D, Mozo A, Navarro P, Rodríguez D, Tobías G. 2009. Orbits and clocks for GLONASS Precise-Point-Positioning. In: Wang J, Feng Y, Wang C, editors. Proceedings of 22nd International Technical Meeting of the Satellite Division of the Institute of Navigation 2009 (ION GNSS 2009); Savannah (GA); p. 2415–2424.
Psimoulis P, Stiros S. 2007. Measurement of deflections and of oscillation frequencies of engineering structures using Robotic Theodolites (RTS). Eng Struct. 29:3312–3324.
Psimoulis PA, Pytharouli S, Karambalis D, Stiros SC. 2008. Potential of global positioning system (GPS) to measure frequencies of oscillations of engineering structures. J Sound Vib. 318:606–623.
Psimoulis PA, Stiros SC. 2008. Experimental assessment of the accuracy of GPS and RTS for the determination of the parameters of oscillation of major structures. Comput-Aided Civil Infrastructure Eng. 23:389–403.
Psimoulis PA, Stiros SC. 2013. Measuring deflections of a short-span railway bridge using a robotic total station. J Bridge Eng. 18:182–185.
Retscher G. 2002. Accuracy performance of virtual reference station (VRS) networks. J Glob Positioning Syst. 1:40–47.
Rizos C, Janssen V, Roberts C, Grinter T. 2012. PPP versus DGNSS. Geomatics World. September/November:18–20.
Roberts GW, Meng X, Dodson A. 2004. Integrating a global positioning system and accelerometers to monitor deflection of bridges. J Surv Eng. 130:65–72.
Savage JC, Gan W, Prescott WH, Svarc JL. 2004. Strain accumulation across the coast ranges at the latitude of San Francisco, 1994–2000. J Geophys Res. 109:1–11.
Shi C, Lou Y, Zhang H, Zhao Q, Geng J, Wang R, Fang R, Liu J. 2010. Seismic deformation of the M_w 8.0 Wenchuan earthquake from high-rate GPS observations. Adv Space Res. 46:228–235.
Shi J, Gao Y. 2014. A troposphere constrain method to improve PPP ambiguity-resolved height solution. J Navig. 67:249–262.
Tamura Y, Matsui M, Pagnini LC, Ishibashi R, Yashida A. 2002. Measurement of wind-induced response of buildings using RTK-GPS. J Wind Eng Ind Aerodynamics. 90:1783–1793.
Wang G, Blume F, Meertens C, Ibanez P, Schulze M. 2011. Performance of high-rate kinematic GPS during strong shaking: observations from shake table tests and the 2010 Chile earthquake. J Geodetic Sci. 2:1–15.
Xu L, Guo JJ, Jiang JJ. 2002. Time-frequency analysis of a suspension bridge based on GPS. J Sound Vib. 254:105–116.
Xu P, Shi C, Fang R, Liu J, Niu X, Zhang Q, Yanagidani T. 2013. High-rate Precise Point Positioning (PPP) to measure seismic wave motions: an experimental comparison of GPS PPP with inertial measurement units. J Geod. 87:361–372.
Yi J, Zhang JW, Li QS. 2013 Dynamic characteristics and wind-induced responses of a super-tall building during typhoons. J Wind Eng Ind Aerodynamics. 121:116–130.
Yi T, Li H, Gu M. 2013a. Experimental assessment of high-rate GPS receivers for deformation monitoring of bridge. Measurement. 46:420–432.
Yi T, Li H, Gu M. 2013b. Recent research and applications of GPS-based monitoring technology for high-rise structures. Struct Control Health Monit. 20:649–670.

Yigit CO, Li X, Inal C, Ge L, Yetkin M. 2010. Preliminary evaluation of precise inclination sensor and GPS for monitoring full-scale dynamic response of a tall reinforced concrete building. J Appl Geod. 4:103–113.
Zhang X, Andersen OB. 2006. Surface ice flow velocity and tide retrieval of the Amery ice shelf using Precise Point Positioning. J Geod. 80:171–176.
Zumberge JF, Heflin MB, Jefferson DC, Watkins MM, Webb FH. 1997. Precise Point Positioning for the efficient and robust analysis of GPS data from large networks. J Geophys Res. 102:5005–5017.

6

De-noising of GPS structural monitoring observation error using wavelet analysis

MOSBEH R. KALOOP†* and DOOKIE KIM‡

†Department of Public Works and Civil Engineering, Faculty of Engineering, Mansoura University, El-Mansoura 35516, Egypt

‡Department of Civil Engineering, Kunsan National University, Kunsan 573-701, Republic of Korea

In the process of the continuous monitoring of the structure's state properties such as static and dynamic responses using Global Positioning System (GPS), there are unavoidable errors in the observation data. These GPS errors and measurement noises have their disadvantages in the precise monitoring applications because these errors cover up the available signals that are needed. The current study aims to apply three methods, which are used widely to mitigate sensor observation errors. The three methods are based on wavelet analysis, namely principal component analysis method, wavelet compressed method, and the de-noised method. These methods are used to de-noise the GPS observation errors and to prove its performance using the GPS measurements which are collected from the short-time monitoring system designed for Mansoura Railway Bridge located in Egypt. The results have shown that GPS errors can effectively be removed, while the full-movement components of the structure can be extracted from the original signals using wavelet analysis.

1. Introduction

Global Positioning System (GPS) has been successfully applied in the short- or long-time structural health monitoring (SHM) of the long- and short-period domains of large-scale civil engineering structures (Meng 2002; Zhong et al. 2008; Im et al. 2011). Yu et al. (2006) summarized the advantages of GPS to monitor the deformation of civil structures. Nevertheless, as any other developing technology, the GPS multipath errors, systematic effects in the position results, are amplified by weak satellite constellations, and shaking noises have their own disadvantages when they are applied in the precise engineering applications (Roberts et al. 2002; Oluropo et al. 2014). A major barrier is the achievable accuracy of GPS positioning solution, which is affected by many factors and restraints. Therefore, noise reduction of GPS observations, improvement of the accuracy of the GPS time series, and detection of deformation epochs are the key issues of movement analysis.

The process of implementing a movement and damage identification strategy for civil and mechanical engineering infrastructures is referred to as SHM (Im et al. 2011). Wong (2007) illustrated the design of the SHM system on Tsing Ma Bridge using multisensors; in between these sensors is GPS. Also, Ko and Ni (2005)

*Corresponding author. Email: mosbeh@mans.edu.eg

summarize the sensors installed on 20 bridges in China for the SHM of these bridges. Many researchers used GPS in SHM system of structures like buildings, bridges, and so on (Meng 2002; Im et al. 2011). In GPS-SHM, the main works should be done as described in the following: analysis of GPS noise; separation of coloured noise from GPS real-time series; accuracy improvement of the GPS real-time series; and reliability improvement of detecting movement epochs. For analysing the behaviour of structures (movement components) in both time and frequency domains from GPS data, signal pre-processing to mitigate noise and extract useful signals should be done first.

Filtering and smoothing in the context of dynamic systems refer to a Bayesian methodology for computing posterior distributions of the latent state based on a history of noisy measurements. This kind of methodology can be found, for example, in navigation, control engineering, robotics, and machine learning (Julier & Uhlmann 2004; Deisenroth et al. 2009). Solutions to filtering (Ko & Fox 2009) and smoothing (Godsill et al. 2004) in linear dynamic systems are well known, and numerous approximations for non-linear systems have been proposed, for both filtering (Ko & Fox 2009) and smoothing (Godsill et al. 2004). Some researchers choose the filtration and smoothed models to mitigate the GPS errors as references (Meng 2002; Yu et al. 2006; Psimoulis et al. 2008; Zhong et al. 2008; Kaloop 2012), and others used multiple antennae to reduce the multipath error and then used filtrations of observation also as references (Meng 2002; Roberts et al. 2002; Danskin et al. 2009; Oluropo et al. 2014).

Wavelet analysis is a strong tool to eliminate GPS noises according to the noise characteristics (Yu et al. 2006; Psimoulis et al. 2008; Pytharouli & Stiros 2010, 2012). The wavelet analysis is one of the smoothed methods, which can be used to de-noise the GPS observations. This study compares three methods based on wavelet analysis, which are principal component analysis (PCA), wavelet compressed, and de-noised methods. The previous filtration and smoothed studies used the wavelet methods separately or used them with other filters. For example, Ogaja et al. (2003) applied the PCA with Haar wavelet analysis to filter and monitor wind-induced responses based on the GPS monitoring observation system; the method used consists of pre-filtering the original GPS solutions via a finite impulse response (FIR) median hybrid (FMH) filter, and applying the PCA to the Haar wavelet transform of the FMH-filtered results. Yu et al. (2006) applied the wden MATLAB function-based wavelet analysis to eliminate the GPS errors and analyse the time and frequency domains; this application is exploited to eliminate noises of one-dimensional (1D) time series in a MATLAB wavelet analysis packet automatically. Wu et al. (2011) studied the de-noising GPS data-based wavelet and Kalman filter; in this study, they used wavelet de-noising based on integration of feature extraction and low-pass filter to reconstruct the filter signal, and then applying the Kalman filter to improve the high-quality filter signals. From this method, they found that these methods have a very important significance in improving the accuracy of the GPS data processing and expanding the application range of the GPS service. Ma et al. (2009) applied the wavelet de-noise method with an improved threshold function to optimize the GPS/INS navigation signals. This study used the translation invariant threshold wavelet noise reduction method to reduce threshold to the signals with noise after translating, and then reverse translation of de-noised signals and get the processed signals. Zhong et al. (2008) applied a method based on the technique of cross validation for automatically identifying wavelet signal layers, which is developed and used for separating noise from signals in data series. This study used the method of cross-validation filter after the dyadic wavelet decomposition to automatically identify the wavelet-decomposed signal levels, and then the filtered values of the observational series are reconstructed based on the wavelet coefficients which obtained from the signal levels determined. Lilong et al. (2010) mitigated the GPS systematic errors using wavelet de-noise method; this study used the db4 mother wavelet and the de-noise method is soft-threshold de-noise method, the double differential observation forming to decompose double difference with the aim of mitigating systematic errors and recovering double difference observation after that used the de-noising bias elimination outlier detection data compression then GPS observation reconstruction is determined. Finally, Yu et al. (2006) and Aminghafaria et al. (2006) summarized the methods of wavelet analysis eliminating noises as follows. First, one is a compulsive that the high-frequency coefficients are processed to be zero in the decomposed signal constructions of wavelet analysis, and some scale or different scale signal components with these coefficients in the data time series are all eliminated. Then, the signals are reconstructed to analyse their spectrum features. Another method is a threshold-eliminating noise processing where a threshold value is defined depending on experience, and used to process the high-frequency coefficients of wavelet analysis, i.e., the coefficients greater than the threshold are reserved, and the coefficients less than the threshold are processed to be zero. The wavelets have found wide use for signal analysis and noise removal in a variety of fields due to their ability to present deterministic features in terms of a small number of relatively large coefficients (Bakshi 1998).

From the previous studies, the limitations of using the wavelet analysis to extract the movement components of structures based on wavelet analysis are shown. However, this study limits its focus on the de-noised GPS movement monitoring observations to extract the long and short periods of movement components of structures based on wavelet analysis and applied design wavelet MATLAB filter models to de-noise Mansoura railway bridge short monitoring GPS time-series data. These models were fast and easy to use and successful to remove most of the multipath GPS errors and observation noises.

2. De-noising model

De-noising models based on wavelet analyses were proposed in this study. These models depend on de-noising, decomposition coefficient, wavelet decomposition or reconstruction, and thresholds with performances. In this study, three models are introduced: multiscale PCA, multisignal wavelet compression, and de-nosing based on wavelet analysis. The model selections are used widely on the mitigation the signal data errors; however, this study is compared between these methods and applied with monitoring GPS observations. Moreover, these models can be used to simplify the mitigation of GPS structure monitoring observation data.

2.1. Wavelet transform

Wavelet analysis is a multiresolution analysis in time and frequency domains. General overview of wavelets and wavelet analysis are found in Bakshi (1998), Chui (1992), and Aminghafaria et al. (2006). For the most practical applications to measure data, the wavelet dilation and translation parameters are discretized dynamically, and the family of wavelets is represented as follows:

$$\Psi_{mk}(t) = 2^{-\frac{m}{2}}\Psi(2^{-m}t - k) \tag{1}$$

where $\Psi(t)$ is the mother wavelet, and m and k are the dilation and translation parameters, respectively. The translation parameter determines the location of the wavelet in the time domain while the dilation parameter determines the location of it in the frequency domain as well as the scale or extent of the time–frequency localization (Sone et al. 1996; Bakshi 1998). Several mother wavelets that have proven to be especially useful are included in MATLAB. Figure 1 shows an example of Symlets mother wavelet.

Almost all practically useful discrete wavelet transforms (DWT) use discrete-time filter bank. These filter banks are called wavelet and scaling confident in wavelet nomenclature. In order to make $\Psi_{mk}(t)$, as equation (1), a complete orthonormal basis, some methods to generate the analysis wavelet $\Psi(t)$ compactly support the time domain and frequency domain proposed by Daubechies (1988) and Meyer (1989).

In the multiresolution analysis, the scaling function $\varphi(t)$ and analysis wavelet $\Psi(t)$ in the central closed subspace V_0 can be written in terms of the orthonormal basis $\varphi_{1,n}$ in V_1 as follows (Daubechies 1988; Sone et al. 1996):

$$\varphi(t) = \sqrt{2}\sum_{n} h_n \varphi(2t - n) \tag{2}$$

$$\Psi(t) = \sqrt{2}\sum_{n} (-1)^n h_{1-n}\varphi(2t - n) \tag{3}$$

where a sequence h_n satisfies the following relations:

$$\left.\begin{array}{ll} h_n = 0 \quad \text{for} \quad n < 0 \quad \text{or} \quad n > 2N; & \text{else} \quad \sum_{n} h_n = \sqrt{2} \\ \sum_{n} (-1)^n h_{1-n}\, n^m = 0 & \text{for} \quad 0 \leq m \leq N-1 \end{array}\right\} \tag{4}$$

The way to ensure real-valued compact support for the analysing wavelet is to choose the scaling function with compact support for the analysing wavelet is to choose the scaling function with (Bakshi 1998). For this reason, the sequence (h_n) in equation (4) is required to be finite real-valued sequence. Therefore, for the arbitrary integer $N \geq 2$, the finite sequence is determined by Daubechies (1988), so that the support of $\varphi(t)$, the moment of Pth order of $\Psi(t)$, and the regularity of $\varphi(t)$ and $\Psi(t)$ can satisfy the following conditions (Sone et al. 1996):

$$\left.\begin{array}{l} \text{supp}\varphi = [0,\ 2N-1] \\ \int_{\infty}^{-\infty} t^P \Psi(t)\, dt \quad ,0 \leq P \leq N-1 \\ \varphi(t),\ \Psi(t) \in C^{\lambda(N)} \end{array}\right\} \tag{5}$$

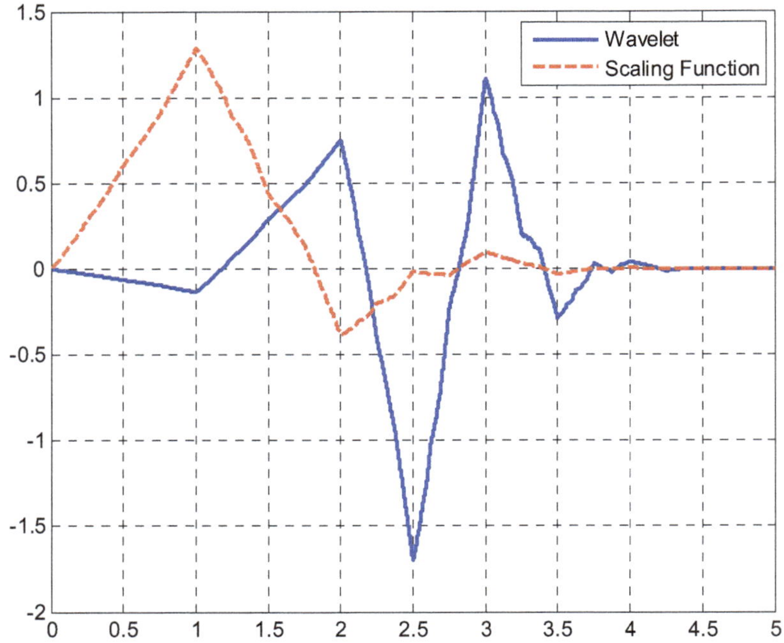

(a)

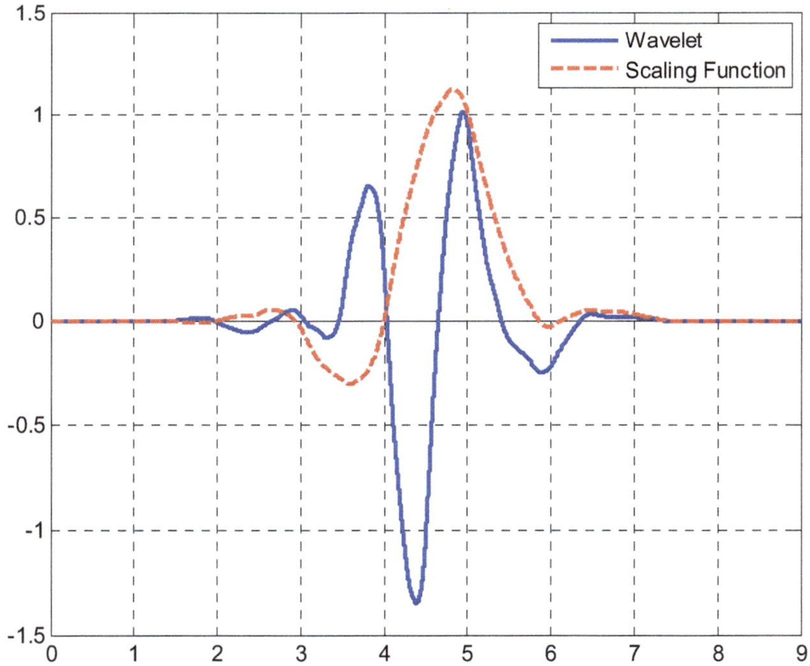

(b)

Figure 1. Scaling function and analysed Symlets wavelet: (a) $N = 3$, (b) $N = 5$.

where $C^{\lambda(N)}$ represents the space consisting of the function that is $\lambda(N)$ times continuously differentiable. For an integer N $\lambda(N)$ is approximated by 0.3485 N (Sone et al. 1996). The example of $\varphi(t)$ and Symlets $\Psi(t)$ is extracted as shown in figure 1; in this figure, integer $N = 3$ and 5. From this example, it can be seen that the regularity of both $\varphi(t)$ and $\Psi(t)$ clearly increases with N.

The low-pass and high-pass filters used in this algorithm are determined according to the mother wavelet in use. The outputs of low-pass filters are referred to as approximation coefficients and the outputs of the high-pass filters are referred to as detail coefficients (Bakshi 1998). Wavelet reconstruction algorithm is the converse of wavelet separation algorithm, whereas the detailed signal that represents high-frequency noise is evaluated as zero. Then, reconstructed function is executed, and the output signal is a de-noised signal (Bakshi 1998).

2.2. *Wavelet compressed signal model*

GPS coordinate time history can be decomposed into its contributions in different regions of the time–frequency space by projection on the corresponding mother wavelet function. The lowest frequency content of the signal is represented on a set of scaling functions, as depicted in figure 1. The number of wavelet and scaling function coefficients decreases dyadically at coarser scales due to dyadic discretization of the dilation and translation parameters. Fast algorithms for computing the wavelet decomposition are based on representing the projection of the signal on the corresponding mother wavelet function as a filtering operation (Mathworks 2008). However, convolution with a filter H represents projection on the scaling function, and convolution with a filter G represents projection on a wavelet. Thus, the coefficients at different scales may be obtained as follows:

$$a_m = Ha_{m-1}, \qquad d_m = Gd_{m-1} \tag{6}$$

where d_m is the vector of wavelet coefficients at scale m, and a_m is the vector of scaling function coefficients. The GPS time-history data are considered to be scaling function coefficients at the finest scale, which means $x = a_0$. Equation (6) may also be represented in terms of the GPS time-history vector, x, as follows:

$$a_m = H_m x, \qquad d_m = G_m x \tag{7}$$

where H_m is obtained by applying the H filter m times, and G_m is obtained by applying the H filter $(m - 1)$ times, and the G filter once. Therefore, the GPS time-history data can be reconstructed exactly and noise can be removed from its wavelet coefficients at all scales, d_m for $m = 1, 2, \ldots, L$, and scaling function coefficients at the coarsest scale a_L, where the wavelet coefficients corresponding to the two step changes are larger than the coefficients corresponding to the uncorrelated stochastic process (Bakshi 1998). The deterministic and stochastic components of the data can be separated by an appropriate threshold (Mathworks 2008). In this section, the design model computes thresholds and performs compression of 1D signals using wavelets. This model returns a compression of the original multisignal-based wavelet decomposition structure. The method of a soft threshold-eliminating noise processing used a fixed form of threshold, and its value is equal to $\hat{\sigma}\sqrt{2\log(p)}$, where

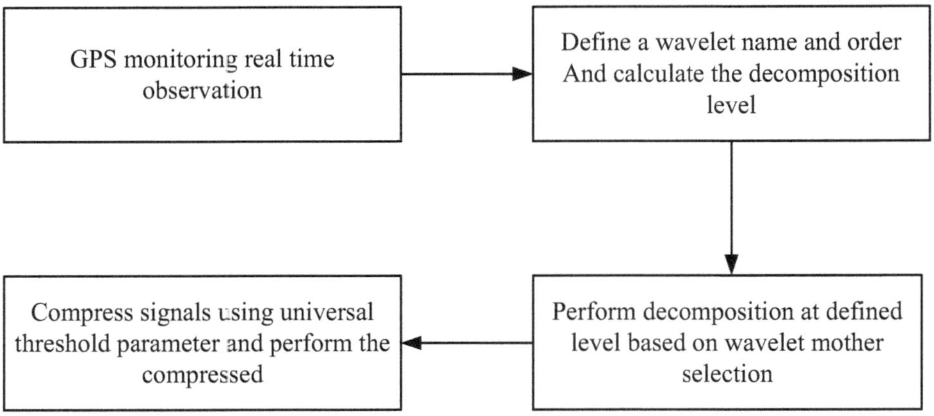

Figure 2. Flow chart diagram of the proposed wavelet compressed model design.

p is the number of observations, and it is used to process the high-frequency coefficients of wavelet analysis, i.e. the coefficients greater than the threshold are reserved, and the coefficients less than it are processed to be zero. The design model is shown in figure 2.

2.3. *Wavelet de-noising model*

Donoho (1995) and Aminghafaria et al. (2006) have proposed a method for reconstructing signals based on de-noised method of the observation data (x) from the correlated noise as follows:

$$x_i = f(t_i) + \sigma z_i \qquad i = 0, \ldots, n-1 \tag{8}$$

where f is deterministic and is the signal to be recovered, $t_i = i/n$, $z_i \sim N(0, 1)$ is a Gaussian whit noise, and σ is a noise level. The interpretation of the term de-noising is that one's goal is to optimize the mean-square error:

$$n^{-1}E\|\hat{f} - f\|_{\ell_n^2}^2 = n^{-1}\sum_{i=0}^{n-1} E(\hat{f}(i/n) - f(i/n))^2 \tag{9}$$

where, with high probability, $\hat{f}$ is at least as smooth as f, the rationale for the side condition (8) is that many statistical techniques simply optimize the mean-squared error. This demands a trade-off between bias and variance which keeps the two terms of about the same order of magnitude. Donoho (1995) proposed three steps for a threshold procedure for recovering signals from noisy observation as follows: apply the interval-adapted pyramidal filtering algorithm of Cohen et al. (1993) to the measured data, obtaining empirical wavelet coefficients. In this study, the method of a soft threshold-eliminating noise processing is also used and some scale or different scale signal components with these coefficients in the GPS data time series are all eliminated. Figure 3 is the proposed wavelet de-noising of the GPS time series.

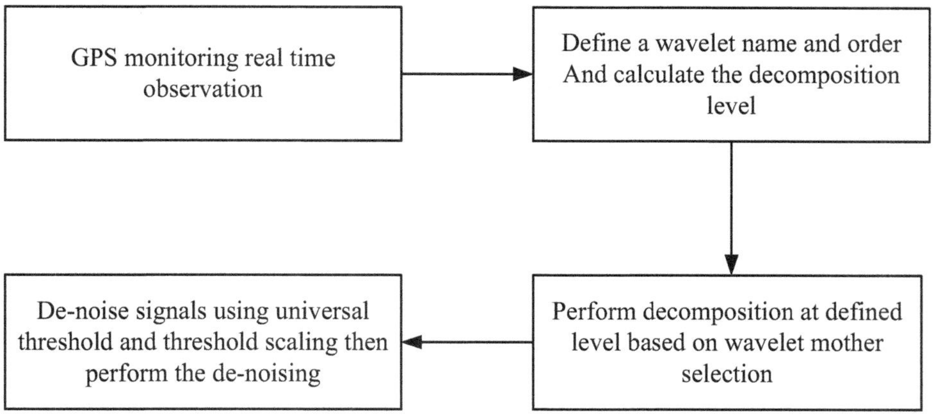

Figure 3. Flow chart diagram of the proposed wavelet de-noising model design.

2.4. *PCA model*

PCA is among the most popular methods for extracting information from the collected data, and it has been applied in a wide range of disciplines (Pytharouli & Stiros 2010, 2012). PCA is suitable for movement monitoring, whereas the correlated variables are being measured simultaneously. The PCA transforms an $(n \times p)$ data matrix, X, by combining the variables as a linear weighted sum as follows:

$$X = TP^T \tag{10}$$

where P is the principal component loadings, T is the principal component scores, n and p are the number of measurements and variables, respectively. In the case of GPS monitoring data analysis, it is assumed that the variables follow a Y-dimensional multivariate (recorded coordinates) normal distribution with mean vector μ' $= (\mu_1, \mu_2, \ldots, \mu_p)$ and covariance matrix Σ, where μ_i is the mean for the ith variable and Σ is a matrix consisting of the variances and covariance of the (Y) variables. The single solution is given by the singular value decomposition (SVD) of Y as follows:

$$Y = U\Lambda V^T \tag{11}$$

where U is the matrix of coordinates of the observations on the principal components, V is the matrix of the loadings constituting these principal components, and Λ is a diagonal matrix such that λ_i^2 is the variance of the ith principal component.

In this method, Aminghafaria et al. (2006) proposed to threshold detail coefficients after they had been projected in the PCA base. In parallel, PCA is performed on approximation coefficients to keep only the most important features of the GPS signals. In this study, the thresholding step of the algorithm was modified by using a heuristic threshold-like in equation (12). This threshold was used for detail coefficients. This empirical modification of the threshold aimed at increasing the threshold value and gave better de-noising results in our study:

$$\delta = \hat{\sigma}\sqrt{2\log(p)\times\max|d_1|} \tag{12}$$

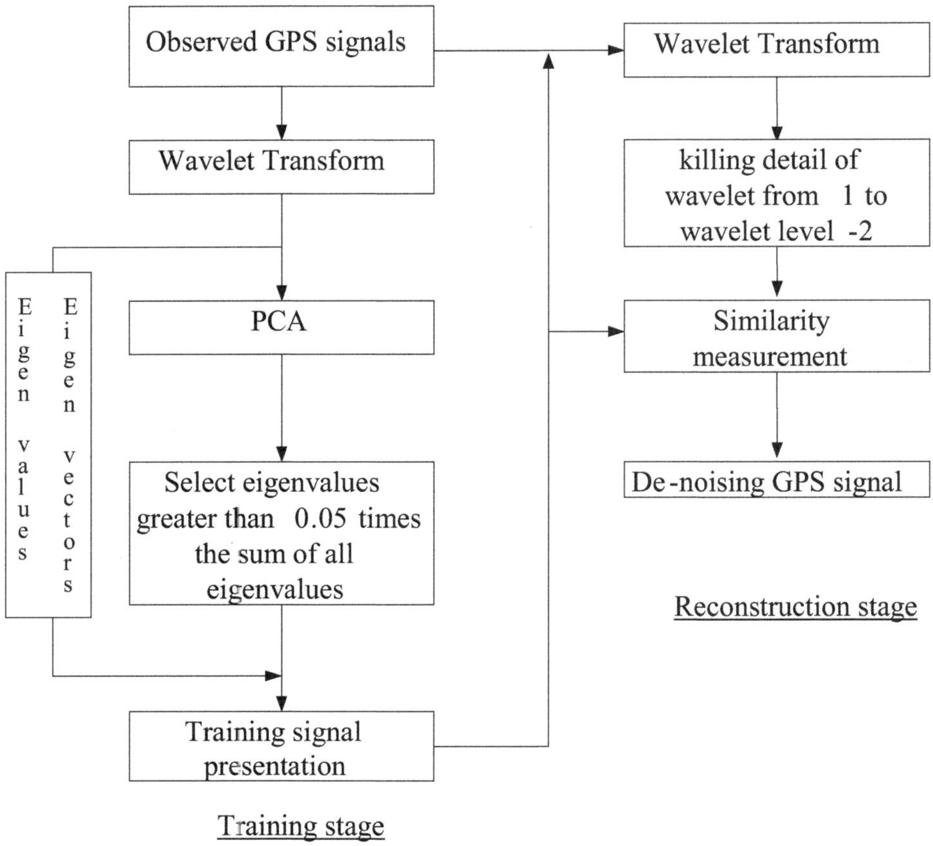

Figure 4. Block diagram of the proposed wavelet PCA model.

where p is the length of the GPS observations and $\hat{\sigma}$ the estimate of the noise standard deviation based on the median absolute deviation (mad) of the wavelet detail coefficients at level 1 (d_1),

$$\hat{\sigma} = \sqrt{2}\text{mad}(d_1)/0.6745 \tag{13}$$

In this study, figure 4 shows the wavelet PCA method stages and process.

The methodology used to eliminate the GPS errors is shown in figure 4. It consists of two stages: training and reconstruction stages. In the training stage, the level of wavelet transformed (DWT) at level J of each observation direction of the GPS observation matrix X with Symlets mother wavelet. The eigenvalues and eigenvectors are calculated from auto-correlation matrix and then arranged the eigenvalues in a descending order to select the eigenvectors with eigenvalues greater than 0.05 times the sum of all eigenvalues of signals. Finally, this stage displays the original and reconstructed signals. In the reconstruction stage, the quality of the reconstructed signals (built from training stage) is checked by calculating the relative mean square errors (should be close to 100%). From the previous stage, the numbers of retained principal components are presented. These results can improve the signals by removing the noise based on killing the wavelet details at selected levels. The correlations between de-noised and original signals were calculated. The correlation returns a value between minus one and one. If it equals one, then the signals are perfectly matched. If it equals minus one, this indicates negative dependency between signals. This model returns a simplified version of the input GPS observation signal obtained from wavelet-based multiscale PCA (Mathworks 2008).

3. Case study: GPS monitoring of bridges

Deformation and movement of bridges are among the problems that widely exist in bridge engineering practice. Therefore, it is very important to monitor and analyse bridge deformation to ensure their safety. In addition, the real-time kinematic GPS (RTK-GPS) mode is an important tool to monitor the continuous movement of natural disasters and structures in short- or long-monitoring time period. This study presents Mansoura-Steel Railway Bridge, Egypt, as shown in figure 5, and was

Figure 5. Mansoura Railway Steel Bridge and GPS monitoring system: (a) bridge view, (b) R2 rover point position, (c) R1 rover point position, (d) base station set-up point, (e) Google Earth View of bridge and GPS monitoring point positions.

applied to prove the performance of the previous methods and movement analysis in response to passing trains. A structural monitoring system is designed based on the RTK-GPS to monitor and assess the bridge behaviour and movements under the effect of trains' loads. This bridge was constructed in 1913 and is the oldest bridge in Mansoura City, Egypt. As shown in figure 5(*a*)–(*c*), the bridge comprises four truss girders and five spans. Each truss span is 70.00 m. This bridge was used for two types of traffic: trains on the middle (double-track) and one vehicle lane on each side of the bridge. This bridge is used to connect the Egyptian railway lines between the East and West Dommieta Nile River branch. In this study, the RTK-GPS (1 Hz) technique is used, for the RTK survey base station is set up over stable ground, as shown in figure 5(*d*) and (*e*), and the radio transmitter is attached. Yeh et al. (2012) concluded that the rover station must be located within ~10 km of the reference station to achieve centimetre-level accuracy. In this study, the GPS base receiver was placed approximately 188.00 m away from point R2 at stable ground, as shown in figure 5 (*e*). Since the base line is short, errors such as ionospheric, tropospheric, and orbit biases are assumed to be essentially zero.

The data analysis in the present paper came from the GPS (rover) receivers clamped at the centres of the top–mid span of the main first span girder (R2) and one at the bridge deck (R1) as shown in figure 5(*b*), 5(*c*), and 5(*e*). Under the mode of the RTK, the reference station serves as a stationary checkpoint of which 3D coordinates have been previously determined by the conventional static GPS method and constantly records the difference between its known position and the position calculated from the satellite data (Berber & Arslan 2013). The detected differences are indicative of the errors from the satellite hardware and, more importantly, lower atmospheric delays with low time required to calculate a rover position. An ultra-high-frequency radio set is then used to send the errors to the rover. The rover, which is the GPS receiver of which the position is being tracked, uses this error information to improve its accuracy. The clock offsets in the receivers, satellites, and atmospheric propagation delays can be ignored because the two receivers are in close proximity, which means that the errors are strongly correlated. In this study, the base GPS, rover GPS, and radio unit are used to collect raw data at rate of 1 Hz, as shown in figure 5(*b*)–(*d*). The measuring condition was favourable: the receiver was free of any obstruction of 15° angle view of the horizon and at least four satellites were tracked continuously. A rover and base-type Trimble-5700 dual-frequency receivers recording at a rate of 1 Hz were used. In addition, the accuracy of GPS instruments used is 1 cm + 1 ppm (length of base line) in horizontal direction and 2 cm + 1 ppm (length of base line) in the vertical direction.

The data collected were post-processed using GPS-Trimble Business Center software. The outputs of the GPS software were the time series of instantaneous Cartesian coordinates of the rover receiver in the WGS84 coordinate system (N, E, H). A local bridge coordinate system (BCS) (X, Y, Z) was established to be used in the analysis and evaluation of the observed data. The coordinates in WGS84 are transformed into those in BCS by 2D similarity transformation (equation (14)). The azimuth (α) of the bridge is 21° 52′ 23.29″, and can be calculated based on the two monitoring points on the bridge girder. Herein, the X-data represent the displacement changes along the longitudinal of the bridge, the Y-data represent the displacement changes along the transverse of the bridge, and the Z-data represent the relative displacement changes along the altitude direction of bridge:

$$\begin{bmatrix} X \\ Y \\ Z \end{bmatrix} = \begin{bmatrix} \cos\alpha & \sin\alpha & 0 \\ -\sin\alpha & \cos\alpha & 0 \\ 0 & 0 & 1 \end{bmatrix} \begin{bmatrix} N \\ E \\ H \end{bmatrix} \tag{14}$$

The receiver coordinates in the three dimensions (X, Y, Z) were transformed into time series of apparent displacements (DX_i, DY_i, DZ_i; $i = 1, 2, \ldots, n$) around a relative zero representing the equilibrium level of the monitoring point. This similar transformation was based on the following equation:.

$$\begin{bmatrix} DX_i \\ DY_i \\ DZ_i \end{bmatrix} = \begin{bmatrix} X_i \\ Y_i \\ Z_i \end{bmatrix} - \frac{1}{n}\sum \begin{bmatrix} X_i \\ Y_i \\ Z_i \end{bmatrix} \tag{15}$$

where n represents the total number of interval recordings.

3.1. Test de-noised models analysis

Figure 6 shows the 500-second unload unfiltered apparent displacement of the three dimensions of the two monitored bridge points. As shown in figure 6, the relative displacements of the two points are calculated based on equation (15) and the errors which are more than three times the standard deviation of the relative movements are removed. Table 1 presents the statistical signals (variance (Var.) and root mean square error (RMSE)) of the apparent displacement presented in figure 6.

The RMSE for the error de-noised movements of X-, Y-, and Z-directions for point R1 is shown in table 2. From the comparison between these results and the RMSE in table 1 for the same point, it can be concluded that the multipath error and observation noise are the main errors that affect the GPS bridge deck observations. Also, from figures 6–8 and tables 1 and 2, it can be shown that the multipath error of the GPS observation of point R1 is higher than that of point R2, which assures the GPS multipath error concept. Also, it contributed by about 30%–50% of the total errors of deck measurements and the remaining signals represent both the observation noises and the dynamic movement component.

Otherwise, figure 7 shows the kernel density estimate (KDE) of residual between GPS time-series observations and PCA de-noised time series of X, Y, and Z movement directions for the bridge deck (D) and top girder (G). From this figure, it is noticeable that the KDE of the girder movement residual in the three directions are higher than the deck movement residual and the kurtosis of Z movement residual of deck and girder. It means that the ambient noise and multipath errors are significant in deck observations more than in girder movements, especially in Z-direction. In addition, it can be seen that the KDE of Z-direction is lower than other directions in both bridge deck and girder. Also, it can be seen that the movement residual of the bridge deck in Z-direction is higher than in other directions, which means that the error component of Z-direction observation is higher than that of other directions. The above assures the GPS observation accuracy.

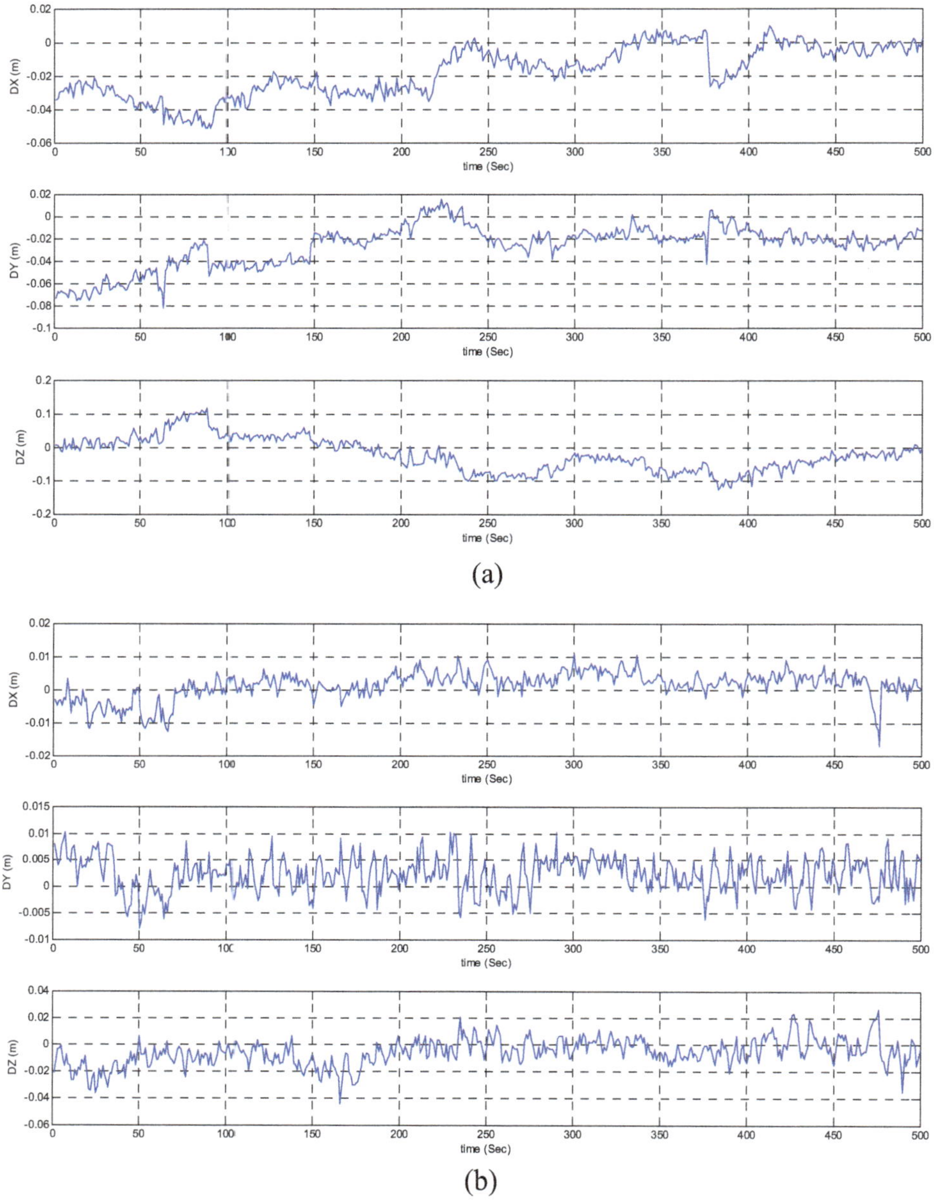

Figure 6. Bridge monitoring movement points of (a) deck and (b) girder.

Table 1. Bridge displacement statistical measurements (unit: mm).

	Bridge deck (R1)		Bridge girder (R2)	
Dir.	Var.	RMSE	Var.	RMSE
X	0.22	22.8	1.72E−2	4.4
Y	0.35	32.2	1.11E−2	4.1
Z	2.42	53.4	0.11	11.8

Table 2. Comparison of apparent GPS displacement error after filter for point R1.

Method		RMSE (mm)	EPP (mm)	ENP (mm)	SNR (dB)
PCA model	*X*	2.9	13.4	−11.0	40.85
	Y	4.4	13.5	−32.0	39.62
	Z	11.2	4.4	−38.0	30.81
Compressed model	*X*	3.0	13.4	−11.0	40.29
	Y	4.5	13.5	−32.0	39.14
	Z	11.2	4.4	−38.0	30.80
De-noised model	*X*	3.0	11.5	−7.8	40.29
	Y	4.5	14.0	−29.3	36.14
	Z	11.2	4.4	−38.0	30.80

The design models, which are mentioned in section de-noising models, were applied on the bridge deck GPS observations as shown in figure 6(*a*). The standard deviation for the real data are 0.015, 0.019, and 0.05 m for *X*-, *Y*-, and *Z*-directions, respectively. The wavelet de-noise results are shown in figure 8. For the compressed model, the GPS observations smoothed return a compressed of the original multisignal GPS original matrix, whose wavelet decomposition structure. The output GPS

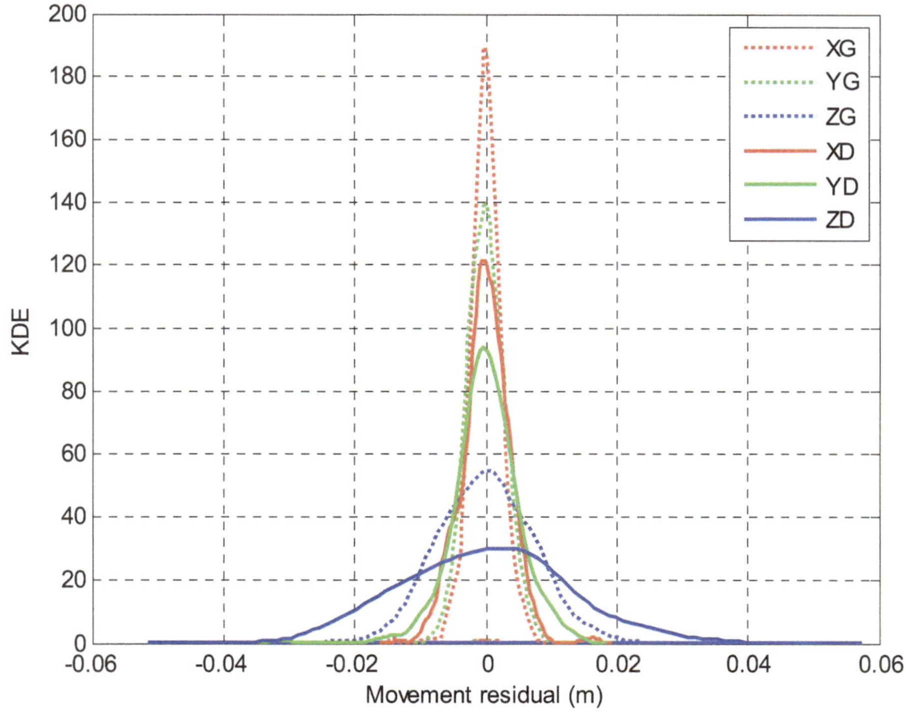

Figure 7. Kernel density estimate of movement residuals of bridge deck (D) and girder (G).

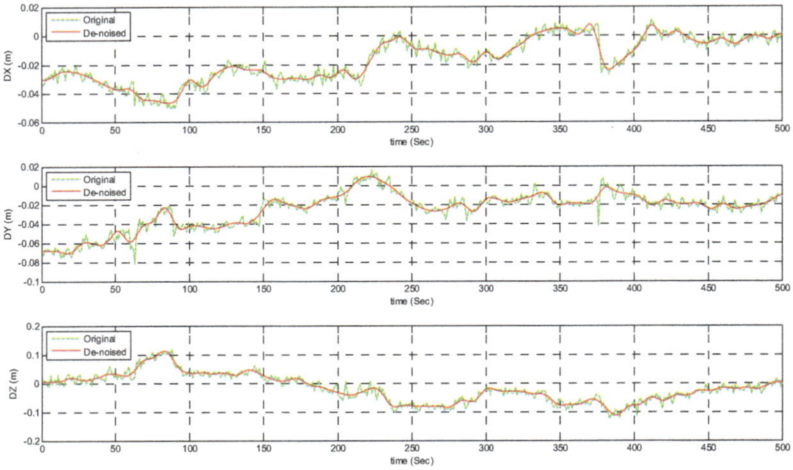

(a) Compressed model

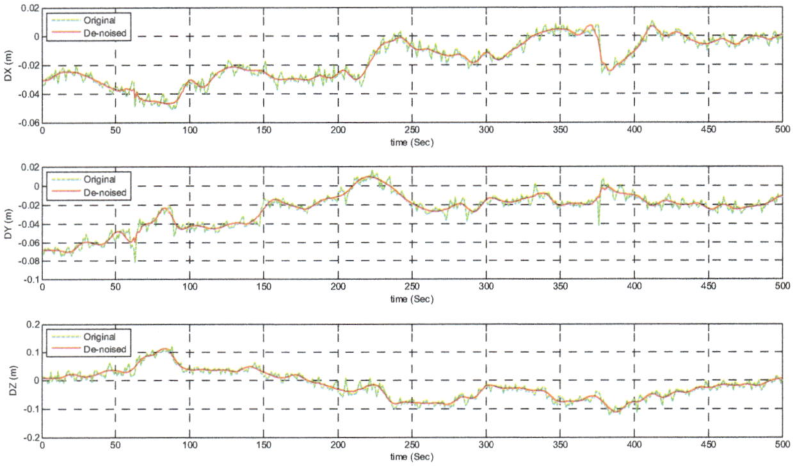

(b) De-noised model

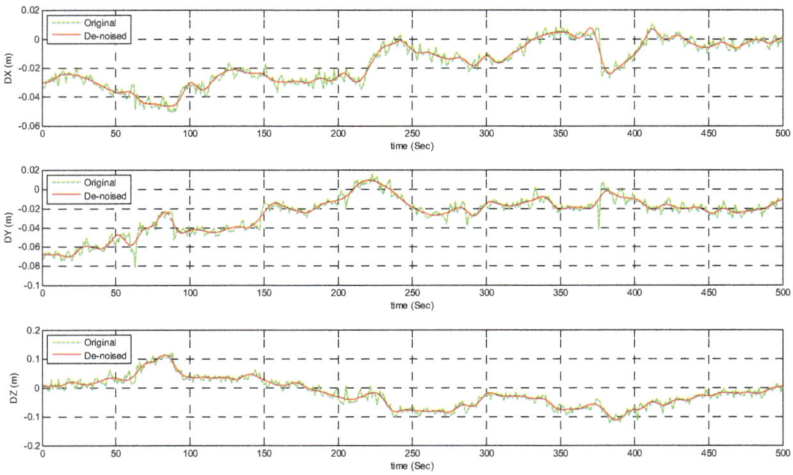

(c) PCA

Figure 8. De-noised GPS deck bridge monitoring movements.

smoothed is obtained by thresholding the wavelet coefficients based on soft thresholding; however; this method returns the wavelet decompressed at level 3 of each GPS direction using wavelet mother Symlet 12, which is the wavelet decomposition associated with the smoothed GPS observation at zero WDT shift. Three approximate coefficients and three cells for the detailed coefficient are used with the decomposition and reconstruction wavelet and scaling function coefficient, refer to figure 1; amplitudes at range are −0.17; 0.76 and −0.46; 0.76, respectively. The threshold value calculations are 0.0174, 0.0221, and 0.0419 for *X*-, *Y*-, and *Z*-directions, respectively. In addition, the balance sparsity-norm method (Mathworks 2008) is used in the compress method. The GPS observations and the associated parameter are not needed to define this method. From figure 8(*a*), it can be shown that the compress method can eliminate the GPS errors and noises, whereas the standard deviation of smoothed signals are 0.014, 0.018, and 0.047 m for *X*-, *Y*-, and *Z*-directions, respectively. In addition, the correlation between the real data and reconstruct GPS observation is 0.98 for three directions, approximately.

For the de-noising wavelet model using DWT at level 3 with Symlets mother wavelet ($N = 12$), the output GPS smoothed is obtained by thresholding the wavelet coefficients based on soft universal threshold with defined associated parameter for the threshold scaling used at level-dependent estimation of level noise (Aminghafaria et al. 2006; Mathworks 2008). In this method, the threshold value calculations are 0.0073, 0.0178, and 0.046 for *X*-, *Y*-, and *Z*-directions, respectively. From figure 8(*b*), it can be shown that the de-noised method can eliminate the GPS errors and noises, whereas the standard deviation of smoothed signals are 0.014, 0.018 and 0.047 m for *X*-, *Y*-, and *Z*-directions, respectively. In addition, the correlation between the real data and reconstruct GPS observation is 0.98 for three directions, approximately.

For the wavelet PCA model using the DWT at level 3 of each spectrum of the observation matrix *X* with Symlets mother wavelet ($N = 12$) and then running the steps of wavelet PCA which are discussed in PCA model. From the results it can be seen that the quality of the reconstructed signals equals 98.25 and 98.04% for *X*- and *Y*-directions, whereas in *Z*- direction it is equal to 95.60%. In addition, the correlation between the real data and reconstruct GPS observation is 0.98 for three directions, approximately. From figure 8(*c*), it can be shown that the wavelet PCA method can eliminate the GPS errors and noises, whereas the standard deviation of smoothed signals are 0.014, 0.018, and 0.047 m for *X*-, *Y*-, and *Z*-directions, respectively.

In addition, table 2 summarizes the statistical (RMSE, error positive peak ($EEP = \max(x_i' - x_i)$), error negative peak ($ENP = \min(x_i' - x_i)$), and signal-to noise ratio (SNR) (equation (16)) of the residuals between the original (GPS time series) and de-noised signals (long-period movement component (Moschas & Stiros 2011)), which is a short-period movement component, based on the design models. The maximum SNR and minimum RMSE are better effects of the de-noising and filter models:

$$SNR = 20\log \frac{\sqrt{\sum_{i=1}^{N} x_i^2}}{\sqrt{\sum_{i=1}^{N} (x_i' - x_i)^2}} \tag{16}$$

where x_i *and* ${}^{i'}_{x}$ are the de-noised and measured apparent displacements, respectively, and N is the number of observations.

From table 1, it can be seen that the RMSE in *Z*-direction is higher than in *X*- and *Y*-directions, which means that the observation noise and GPS multipath errors of the bridge deck are high in this direction. Also, the concept accuracy of GPS in *Z*-direction is lower than in *X*- and *Y*-directions. From figure 8, it can be seen that the de-noised signal and the apparent displacement shapes are highly correlated. In addition, the wavelet models simulated the movement from GPS time-series observations without noise in the three-direction movement observations. Also, it can be shown that the signal noises are affected in the de-noised signal for the de-noised models. From table 2, it can be seen that the three models can be used in the GPS de-noise signal to extract the long-period bridge movement components with high accuracy, whereas the RMSE and SNR values for three models are much closed. Furthermore, the best model in *X*- and *Y*-directions is the PCA model, while in *Z*-direction the three models can be used. The short-period movement component can be calculated after subtracting the long-period time series from the apparent movements. Now, the dynamic displacement of bridge can be extracted after the signal is de-noised based on applying the band-pass filter on the short-period movement component (Moschas & Stiros 2011).

3.2. Monitoring bridge during passing trains

Figure 9 shows the short-time 3D time-series movement observation for point R2 during the passing of three trains on the bridge in both directions. The observed time when the trains passed on the bridge were from 839 to 1649 seconds. Moschas and Stiros (2011) concluded that a simple high-pass filter permits us to separate apparent displacement into a long-period and a short-period components. The long-period component is sometimes identified with the background or ambient noise (Moschas & Stiros 2011, 2013). In this section, the PCA filter model is used to de-noise the GPS time history observation and extract the long-period component displacement of the bridge girder. In our case, it contains both the ambient noise and the semi-static displacement.

From figure 9(*a*), it can be seen that the period of signals is shown clear from 810 to 2100 seconds due to the passing of trains at this time period. In addition, the correlation between recorded and filtered data is very high with no displacement information losses. Accordingly, the PCA model filter is suitable to extract the long-period component displacement from GPS-recorded observations. Also, it can be seen that the *X*- and *Y*-direction movements occur in the same direction and *Z*-direction displacement occurs in the opposite direction at the same time interval of the passing trains. It is noticed that the maximum long-period component displacements are 17.0, 23.4, and 76.5 mm; the mean displacements deviation are 3.85, 4.74, and 13.92 mm; and the standard deviation displacements are 4.6, 6.2, and 19.1 mm in the *X*-, *Y*-, and *Z*-directions, respectively. The correlations between the 3D displacements are 0.82 (*X* and *Y*), −0.54 (*X* and *Z*), and −0.72 (*Y* and *Z*).

These results indicate that the correlation between *X*- and *Y*-directions is a very high positive correlation, while the correlations between *X*- and *Y*-directions and *Z*-direction are low and negative due to the passing trains. Accordingly, it can be concluded that the performance of the GPS observations is highly sensitive when trains are passing on the bridge, and the long-period component bridge behaviour under train passes is safe. In addition, the correlation between the two direction (*X* and *Y*) movements and *Z*-direction movement are strongly influenced.

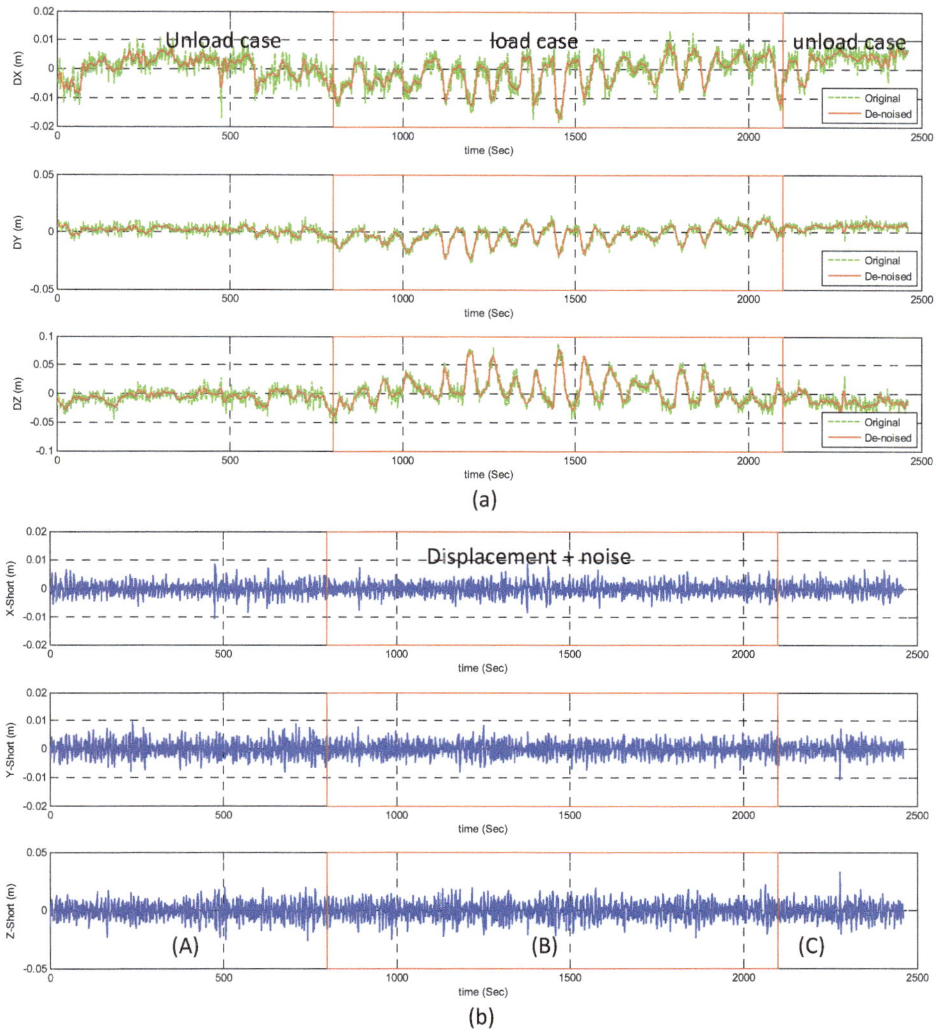

Figure 9. Girder bridge movement components during train's passes: (a) apparent and extracted long-period component displacement, (b) short-period component of the apparent bridge displacement.

From figure 9(*b*), it is shown that the short-period component, which can be computed based on the supervised long-period component from apparent displacement, contains a dynamic displacement and remnant noise (Moschas & Stiros 2013). The extracted short-period component can be used to examine the quality of the filter used and guide understanding of the behaviour of structures under affecting loads. From figure 9(*b*), it can be seen that the short-period component ranges are 19.4, 20.1, and 59.2 mm, and the standard deviations are 2.1, 2.5, and 7.2 mm in the *X*-, *Y*-, and *Z*-directions, respectively. This result indicates that the maximum short-period component occurs in the *Z*- and *Y*-directions, and then in the *X*-direction. These results occurred due to the GPS accuracy and dynamic components for these directions. In addition, as per the observed conditions for the monitoring R2 point, it is

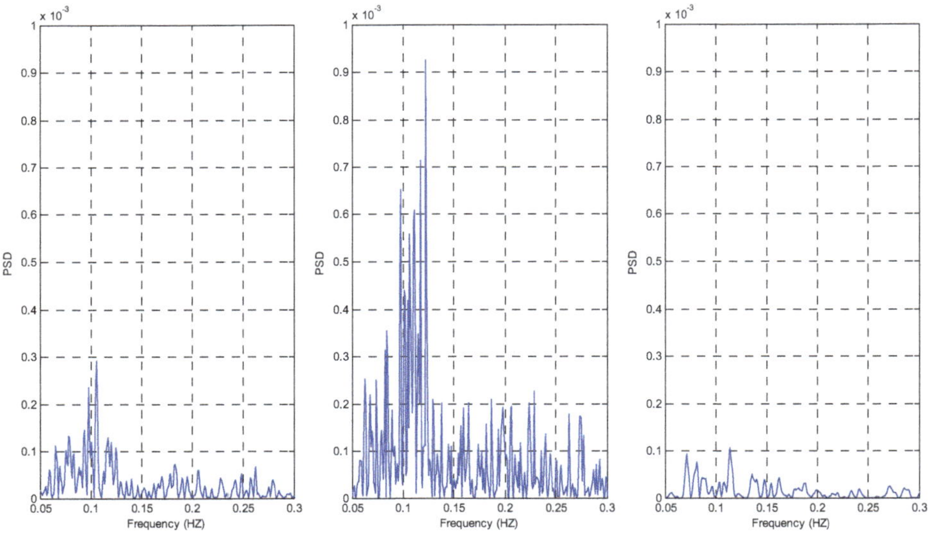

Figure 10. Spectral analysis results for the three intervals A, B, and C of the short-period component for the Z-direction as shown in figure 7(b).

concluded that the dominant noise for the short-period component is observation noise for the monitoring point. In addition, figure 10 shows the spectral analysis for the three interval cases: A (before the train passes), B (when the train passes), and C (after the train passes) (as shown in figure 9(b)) for the Z-direction short-period component.

The bridge frequency was determined by using the fast Fourier transformation based on Hamming window and band-pass filter for the frequency range 0.02–0.3 Hz. From this figure, it can be seen that the high power spectral density of frequency occurred during the train passes. The interval B contains a peak in the frequency range 0.097–0.122 Hz, while intervals A and C contain only white noise and low-frequency, coloured noise characterizing GPS measurements. The results of the spectral analysis, therefore, permit us to confirm that the interval B contains the vibration signal. Finally, from this study result and the previous de-noising GPS structural monitoring limited studies (Ogaja et al. 2003; Yu et al. 2006), it can be concluded that the wavelet analysis can mitigate the errors for GPS time-series monitoring data. In addition, from this study it can be seen that the wavelet de-noising models can be used to analyse the behaviour of structure components without using any other filters. In addition, this method is simplified and used in the process of structural health monitoring systems.

4. Summary

Recently, GPS is used widely in structural health monitoring in monitoring the structures' deformations under different loads or in the construction of important structures. For analysing, pre-processing of the GPS signals should be done first to de-noise the GPS errors and measurements noises, which are the major barrier in achieving high accuracy of a GPS positioning solution that is affected by many factors and restraints. Therefore, noise reduction of GPS observations, improvement of the accuracy of the GPS time series, and detection of deformation epochs are the key issues of long- and short-movement period component analysis.

Wavelet smoothing has been extensively used in signal de-noising and extraction due to its good localized time–frequency features. Eliminating the signal noise must be processed into three steps: first, to compute the wavelet decomposition of the observed signal up to level J; second, to threshold conveniently the wavelet detail coefficients; and third to reconstruct a de-noised version of the original signal, from the threshold detail coefficients and the approximation coefficients, using the inverse wavelet transform.

The case study shows the data-sets which have been de-noised by using the three wavelet models. The conclusions of the case study can be summarized as follows:

- The wavelet methods are simple and fast tools in extracting the de-noised GPS observations in structural health monitoring systems.
- The GPS measurements can be used as a trustworthy tool for characterizing the dynamic behaviour of bridges in both time and frequency domains.
- The GPS multipath errors can be contributed to about 30%–50% of the total errors.
- The de-noised signal represents the long-period movement component of the structure, while the short-period movement component is represented by subtracting the long-period component from the apparent displacement.
- The spectral analysis can be confirmed with the vibration signal intervals.

Acknowledgements

The authors would like to thank the reviewers for their valuable comments and suggestions to improve the quality of the paper. They are also grateful to Mr. Mohamed Sayed and Mr. Mohamed Tharwat for their technical support and collection data.

Funding

This work was supported by the National Research Foundation of Korea Grant funded by the Korean Government [grant number NRF-2012K2A4A1034757] and Mansoura University, Egypt.

References

Aminghafaria M, Chezea N, Poggia J. 2006. Multivariate denoising using wavelets and principal component analysis. Comput Stat Data Anal. 50:2381–2398.

Bakshi B. 1998. Multiscale PCA with application to multivariate statistical process monitoring. AIChE J. 44:1596–1610.

Berber M, Arslan N. 2013. Network RTK: a case study in Florida. J Int Measurement Confed. 46:2798–2806.

Chui CK. 1992. Introduction to wavelets. San Diego (CA): Academic Press.

Cohen A, Daubechies I, Jawerth B, Vial P. 1993. Multiresolution analysis, wavelets, and fast algorithms on an interval. Comput Rendus Acad Paris. 316:417–421.

Danskin S, Bettinger P, Jordan T. 2009. Multipath mitigation under forest canopies: a choke ring antenna solution. J Forest Sci. 55:109–116.

Daubechies I. 1988. Orthonormal bases of compactly supported wavelets. Commun Pure Appl Math. 41:909–996.

Deisenroth MP, Huber MF, Hanebeck UD. 2009. Analytic moment-based Gaussian process filtering. In Danyluk A, College W, editors. Proceedings of the 26th Annual International Conference on Machine Learning; 2009 June 14–18; Montreal, QC, Canada; p. 225–232.

Donoho L. 1995. De-noising by soft-thresholding. IEEE Trans Inf Theory. 41:613–627.

Godsill SJ, Doucet A, West M. 2004. Monte Carlo smoothing for nonlinear time series. J Am Stat Assoc. 99:438–449.

Im S, Hurlebaus S, Kang Y. 2011. A summary review of GPS technology for structural health monitoring. J Struct Eng. 139:1653–1664.

Julier SJ, Uhlmann JK. 2004. Unscented filtering and nonlinear estimation. Proc IEEE. 92:401–422.

Kaloop M. 2012. Bridge safety monitoring based-GPS technique: case study Zhujiang Huangpu Bridge. J Smart Struct Syst. 9:473–487.

Ko J, Fox D. 2009. GP-Bayes filters: Bayesian filtering using Gaussian process prediction and observation models. Auton Robots. 27:75–90.

Ko J, Ni Y. 2005. Technology developments in structural health monitoring of large-scale bridges. J Eng Struct. 27:1715–1725.

Lilong L, Hongyan W, Bin L. 2010. Mitigation of systematic errors of GPS positioning based on wavelet denoise systems. In: Huang X, Xu LD, Zhou ZD, Fan Z, Gupta MM, editors. 2010 Second WRI Global Congress on Intelligent Systems, Wuhan, China, 16–17 Dec. 2010, IEEE; p. 253–255.

Ma H, Hu Y, Wu J, Wang J, Guo W. 2009. Application of lifting scheme translation-invariant wavelet de-noising method in GPS/INS integrated navigation. In: Proceedings of the 2009 International Conference on Wavelet Analysis and Pattern Recognition; 12–15 July; Baoding. p. 303–307.

Mathworks. 2008. Matlab, Release 12. Natick, MA: Mathworks, Inc.

Meng X. 2002. Real-time deformation monitoring of bridges using GPS/accelerometers [PhD thesis]. The University of Nottingham.

Meyer Y. 1989. Orthonormal wavelet, in wavelets. New York (NY): Springer.

Moschas F, Stiros S. 2011. Measurement of the dynamic displacements and of the modal frequencies of a short-span pedestrian bridge using GPS and accelerometer. Eng Struct J. 33:10–17.

Moschas F, Stiros S. 2013. Noise characteristics of high-frequency, short-duration GPS records from analysis of identical, collocated instruments. Measurement. 46:1488–1506.

Ogaja C, Wang J, Rizos C. 2003. Detection of wind-induced response by wavelet transformed GPS solutions. J Surv Eng. 129:99–104.

Oluropo O, Gethin W, Christopher J. 2014. GPS monitoring of a steel box girder viaduct. J Struct Infrastruct Eng. 10(1):25–40.

Psimoulis P, Pytharouli S, Karambalis D, Stiros S. 2008. Potential of Global Positioning System (GPS) to measure frequencies of oscillation of engineering structures. J Sound Vibr. 318:606–623.

Pytharouli S, Stiros S. 2010. Kinematics and rheology of a major landslide based on signal analysis techniques. Geotechnique. 60:201–222.

Pytharouli S, Stiros S. 2012. Analysis of short and discontinuous tidal data: a case study from the Aegean Sea. Surv Rev. 44:239–246.

Roberts G, Meng X, Dodson A, Cosser E. 2002. Multipath mitigation for bridge deformation monitoring. J Glob Position Syst. 1:25–33.

Sone A, Yamatoto S, Masuda A, Nakaoka A. 1996. Health monitoring system of building by using wavelet analysis. Proceedings of the Eleventh World Conference on Earthquake Engineering, Acapulco, Mexico, 23–28 June 1996. (Paper No. 235).

Wong K. 2007. Design of a structural health monitoring system for long-span bridges. J Struct Infrastruct Eng. 3(2):169–185.

Wu L, Ma H, Ding W, Hu Q, Zhang G, Lu D. 2011. Study of GPS data De-noising method based on Wavelet and Kalman filtering. In: Li J, editor. Circuits, Communications and System (PACCS), Third Pacific-Asia Conference, July 17–18, Wuhan, China. IEEE; p. 1–3.

Yeh TK, Chao BF, Chen CS, Chen CH, Lee ZY. 2012. Performance improvement of network based RTK GPS positioning in Taiwan. Surv Rev J. 44:3–8.

Yu M, Guo H, Zou C. 2006. Application of wavelet analysis to GPS deformation monitoring. In: Proceeding of Position, Location, and Navigation Symposium, 24–27 Apr. 2006; Coronado, CA. IEEE; p. 670–676.

Zhong P, Ding X, Zheng D, Chen W, Huang D. 2008. Adaptive wavelet transform based on cross-validation method and its application to GPS multipath mitigation. J GPS Solut. 12:109–117.

7

Landslide hazard mapping considering rainfall probability in Inje, Korea

M.J. LEE†, I. PARK‡, J.S. WON§ and S. LEE*¶#

†Korea Adaptation Center for Climate Change, Korea Environment Institute (KEI), Seoul 122-706, Korea

‡Department of Geoinformatics, University of Seoul, Seoul 130-743, Korea

§Department of Earth System Sciences, Yonsei University, Seoul 120-749, Korea

¶Geological Research Division, Korea Institute of Geoscience & Mineral Resources (KIGAM), Daejeon 305-350, Korea

This study evaluated the landslide hazard at Inje, Korea, using a geographic information system (GIS) and rainfall probabilistic data. The locations of landslides were identified in the study area by aerial photograph interpretation and field surveys. Data about rainfall probability, topography, and geology were collected, processed, and compiled in a spatial database using GIS. Then, the probability of landslides in the study area in recurrence interval years in the future was calculated assuming that landslides are triggered by a daily rainfall of 202 mm or a three-day cumulative rainfall of 449 mm. Twelve factors that influence landslide occurrence were chosen from a database of topography, soil, and forest cover. Landslide susceptibilities were analysed and mapped according to these landslide-occurrence factors, employing the frequency ratio method. Of the total landslide locations, 50% were used for hazard analysis and the remaining 50% were used for model validation. Validation results for the daily rainfall of 202 mm and three-day cumulative rainfall of 449 mm for recurrence interval years were from 89.22% to 91.80% and from 89.38% to 93.80%, respectively. This analysis of landslide hazards took rainfall probability into account. Rainfall, including heavy rainfall, is expected to increase in the future.

1. Introduction

Korea experiences frequent landslides. Notable landslide events occurred in 1996, 1998, 1999, 2002, and 2006, and increased landslide frequency has been noted in recent years (http: //sansatai.forest.go.kr). The Fourth Report of Climate Change Assessment, published by the Intergovernmental Panel on Climate Change (IPCC) in 2007, predicted climate change would persist for several more centuries due to the greenhouse gases that have already been discharged, even though they could be mitigated by efforts such as reduction in greenhouse gas emissions. The abnormal climate and localized torrential rain caused by climate change may lead to higher landslide rates (Evans & Clague 1994; Blijenberg 1998; Van Steijn 1999; Jomelli et al. 2007; Floris et al. 2010). An analysis of the causes of landslides, with a multilateral, holistic

*Corresponding author. Email: leesaro@kigam.re.kr

#Present address: Korea University of Science and Technology, Daejeon 305-350, Korea view, is needed, given this expected increase in landslides is due to climate change. Damage from landslides can be minimized by assessing landslide susceptibility and hazards (IPCC 2007). Landslides are major natural geological hazards that annually cause enormous property damage and incur both direct and indirect costs. One of the main triggers of landslides is the concentration of heavy rainfall within a short time interval. Rainfall probability analysis suggests that intense heavy rainfall events will become even more frequent in the future. To reduce the damage from landslides associated with heavy rainfall, we must develop a scientific methodology to investigate landslide occurrence in the context of potential changes in future climate. Understanding current and past relationships between rainfall and landslides is central to predicting future landslides. At the same time, a method to quantitatively predict changes in rainfall due to climate change is also needed.

Fell et al. (2008) defined landslide susceptibility, hazard, and risk. Landslide susceptibility is a quantitative or qualitative assessment of the classification, volume (or area), and spatial distribution of landslides which exist or potentially may occur in an area. Susceptibility may also include a description of the velocity and intensity of the existing or potential landsliding. Although it is expected that landsliding will occur more frequently in the most susceptible areas, in the susceptibility analysis, the time frame is not explicitly taken into account. Landslide susceptibility includes landslides which have their source in the area, or may have their source outside the area but may travel onto or regress into the area. A hazard is a condition with the potential for causing an undesirable consequence. Descriptions of landslide hazards should include the location, volume (or area), classification, and velocity of the potential landslides and any resultant detached material, and the probability of their occurrence within a given period of time. Risk is a measure of the probability and severity of an adverse effect to health, property, or the environment. Risk is often estimated by the product of the probability of a phenomenon of a given magnitude times the consequences. However, a more general interpretation of risk involves a comparison of the probability and consequences in a non-product form. For quantitative risk assessment, the use of the landslide intensity is recommended.

Many methods have been proposed for assessing landslide susceptibility, with increasing use of geographic information systems (GIS) and various models. Many of these studies have applied probabilistic models such as the frequency ratio and weight of evidence approaches (Lee & Min 2001; Lee & Pradhan 2006; Oh et al. 2009; Ozdemir 2009; Vahidnia et al. 2009; Regmi et al. 2010; Yilmaz 2010; Lee et al. 2013). Among statistical models, the logistic regression model has also been applied to landslide susceptibility mapping (Lee 2005, 2007; Dong et al. 2009; Nandi & Shakoor 2010; Oh & Lee 2010; Pradhan & Lee 2010a; Bai et al. 2011). More sophisticated assessments have involved fuzzy logic, artificial neural network, analytic hierarchy process, Dempster–Shapfer theory of evidence, and Monte Carlo methods (Lee et al. 2004; Yalcin 2008; Shou et al. 2009; Tangestani 2009; Yilmaz 2009; Pradhan & Lee 2010b; Park 2011; Pradhan 2011). Many research works developed landslide susceptibility maps without considering rainfall. These susceptibility maps have a limit as to predict the hazard area according to rainfall in a real rainfall situation, because susceptibility maps do not reflect the correlation between landslide and rainfall. Thus, research on mapping landslide hazard maps is necessary by considering rainfall data that has a major influence on landslide occurrence. Research on the calculation of rainfall probability has primarily been limited to improving the accuracy of rainfall probability predictions and to studies focused on water resources

(Borga 2002; Herr & Krzysztofowicz 2005, Fell et al. 2008). However, some cases show rainfall probability can be used in various areas. For example, rainfall probability was used for analysing susceptibilities of landslides (Lee et al. 2013). This research subdivides the results and diversifies data by using time series of various recurrence interval years. In this study, landslide hazard maps were produced considering rainfall in the future through temporal analysis.

This study used rainfall probability to evaluate future landslide hazards. First, a GIS database was established by calculating landslide-related items in the study area. Second, the landslide location and frequency ratio of the related items were calculated. Third, a susceptibility map of the study area was created based on the frequency ratio of related items. Fourth, the rainfall probability of the study area was calculated. Fifth, a landslide hazard map was created taking into account the rainfall probability for future recurrence interval years. Finally, the susceptibility and hazard maps of the study area were validated. ArcGIS 10.0 (ESRI, Redlands, CA, USA) was used as the basic analysis tool for spatial management and data manipulation.

The study area was the Inje region (altitude above sea level: max 1220 m and min 190 m), composed of a valley in a mountainous terrain with a river flowing through the middle. A road follows the flow of the river (east–west) and houses can be found within 30 m of the road. Geologically, the area consists mainly of banded gneiss and granite. Soils are sandy loam, rocky loam, rocky sandy loam, and very rocky loam in order of decreasing abundance. In the period 11–15 July 2006, typhoon Ewiniar struck the Inje area with heavy storms and heavy rainfall. Daily rainfall reached 202 mm, with a cumulative rainfall over three days of 447 mm. For the form of the landslide in the study area, the total length was on average about 10 m. The typhoon Ewiniar also caused landslide, which is inactivated until now, during the period 11–15 July 2006. In the 37 km^2 study area, 693 landslides were recorded (figures 1 and 2). The typhoon took lives of 17 people and caused property damage of

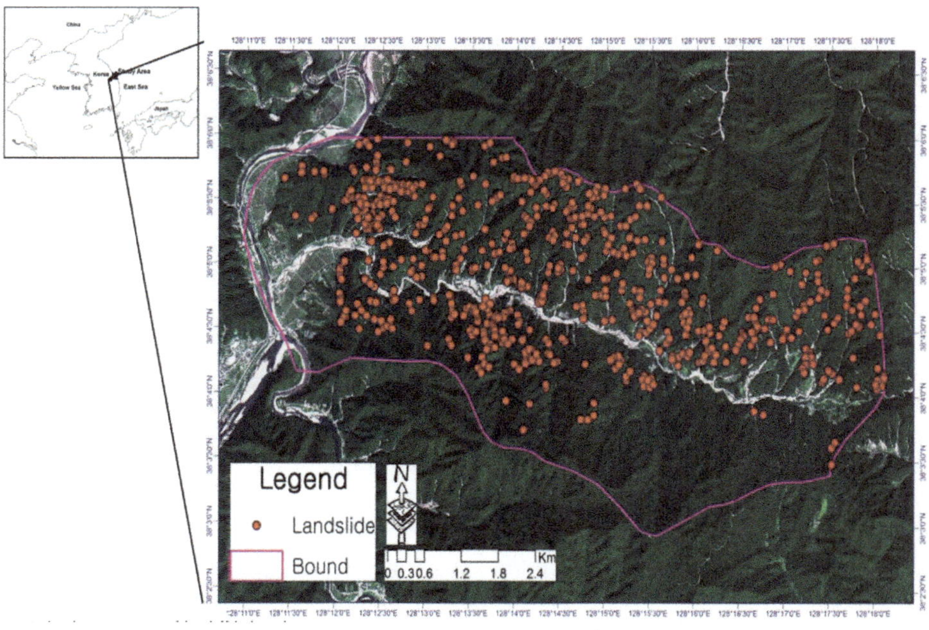

Figure 1. Study area map and landslide location map.

Figure 2. Photographs of landslide in the study area (above : near road in the field).

approximately $100 million. There were total 81 casualties, specifically 17 fatalities and 64 injuries. It was the first large-scale landslide occurrence in the study area. In addition, the rainfall of 2006, when the landslide occurred, was the highest in the history. Thus, this research has only considered damage from typhoon Ewiniar. The extensive damage from rainfall-induced landslides in the Inje area can be partly attributed to the lack of landslide assessments and prediction and thus a lack of measures to minimize damage and injury. Through scientific analysis of landslides, we can assess and predict areas with high landslide hazards and implement preparative measures to decrease the landslide damage. Toward this end, we created a landslide hazard map using rainfall probability analysis techniques based on a frequency ratio model. Rainfall probability analysis suggested that the frequency of concentrated rainfall will increase with climate change and alteration of future rain patterns. Figure 3 presents a flowchart of the entire method employed in this research.

2. Data

Accurate detection of landslide location is crucial for probabilistic landslide hazard analysis. Data from remote sensing methods can be used as a reliable and cost-effective source of information on landslides. In this study, 1:10,000–1:50,000 scale aerial photographs, taken during 2009, were examined to detect landslide locations, which were validated through fieldwork. Recent landslides were detected in the aerial photographs from breaks in the forest canopy, bare soil, or geomorphic characteristics typical of landslide scars, such as head and side scarps, flow tracks, and soil and debris deposits below a scar. To form a database to assess the surface area and number of landslides in the study area, 693 landslides were mapped in the 37 km^2 study area. To apply the probabilistic method, a spatial database that considers landslide-related factors was designed and constructed.

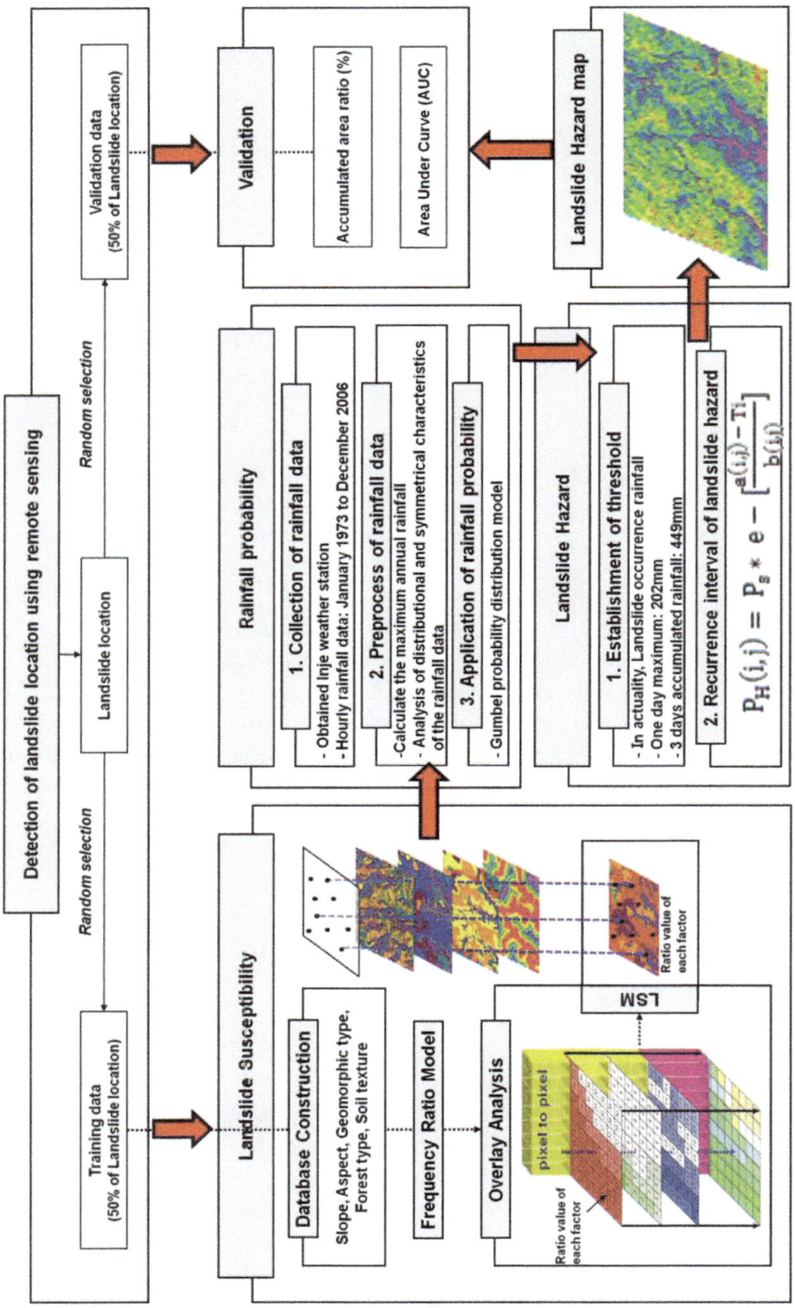

Figure 3. Study flowchart.

Relevant topical maps, including those of topography, geology, and soil and forest cover, are available either in print or in electronic form, at different scales and formats, in Korean archives. This study used a landslide map showing point coverage, a topographic map in the form of line and point coverage with a scale of 1:5,000, a soil map with polygon coverage and 1:25,000 scale, and also a forest map with polygon coverage and 1:25,000 scale. The map data used were compiled by national organizations, such as the National Geographic Institute in the case of the topographic map, the Korea Institute of Geoscience and Mineral Resources for the geological map, the National Institute of Agricultural Science and Technology for the soil map, and the National Forest Research Institute for the forest map.

Twelve factors related to landslides were considered in calculating the landslide probability. These factors were extracted from the constructed spatial database and transformed into a vector-type spatial database using GIS. Using the topographic database, a digital elevation model (DEM) was created first. Contour and survey base points were extracted from the 1:5,000 scale topographic maps, and a DEM was created with a resolution of 5 m. Using this DEM, the slope angle, slope aspect, and slope curvature were calculated. Topography, texture, drainage, material, and thickness of soil were acquired from the soil map and the type, diameter, age, and density of timber were obtained from forest maps (figure 4). To calculate the rainfall probability accurately, understanding the accuracy of past weather data is important. Rainfall probability in the study area was analysed along with hourly rainfall data taken over 33 years (from 1 January 1973 to 31 December 2006). For more accurate analysis of future rainfall probabilities, the quality and quantity of past rainfall data are a significant influence and may serve to enhance accuracy. Therefore, acquiring and analysing past rainfall data should be done in advance. To measure the rainfall probability in the study area, hourly weather data were obtained from the Korea Meteorological Administration. Hourly rainfall data were acquired for the climate observation spots in the study area. The data form a time series, allowing for consideration of both rainfall and time. Four primary objectives for analysing the time series data were to better describe, forecast, explain, and obtain better controls on the data, with the main goal of aiding forecasting. By analysis of past time series, we can predict how past patterns will continue into the future. Therefore, accurate analysis of the accumulated historical data is important. In this study, the rainfall intensity and recurrence interval (mm /hr) were measured by incorporating research from previous studies. In addition, the return period and rainfall intensity parameters were used to calculate the probability of future landslides.

Whole data were converted to raster from vector, and the 12 factors that were calculated and extracted were converted to a 5 × 5 m grid (ARC /INFO GRID-type). For the study area, the total grid number was 1,466,978 and the selected landslide-occurrence location grid was 693.

3. Method

3.1. Landslide hazard map mapping using frequency ratio

In general, to predict landslides, it is necessary to assume that landslide occurrence is determined by landslide-related factors, and that future landslides will occur under the same conditions as past landslides. On this basis, the relationship between areas where a landslide has occurred and landslide-related factors can be distinguished

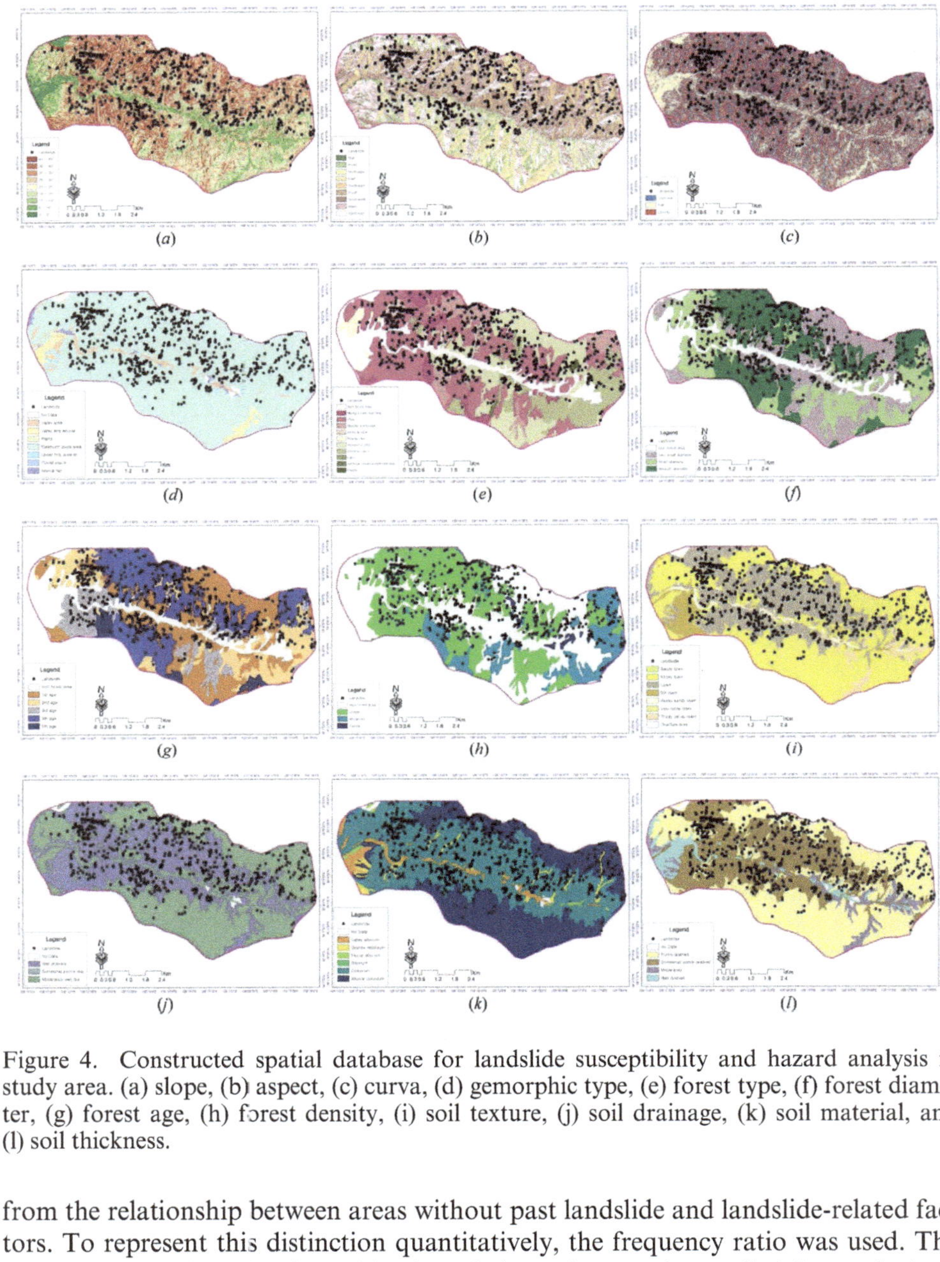

Figure 4. Constructed spatial database for landslide susceptibility and hazard analysis in study area. (a) slope, (b) aspect, (c) curva, (d) gemorphic type, (e) forest type, (f) forest diameter, (g) forest age, (h) forest density, (i) soil texture, (j) soil drainage, (k) soil material, and (l) soil thickness.

from the relationship between areas without past landslide and landslide-related factors. To represent this distinction quantitatively, the frequency ratio was used. The frequency ratio is a simple and basic technique that can be applied for analysis to prepare a landslide susceptibility map, and it is highly compatible with GIS. The frequency ratio is the ratio of the area where the landslide occurred to the total study area, and also, is the ratio of the probabilities of a landslide occurrence to a non-occurrence for a given attribute. The frequency ratio is the ratio of the area in which landslide occurred in class or type for a given factor of the study area. In the case of landslide occurrence, the area ratio of landslide occurrence to non-occurrence is calculated for the class or type of each factor. In the following equation, $P(P)$ denotes the area ratio for the class or type for a given number of unit cells containing a percentage of pixels in the domain for the class, and $P(O)$ denotes the percentage of occurrence in the total event. The frequency ratio of each factor's type or class, C, is then expressed by

$$\mathrm{FR}_C = \frac{P_C(O)}{P_C(P)}. \tag{1}$$

Therefore, the greater this ratio is above unity, the stronger the relationship between landslide occurrence and the given factor's attribute. The lower the ratio is below unity, the lesser the relationship is between landslide occurrence and the given factor's attribute. To calculate the frequency ratio, the cross tabulation in ArcGIS 10 was used. Then, the frequency ratios for each class of a reclassified factor or categorical factor (i.e. geology and land use) were calculated (table 1) and the ratio was assigned to each factor class. Finally, the ratios were used for calculating the landslide susceptibility index and mapping. The ratios for each factor type or range were added to calculate the landslide susceptibility index (LSI), as shown in the following equation:

$$\mathrm{LSI} = \Sigma Fr, \tag{2}$$

where Fr is the rating of each factor's type or range.

3.2. *Landslide hazard map mapping using rainfall probability*

It is also well known that the main triggering factor for landslides in the region is short-lived heavy rainfalls (Frattini et al. 2009). Therefore, to evaluate the landslide occurrence hazard in the study area, the temporal behaviour of the responsible triggering factor, short-lived heavy rainfalls, was considered. For this purpose, rainfall data obtained from the available meteorological stations were evaluated. To measure the rainfall probability in the study area, hourly rainfall data from January 1973 to December 2006 for the study area were obtained from the Inje station, which showed high consistency with the critical dates of the events that occurred in the study area. Analysis of rainfall data from July and August 2006, covering the period of typhoon Ewiniar (landslide activity occurred from 11 to 14 July 2006). To simulate the behaviour of the responsible triggering factor in time, the data measured at the Inje station were used along with the Gumbel probability distribution model (Zezere et al. 2008).

For calculating rainfall, data from local weather stations were first used to calculate the maximum annual rainfall. Second, the distributional and symmetrical characteristics of the climate data were analysed. The average, standard deviation, variance, and coefficient of variation of the data were calculated and primary statistical and preliminary analysis was performed. Third, the probability distribution to be applied in the frequency analysis was selected. The Gumbel distribution, often used in other frequency analyses of hydrological data, was selected in this case. Fourth, the parameter of probability distribution, which is adjusted according to the recurrence interval, was calculated. The method of moment was used to estimate the parameters of the probability distribution. Finally, when the proper optimal probability distribution was selected, the recurrence interval and future rainfall levels were calculated.

Table 1. Frequency ratio of factors to landslide occurrences.

Factors	Class	Number of pixels in domain[a]	Percentage of domain	Number of landslide[b]	Percentage of landslide	Frequency ratio
Slope	0°–5°	65,606	4.47	3	0.43	0.10
	6°–10°	84,813	5.78	4	0.58	0.10
	11°–15°	123,479	8.42	15	2.16	0.26
	16°–20°	166,365	11.34	23	3.32	0.29
	21°–25°	235,666	16.06	76	10.97	0.68
	26°–30°	255,125	17.39	113	16.31	0.94
	31°–35°	239,647	16.34	200	28.86	1.77
	36°–40°	171,456	11.69	129	18.61	1.59
	41°–90°	124,821	8.51	130	18.76	2.20
Aspect	Flat	12,418	0.85	0	0.00	0.00
	North	178,128	12.14	47	6.78	0.56
	Northeast	144,479	9.85	38	5.48	0.56
	East	129,251	8.81	114	16.45	1.87
	Southeast	169,033	11.52	166	23.95	2.08
	South	170,758	11.64	86	12.41	1.07
	Southwest	196,686	13.41	86	12.41	0.93
	West	231,150	15.76	83	11.98	0.76
	Northwest	235,075	16.02	73	10.53	0.66
Curvature	Concave	505,504	34.46	279	40.26	1.17
	Flat	464,390	31.66	138	19.91	0.63
	Convex	497,084	33.88	276	39.83	1.18
Geomorphic type	No data	7524	0.51	0	0.00	0.00
	Valley area	48,501	3.31	6	0.87	0.26
	Valley and alluvial	36,891	2.51	0	0.00	0.00
	Plains	1659	0.11	0	0.00	0.00
	Piedmont slope area	1,348,326	91.91	684	98.70	1.07
	Lower hilly area an	12,405	0.85	2	0.29	0.34
	Fluvial plains	268	0.02	0	0.00	0.00
	Alluvial fan	11,404	0.78	1	0.14	0.19
Forest type	Non-forest area	166,668	11.36	9	1.30	0.11
	Mixed broad-leaf tree	86,092	5.87	27	3.90	0.66
	Pine	528,932	36.06	305	44.01	1.22
	Needle and broad	126,759	8.64	81	11.69	1.35
	Artificial pine	17,463	1.19	26	3.75	3.15
	Rigida pine	25,395	1.73	4	0.58	0.33
	Korea nut pine	310,110	21.14	170	24.53	1.16
	Artificial larch	197,038	13.43	68	9.81	0.73
	larch	1569	0.11	0	0.00	0.00
	Artificial mixed broad-leaf tree	2849	0.19	0	0.00	0.00
	Poplat	4101	0.28	3	0.43	1.55
Forest diameter	Non-forest area	193,632	13.20	13	1.88	0.14

(continued)

Table 1. (*Continued*)

Factors	Class	Number of pixels in domain[a]	Percentage of domain	Number of landslide[b]	Percentage of landslide	Frequency ratio
	Very small diameter	411,635	28.06	218	31.46	1.12
	Small diameter	422,539	28.80	212	30.59	1.06
	Medium diameter	439,170	29.94	250	36.08	1.21
Forest age	Non-forest area	193,632	13.20	13	1.88	0.14
	First age	411,635	28.06	218	31.46	1.12
	Second age	237,161	16.17	120	17.32	1.07
	Third age	175,273	11.95	92	13.28	1.11
	Fourth age	406,194	27.69	246	35.5	1.28
	Fifth age	43,081	2.94	4	0.58	0.20
Forest density	Non-forest area	605,267	41.26	231	33.33	0.81
	Loose	604,447	41.2	357	51.52	1.25
	Moderate	241,453	16.46	102	14.72	0.89
	Dense	15,809	1.08	3	0.43	0.40
Soil texture	Sandy loam	3640	0.25	0	0.00	0.00
	Rocky loam	766,584	52.26	261	37.66	0.72
	Loam	540,093	36.82	404	58.30	1.58
	Silt loam	68,905	4.70	18	2.60	0.55
	Rocky sandy loam	83,604	5.70	10	1.44	0.48
	Very rocky loam	268	0.02	0	0.00	0.00
	Overflow area	3884	0.26	0	0.00	0.00
Soil drainage	No data	7524	0.51	0	0.00	0.00
	Well drained	686,189	46.78	442	63.78	1.36
	Somewhat poorly drained	1836	0.13	0	0.00	0.00
	Moderately well drained	771,429	52.59	251	36.22	0.69
Soil material	No data	7524	0.51	0	0.00	0.00
	Valley alluvium	61,597	4.20	4	0.58	0.14
	Granite residuum	34,112	2.33	10	1.44	0.62
	Fluvial alluvium	7225	0.49	0	0.00	0.00
	Diluvium	268	0.02	0	0.00	0.00
	Colluvium	809,966	55.21	518	74.75	1.35
	Alluvial colluvium	546,286	37.24	161	23.23	0.62
Soil thickness	No data	7524	0.51	0	0	0.00
	Prooly shallow	712,174	48.55	248	35.79	0.74
	Shallow	590,636	40.26	430	62.05	1.54
	Normal	90,499	6.17	8	1.15	0.19
	Deep	66,145	4.51	7	1.01	0.22

Note: Domain: pixels in study area.
Domain (%): domain / total pixels in study area.
Landslide: number of landslide occurrences.
Landslide (%): landslide / total number of landslide occurrences.
[a] Number of total grids in study area: 1,466,978 (without no data)
[b] Number of landslide occurrence grids: 693.

The values of landslide susceptibility and the probability of a given rainfall threshold are completely independent. The product of the probability of a given rainfall threshold and the landslide commend susceptibility in each terrain unit provides spatially distributed landslide commend hazard values within the study area. However, in this case, two important aspects are considered: (1) What is the critical rainfall threshold in the study area or how should be it decided? (2) In how many years is this critical threshold value expected to occur? As mentioned in section 3, one of the possible solutions for the first question is to apply a accumulate rainfall. Another possible way to select an appropriate threshold is to evaluate a known scenario that has occurred in the region (Nefeslioglu et al. 2011).

This research has calculated landslide critical value of rainfall by applying previous researches and characters of study area. To accurately calculate the landslide critical value of rainfall, understanding of past weather data is important. Rainfall probability in the study area was analysed along with hourly rainfall data taken over a period of 33 years (from 1973 to 2006). By analysing the past time series, predictions of how past patterns will continue into the future may be forecast. Therefore, accurate analysis of the accumulated historical data is important. In this study, the rainfall intensity and recurrence interval (mm /hr) were measured by incorporating research from previous studies. In addition, the return period and rainfall intensity parameters were used to calculate the probability of future landslides (Frattini et al. 2009).

To calculate the landslide critical value of rainfall in the study area, hourly weather data obtained from the Korea Meteorological Administration were analysed. Analysis of precipitation data from July and August 2006, covering the period of typhoon Ewiniar (landslide activity occurred from 11 to 14 July), showed that the highest daily rainfall of 202 mm occurred on 14 July, with a cumulative rainfall of 449 mm over the three-day period. Rainfall across the whole study area was the same during this period. Because the study areas share a single weather station, their areas (Deokeokri: 36 km^2 and Karisanri: 38 km^2) were used to calculate cumulative rainfall which was then considered with respect to landslide occurrence over the 20 years of data. The data were analysed in terms of the date and location of a landslide, the nature of the landslide and the area in which it occurred, and the meteorological data available from the local station for the previous five days. In figures 5–7, the horizontal (x) axis indicates rainfall on the day calculated to be the day of the landslide, and the vertical (y) axis indicates rainfall accumulation over one, three, and five days prior to the landslide (figures 5–7 and table 2). Landslides were considered to be triggered by rainfall on the day of occurrence when data plotted to a region over the 45° central border line and to be caused by cumulative rainfall when data plotted to a region under the line. Table 2 is a summary of the number of landslide locations over and under the 45° central border line. Looking at landslides in the 2000 s (figure 5– showing rainfall on the day of a landslide and the cumulative rainfall for one day before), there is a clear separation between those plotted over and under the 45° central border line. There are 162 data points over and 24 under the central border lines (table 2). Therefore, landslides are estimated to be caused mainly by rainfall on the day of occurrence. In figure 6, which includes rainfall on the day of the landslide and cumulative rainfall for three days before the landslide, the data points are equally distributed over and under the 45° central border line, indicating that landslides were generated equally by rainfall on the day of landsliding and by cumulative three-day-prior rainfall (86 data points over the border and 80 under the border; table 2). Figure 7 compares rainfall on the day of the landslide with

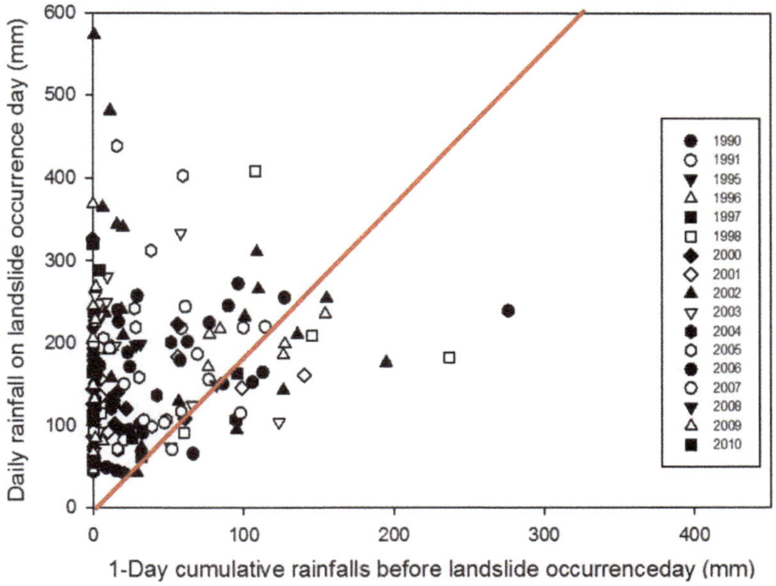

Figure 5. One-day cumulative rainfall before the failure day. Landslides were considered to be triggered by rainfall on the day of occurrence when data plotted to a region over the 45° central border line and to be caused by cumulative rainfall when data plotted to a region under the line.

cumulative rainfall for five days prior. More landslides plotted to the area under rather than over the 45° central border line (56 data points over the border and 130 under the border; table 2), indicating that these landslides were caused mainly by five-day cumulative rainfall, rather than by rainfall on the day of occurrence.

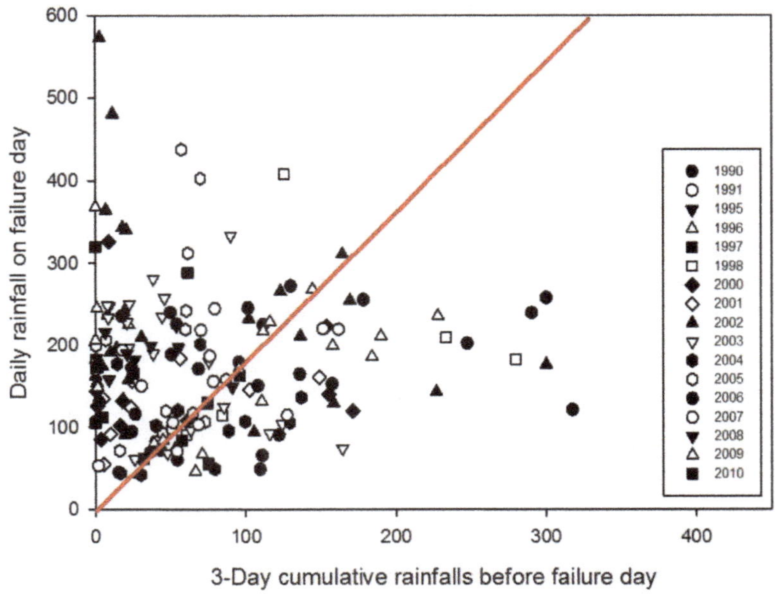

Figure 6. Three-day cumulative rainfall before the failure day.

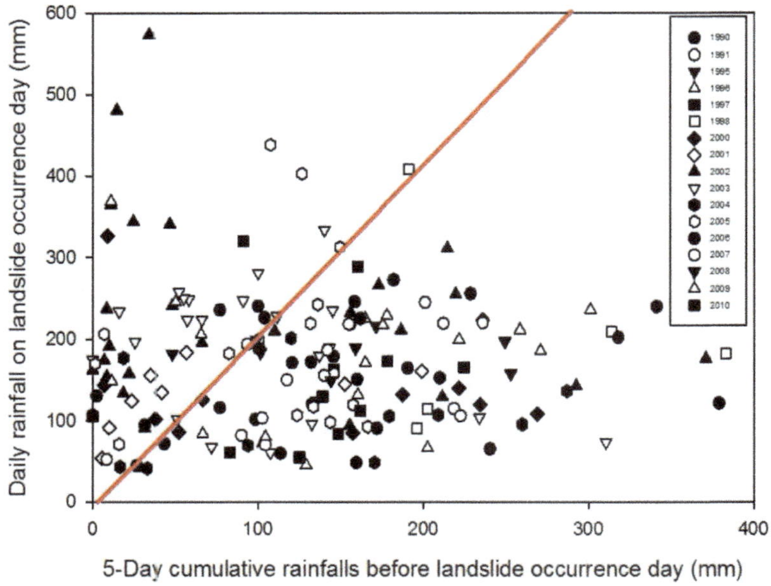

Figure 7. Five-day cumulative precipitation before the landslide occurrence day.

Three-day cumulative rainfall may therefore be useful for identifying whether landslides are caused by rainfall on the day of occurrence or by cumulative rainfall. As seen in figures 5–7, many landslides in the 2000s were closely related to three-day cumulative rainfall, and thus landslides in the 2000s may have been more influenced by cumulative rainfall than by rainfall on the day of occurrence. As a result, one-day rainfall and three-day cumulative rainfall, rather than cumulative rainfall for other durations, had a higher correlation with landslide occurrence.

The recurrence interval from 2 to 100 years in the future, with one day and three days cumulative rainfall amounts used to infer the occurrence of landslides. As shown in table 3, the frequency of 132.9 mm of rainfall falling over the course of one day in the study area is once about every two years. The frequency of breaking the one-day and three-day thresholds of cumulative rainfall will increase with time. For example, the probability of one-day rainfall of 300 mm is once every 80 years and will increase in the future. The probability that 449 mm of rainfall will occur cumulatively over three days is once every 100 years and will also increase in each future recurrence interval year. In this study, thresholds of 202 mm and 449 mm were used. According to this study, the frequency of 202 mm occurring on one day is once every 10 years, while that of 449 mm of rainfall over three days is once every 100 years in

Table 2. Summary of landslide spot counts over and under the 45° central border line (number of spots).

Classification	Over	Under
Daily rainfall and one-day cumulative rainfalls before failure data	162	24
Daily rainfall and three-day cumulative rainfalls before failure data	86	80
Daily rainfall and five-day cumulative rainfalls before failure data	56	130

Table 3. Recurrence interval of probability rainfall.

Recurrence interval	One day	Three days
2 years	132.9	163.5
3 years	156.9	199.5
5 years	183.7	239.6
10 years	217.4	290.0
20 years	249.7	338.3
30 years	268.3	366.1
50 years	291.5	400.9
70 years	306.7	423.7
80 years	312.8	432.7
100 years	322.9	447.8

the recurrence interval years. Thus landslide-causing events of 202 mm of rainfall over one day may occur every 10 years and events of 449 mm over three days may occur every 100 years.

Future rainfall levels can vary depending on the recurrence interval, duration time, input data accuracy, and shape. The data in table 3 were generated by applying equation (3) and using data from table 4, which includes rainfall levels cumulative over one and three days:

$$P_{H_{(i,J)}} = P_S \times e^{-\left[\frac{a_{(i,t)} - T_i}{b_{(i,t)}}\right]}, \tag{3}$$

where P_H is the probability of hazard and P_S is the probability of susceptibility. T_i is the threshold of rainfall before the breakout of a landslide, and a and b are given as the time constants. In order to acquire the time constants given in the equation, the method is as follows: (1) calculation of the Gumbel probability of occurrence of daily rainfall data in the periods of observation, (2) plotting the probabilities on a semi-logarithmic graph with respect to daily rainfall data, (3) construction of the logarithmic function which represents the behaviour of the triggering factor for a given return period, and (4) time constants are acquired from the function written in the

Table 4. One-day and three-day accumulation recurrence interval of parameters.

One-day parameter	*A*	*b*	Three-day parameter	*a*	*b*
2 years	38.95	69.74	2 years	28.85	56.28
3 years	47.95	81.51	3 years	33.71	65.17
5 years	58.04	94.59	5 years	39.16	75.07
10 years	70.79	110.98	10 years	46.06	87.49
20 years	83.06	126.67	20 years	52.70	99.37
30 years	90.13	135.71	30 years	56.50	106.23
50 years	98.99	146.96	50 years	61.28	114.76
70 years	104.80	154.36	70 years	64.42	120.36
80 years	107.09	157.30	80 years	65.67	122.60
100 years	110.94	162.19	100 years	67.74	126.30

form of equation (4) (Nefeslioglu et al. 2011). $f(x) = b \ln(x)$ is a real-valued function. a is given as the time constants.

$$f(x) = -b \ln(x) + a. \tag{4}$$

3.3. Validation

Using the LSI, a landslide susceptibility and hazard map was constructed and validated based on the locations of existing landslides. For the validation, the calculated LSI values for all cells in the study area were sorted in descending order. Then, the landslides (%) were divided into classes of the cumulative area ratio (%) according to the LSI value and hazard value of rainfall probability. To compare the results quantitatively, the areas under the curve (AUC) were re-calculated with the total area indicating perfect prediction accuracy (Lee & Dan 2005; Lee & Sambath 2006). Hence, the AUC can be used to assess the prediction accuracy qualitatively.

4. Result

Using the frequency-ratio model, the frequency ratio of each factor in each class range was calculated (table 1). From the frequency ratio, the relationships between landslides and the examined factors were as follows. The ratio for slopes of 41°–90° was greater than 2, indicating the highest probability of landslides, but that for classes of slopes from 0° to 30° was less than 1.0, indicating lower probability. Landslides were found to be most abundant on southeast-facing hillslope, indicating that this aspect makes slopes highly affected by. Curvature values represent the morphology

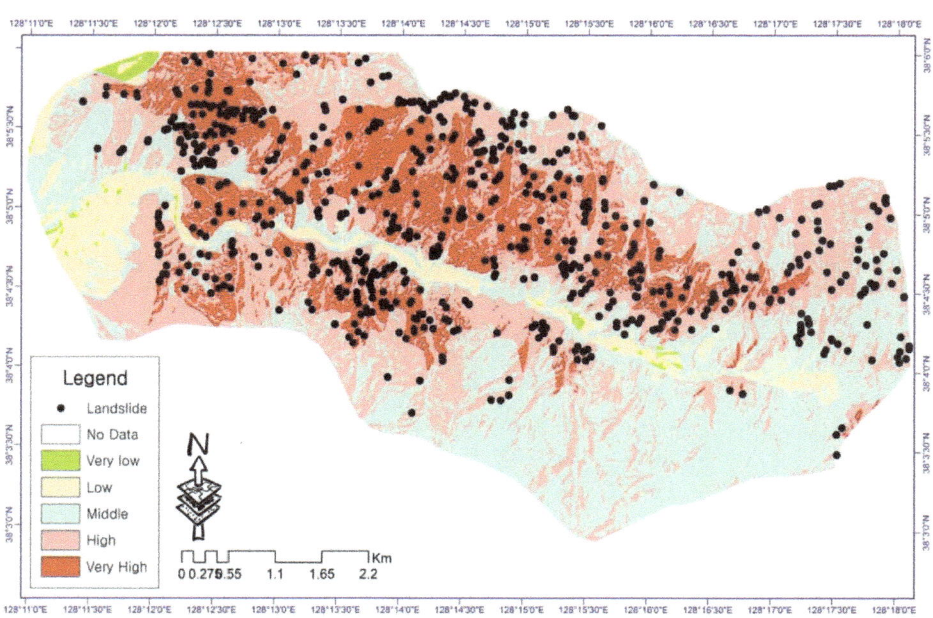

Figure 8. Landslide susceptibility map.

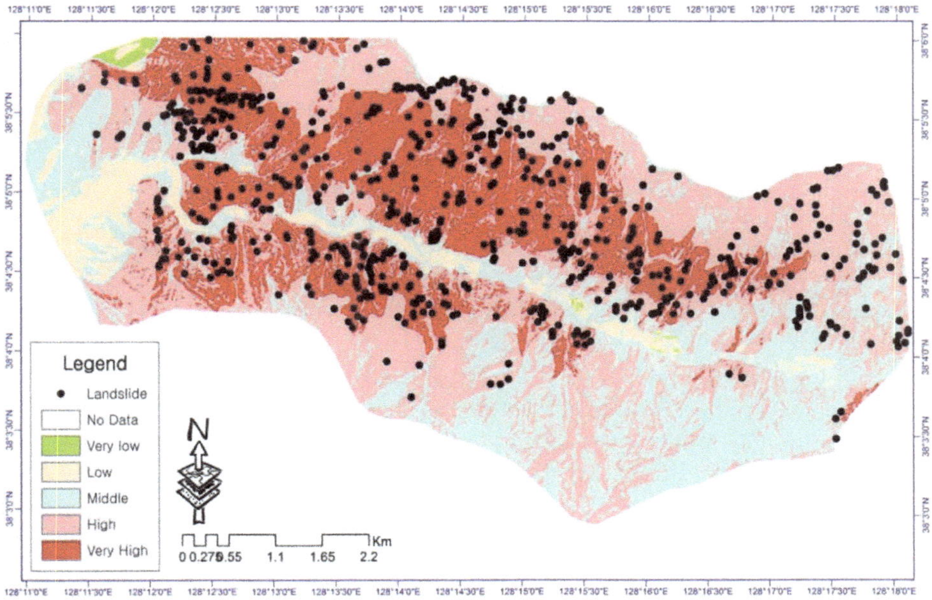

(*a*)

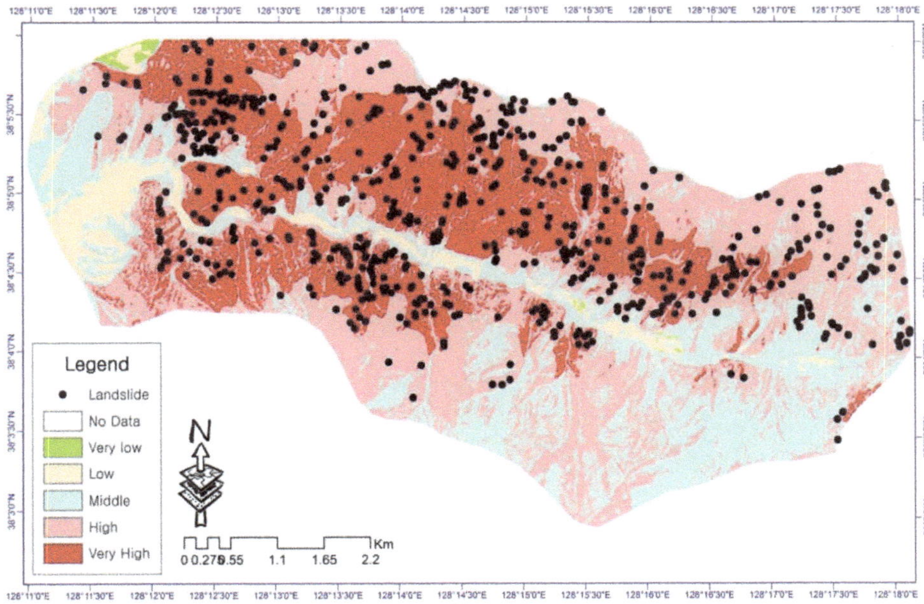

(*b*)

Figure 9. Landslide susceptibility map and hazard map based on frequency ratio, using a threshold 202 mm at time series: (a) landslide hazard map after five years, (b) landslide hazard map after 10 years, (c) landslide hazard map after 50 years, and (d) landslide hazard map after 100 years.

of topography. Positive curvature indicates that the surface is upwardly convex at that cell. Negative curvature indicates that the surface is upwardly concave. The reason is that in concave part of the surface as catchment area increases the corresponding slope decreases (Gangodagamage et al. 2011). A zero value indicates that the

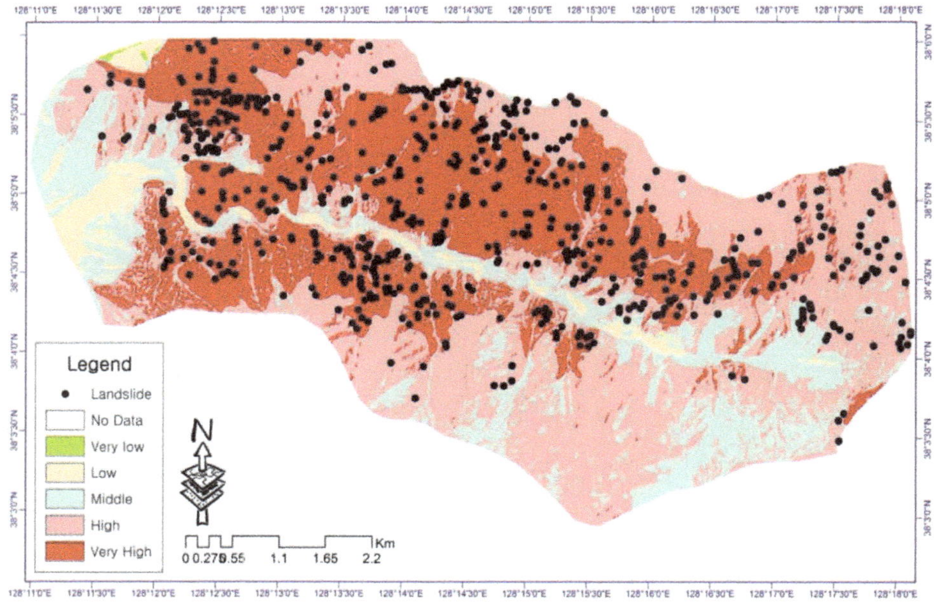

(*c*)

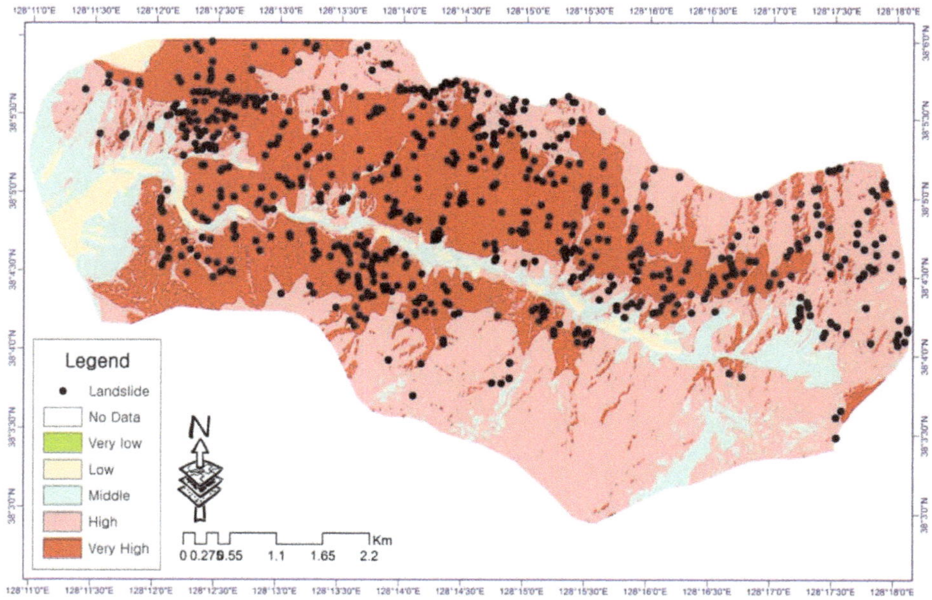

(*d*)

Figure 9. (Continued)

surface is planar. The more negative or positive the value is, the higher the probability of landslide occurrence. Planar areas had a low value of 0.63. In the case of geomorphic type, the piedmont slope area had a ratio of 1.07, indicating a high probability of landslide occurrence. Regarding forest type, areas of artificial pine had a ratio of 3.15, indicating a high probability.

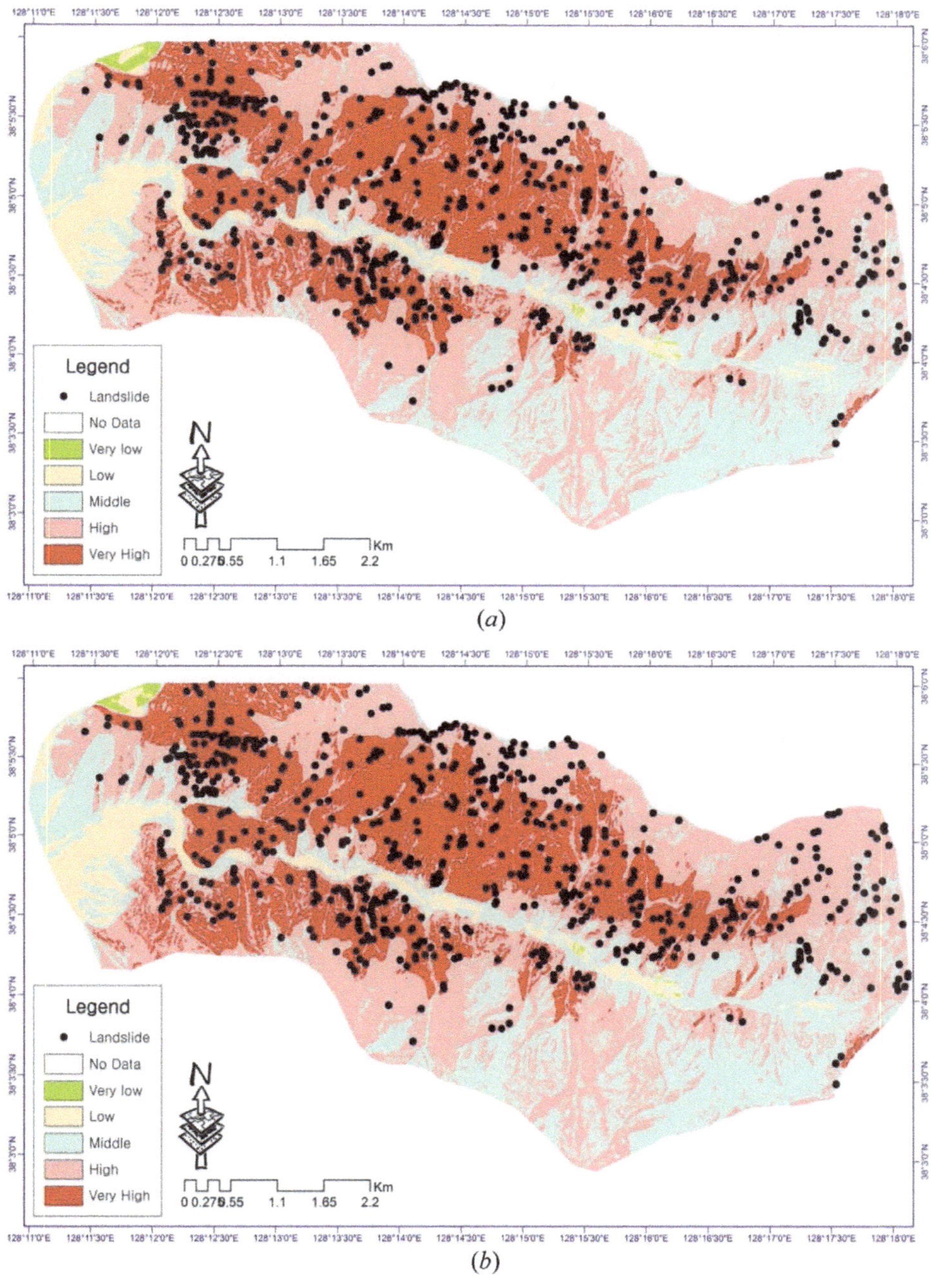

Figure 10. Landslide susceptibility map and hazard map based on frequency ratio, using a threshold 449 mm at time series: (a) landslide hazard map after 5 year, (b) landslide hazard map after 10 year, (c) landslide hazard map after 50 year, (d) landslide hazard map after 100 year.

Using the frequency ratio and equation (1), the LSI values were calculated. Then LSI values converted to probability of susceptibility values. Also using the probability landslide susceptibility values and equation (3), the probability of hazard values was computed using various periods. The computed probability susceptibility and

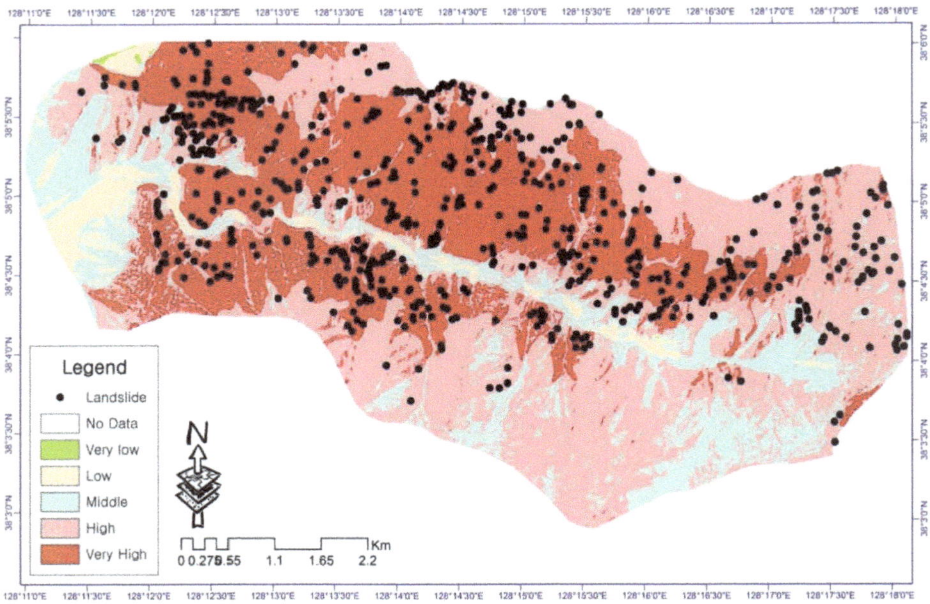

(*c*)

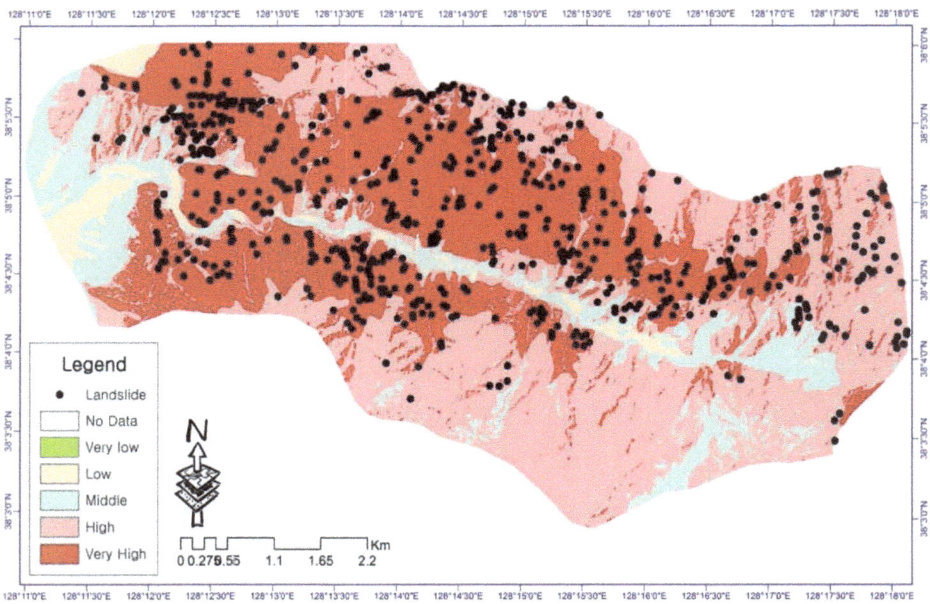

(*d*)

Figure 10. (Continued)

hazard values were mapped to landslide susceptibility map and hazard maps respectively, to allow interpretation such as that illustrated in figures 8–10. The susceptibility and hazard values were classified into five classes (Very high, High, Middle, Low, Very low) based on area to ease visual interpretation. If the susceptibility and hazard value are high, there is a higher landslide hazard; a lower value indicates a lower

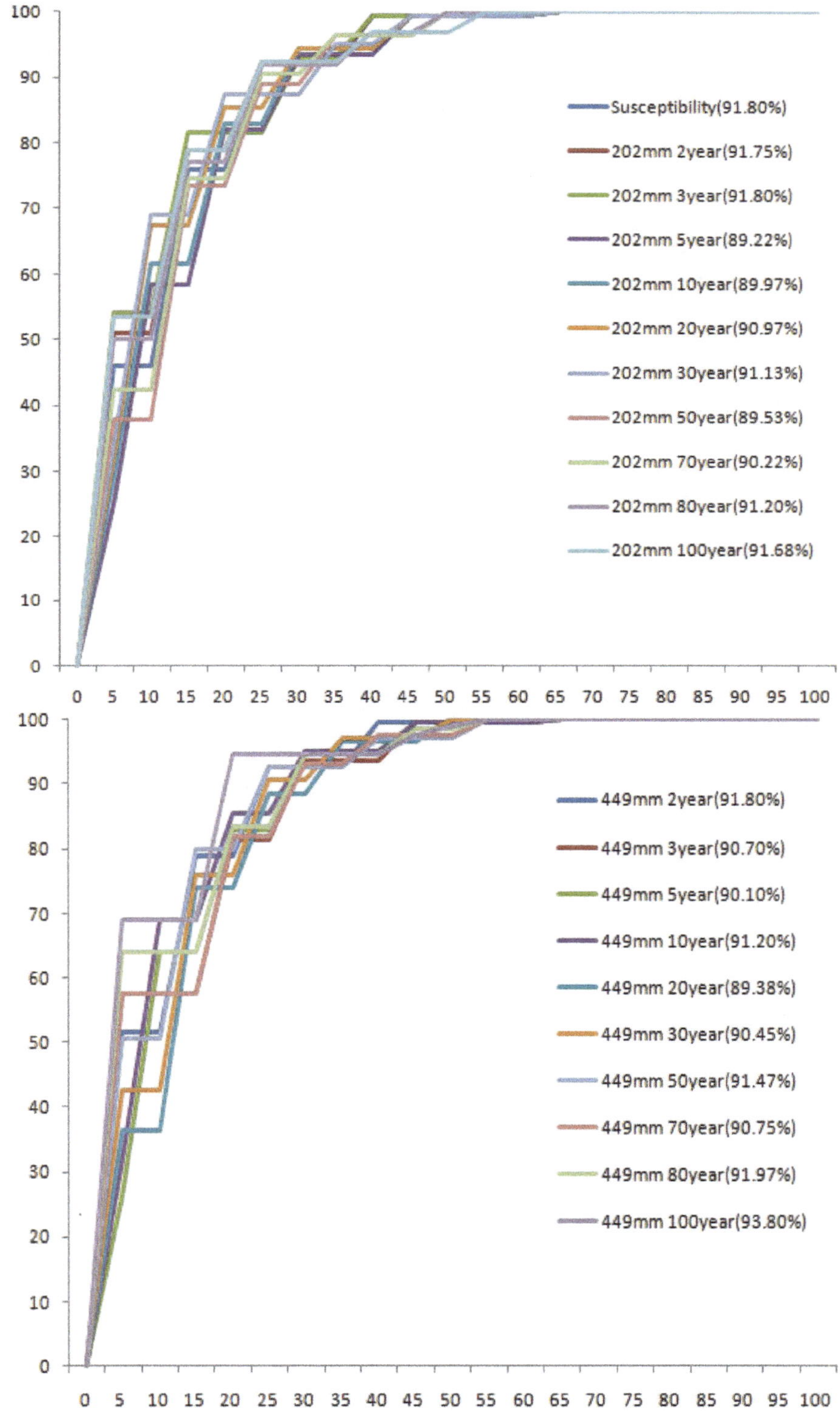

Figure 11. Cumulative frequency diagram showing landslide susceptibility and hazard rank (*x*-axis) occurring in cumulative per cent of landslide occurrence (*y*-axis).

hazard. Figures 9 and 10 show the final landslide hazard map considering the recurrence interval years refer to 2–100 years in the future.

The results of the landslide hazard map were validated using the landslide locations. The total number of landslide locations is 693. About 50% (347) of the total landslide locations were used for frequency ratio and hazard analysis. The remaining 50% (346) were used for validation.

To validate landslide susceptibility and hazard calculation methods, two basic assumptions were needed. One is that landslides can be related to spatial information (such as topography, soil and forest) and the other is that future landslides will be triggered by a specific impact factor, such as rainfall or an earthquake. In this study, these two assumptions are satisfied because the landslides are related to the spatial information and all the landslides were triggered by a single cause, namely, heavy rainfall in the Inje area. The results show that the 100% susceptibility class with the highest probability of a landslide contains 91.80% of the predicted landslides (figure 11). The 202-mm daily rainfall threshold was validated by the AUC method for the recurrence interval years of 2, 3, 5, 10, 20, 30, 50, 70, 80, and 100 years, with results of 91.75%, 91.80%, 89.22%, 89.97%, 90.97%, 91.13%, 89.53%, 90.22%, 91.20%, and 91.68%, respectively. For the 449-mm three-day rainfall threshold, the validation results for the same recurrence interval years were 91.80%, 90.70%, 90.10%, 91.20%, 89.38%, 90.45%, 91.47%, 90.75%, 91.97%, and 93.80%, respectively (figure 11).

5. Discussion and conclusions

Landslides caused by heavy rainfall may become more frequent if climate change causes increased rainfall. This study attempted to identify a relation between nearly rainfall and landslides with the case of typhoon Ewiniar in 2006, and predict the future rainfall probability caused by climate change. It finally drew the recurrence interval and future rainfall levels were constructed by applying the landslide hazard map. Government and research institutions worldwide have attempted to assess landslide hazards and risk, and to show their spatial distribution. Here, landslide susceptibility and hazard maps were constructed using the relationship between landslides and causal factors. Frequency ratios of landslide-related items were calculated using GIS and a susceptibility map was created. Rainfall probability was also calculated, and landslide hazards were analysed for several future recurrence interval years.

In conclusion, the validated landslide hazard map, constructed by considering rainfall probability, showed that the probabilities of reaching the two thresholds would increase in each future target year. Thus, if current conditions in the study area are maintained without additional efforts to prevent future landslides, the risk posed by landslides will increase with time. In addition, heavy rainfall is also likely to increase. The study results, showing landslide hazards in future years, can be used to help develop landslide management plans.

The analysis of landslide hazards, based on past and present data, is important for predicting and minimizing the damage of future landslides. This study incorporated rainfall probability into landslide hazard assessment, providing new insight into hazard evaluation. The results are useful for establishing more efficient management plans and as baseline information for applications in road, rail, and linear industries.

Funding
This research was supported by the Basic Research Project of the Korea Institute of Geoscience and Mineral Resources (KIGAM) funded by the Ministry of Science, ICT & Future Planning, and the Development of Time Series Disaster Mapping Technologies through Natural Disaster Factor Spatio-Temporal Correlation Analysis and Integration: Focused on the Urban Disasters by Extreme Climate Change by National Research Foundation of Korea.

References

Bai S, Lü G, Wang J, Zhou P, Ding L. 2011. GIS-based rare events logistic regression for landslide-susceptibility mapping of Lianyungang, China. Environ Earth Sci. 62:139–149.
Blijenberg HM. 1998. Rolling stones? Triggering and frequency of hill slope debris flows in the Bachelard Valley, Southern French Alps [Ph.D. thesis]. Utrecht University, p. 233.
Borga M. 2002. Accuracy of radar rainfall estimates for streamflow simulation. J Hydrol. 267:26–39.
Dong JJ, Lee CT, Tung YH, Liu CN, Lin KP, Lee JF. 2009. The role of the sediment budget in understanding debris flow susceptibility. Earth Surf Proc Land. 34:1612–1624.
Evans SG, Clague JJ. 1994. Recent climatic change and catastrophicgeomorphic processes in mountain environments. Geomorphology. 10:107–128.
Fell R, Corominas J, Bonnard C, Cascini L, Leroi E, Savage WZ. 2008. Guidelines for landslide susceptibility, hazard and risk zoning for land-use planning. Eng Geol. 102:99–111.
Floris M, D'Alpaos A, Squarzoni C, Genevois R, Marani M. 2010. Recent changes in rainfall characteristics and their influence on thresholds for debris flow triggering in the Dolomitic area of Cortina d'Ampezzo, north-eastern Italian Alps. Nat Hazards Earth Syst Sci. 10:571–580.
Frattini P, Crosta G, Sosio R. 2009. Approaches for defining thresholds and return periods for rainfall-triggered shallow landslides. Hydrol Process. 23:1444–1460.
Gangodagamage C, Belmont P, Foufoula-Georgiou E. 2011. Revisiting scaling laws in river basins: New considerations across hillslope and fluvial regimes. Water Resour Res. 47: W07508.
Herr HD, Krzysztofowicz R. 2005. Generic probability distribution of rainfall in space: the bivariate model. J Hydrol. 306:234–263.
IPCC. 2007. Climate Change 2007 – IPCC 4th Assessment Report, Working Group 1 – The physical science basics. Geneva: IPCC.
Jomelli V, Brunstein D, Grancher D, Pech P. 2007. Is the response of hill slope debris flows to recent climate change univocal? A case study in the Massif des Ecrins (French Alps), Clim Change. 85:119–137.
Lee S. 2005. Application of logistic regression model and its validation for landslide susceptibility mapping using GIS and remote sensing data. Int J Remote Sensing. 26:1477–1491.
Lee S. 2007. Comparison of landslide susceptibility maps generated through multiple logistic regression for three test areas in Korea. Earth Surf Proc Land. 32:2133–2148.
Lee S, Dan NT. 2005. Probabilistic landslide susceptibility mapping in the Lai Chau province of Vietnam: focus on the relationship between tectonic fractures and landslides. Environ Geol. 48:778–787.
Lee MJ, Lee S, Jeon SW, Kim GH. 2013. Landslide vulnerability mapping considering GCI (Geospatial Correlative Integration) and rainfall probability in Inje. J Environ Policy. 38:21–47.
Lee S, Min K. 2001. Statistical analysis of landslide susceptibility at Yongin, Korea. Environ Geol. 40:1095–1113.
Lee S, Pradhan B. 2006, Probabilistic landslide hazards and risk mapping on Penang Island, Malaysia. J Earth Syst Sci. 115:661–672.

Lee S, Ryu JH, Won JS, Park HJ. 2004. Determination and application of the weights for landslide susceptibility mapping using an artificial neural network. Eng Geol. 71:289–302.

Lee S, Sambath T. 2006. Landslide susceptibility mapping in the Damrei Romel area, Cambodia using frequency ratio and logistic regression models. Environ Geol. 50:847–855.

Nandi A, Shakoor A. 2010. A GIS-based landslide susceptibility evaluation using bivariate and multivariate statistical analyses. Eng Geol. 110:11–20.

Nefeslioglu HN, Gokceoglu C, Sonmez H, Gorum T. 2011. Medium-Scale hazard mapping for shallow landslide initiation: the Buyukkoy catchment area (Cayeli, Rize, Turkey). Landslide. DOI 10.1007/s10346-011-0267-7

Oh HJ, Lee S. 2010. Cross-validation of logistic regression model for landslide susceptibility mapping at Geneoung areas, Korea. Disaster Adv. 3:44–55.

Oh HJ, Lee S, Chotikasathien W, Kim C, Kwon J. 2009. Predictive landslide susceptibility mapping using spatial information in the Pechabun area of Thailand. Environ Geol. 57:641–651.

Ozdemir A. 2009. Landslide susceptibility mapping of vicinity of Yaka Landslide (Gelendost, Turkey) using conditional probability approach in GIS. Environ Geol. 57:1675–1686.

Park NW. 2011. Application of Dempster-Shafer theory of evidence to GIS-based landslide susceptibility analysis. Environ Earth Sci. 62:367–376.

Pradhan B. 2011. Manifestation of an advanced fuzzy logic model coupled with Geo-information techniques to landslide susceptibility mapping and their comparison with logistic regression modelling. Environ Ecol Stat. 18:471–493.

Pradhan B, Lee S. 2010a. Delineation of landslide hazard areas on Penang Island, Malaysia, by using frequency ratio, logistic regression, and artificial neural network models. Environ Earth Sci. 60:1037–1054.

Pradhan B, Lee S. 2010b. Regional landslide susceptibility analysis using back-propagation neural network model at Cameron Highland, Malaysia. Landslides. 7:13–30.

Regmi NR, Giardino JR, Vitek JD. 2010. Modeling susceptibility to landslides using the weight of evidence approach: Western Colorado, USA. Geomorphology. 115:172–187.

Shou K, Chen Y, and Liu H. 2009. Hazard analysis of Li-shan landslide in Taiwan. Geomorphology. 103:143–153.

Tangestani MH. 2009. A comparative study of Dempster-Shafer and fuzzy models for landslide susceptibility mapping using a GIS: An experience from Zagros Mountains, SW Iran. J Asian Earth Sci. 35:66–73.

Vahidnia MH, Alesheikh AA, Alimohammadi A, Hosseinali F. 2009. Landslide hazard zonation using quantitative methods in GIS. Int J Civil Eng. 7:176–189.

Van Steijn H. 1999. Frequency of hillslope debris flows in the Bachelard valley (French Alps). In: Panizza M, Soldati M, Bertacchini M, Van Asch W, Malmusi S, editors. Programme in geomorphology. London: Arnold; p. 15–24.

Yalcin A. 2008. GIS-based landslide susceptibility mapping using analytical hierarchy process and bivariate statistics in Ardesen (Turkey): comparisons of results and confirmations. Catena. 72:1–12.

Yilmaz I. 2009. A case study from Koyulhisar (Sivas-Turkey) for landslide susceptibility mapping by artificial neural networks. Bull Eng Geol Environ. 68:297–306.

Yilmaz I. 2010. The effect of the sampling strategies on the landslide susceptibility mapping by conditional probability and artificial neural networks. Environ Earth Sci. 60:505–519.

Zezere JL, Garcia RAC, Oliveira SC, Reis E. 2008. Probabilistic landslide risk analysis considering direct costs in the area north of Lisbon (Portugal). Geomorphology. 94:467–495.

8

Emergency preparedness in the case of Makran tsunami: a case study on tsunami risk visualization for the western parts of Gujarat, India

V. M. PATEL†‡*, M. B. DHOLAKIA§ and A. P. SINGH¶
†Civil Engineering Department, K.D. Polytechnic, Patan 384265, Gujarat, India
‡ Kadi Sarva Vishwavidyalaya, Gandhinagar 382016, Gujarat, India
§L.D. College of Engineering, Ahmedabad 380015, Gujarat, India
¶Institute of Seismological Research, Raisan, Gandhinagar 382009, Gujarat, India

The west coast of India is affected by tsunamigenic earthquake along the Makran subduction zone. On 28 November 1945 at 21:56 coordinated universal time (UTC), a massive Makran earthquake (M8.0) generated a destructive tsunami that propagated across the Northern Arabian Sea and the Indian Ocean. This tsunamigenic earthquake was responsible for the loss of life and great destruction along the coasts of India, Pakistan, Iran and Oman. Modelling of tsunami stages has been made for the coasts of Pakistan, Iran, India and Oman using NAMI-DANCE computer code. The fault parameters of the earthquakes for the generation of tsunami are epicentre (25.15° N, 63.48° E), fault area (200 km length and 100 km width), angle of strike, dip and rake (246°, 7° and 90°), focal depth (15 km), slip magnitude (7 m). The bathymetry data are taken from General Bathymetric Chart of the Oceans (GEBCO) and land topography data were collected using Shuttle Radar Topography Mission (SRTM). The present simulation is carried out for a duration of 360 min. It is observed that the maximum calculated tsunami run-ups were about 0.7–1.1 m along the coast of Oman, 0.5 m near Muscat, 0.1 m near Sur, 0.7–1.35 m along the western coast of India, 0.5–2.3 m along the southern coast of Iran and 1.2–5.8 m along the southern coast of Pakistan. After the tsunamigenic earthquake, the tsunami wave reached the Gulf of Kachchh in about 240 min, Okha in about 185 min, Dwarka in about 150 min, Porbandar in about 155 min, Mumbai in about 300 min and Goa in about 210 min. The calculated 2-hr tsunami travel time to the Indian coast is in good agreement with the available reports and published data. If the tsunami strikes during high tide, we should expect more serious hazards which would impact local coastal communities. The results obtained in this study are converted to be compatible with the geographic information system based applications for display and spatial analysis of modelling results. The paper also presents classification of tsunami risk zones based on elevation vulnerability. We expect that the results presented here will be supportive to the tsunami emergency response system and useful in planning the protection measures due to tsunami.

*Corresponding author. Email: principalkdp@yahoo.com
Mentor author: Dr A. P. Singh (apsingh07@gmail.com)

1. Introduction

Tsunami is a phenomenon of gravity waves produced in consequence of movement of the ocean floor. The giant tsunami in the Indian Ocean on 26 December 2004, claiming more than 225,000 lives (Titov et al. 2005; Geist et al. 2006; Okal & Synolakis 2008, Singh et al. 2012), has emphasized the urgent need for tsunami emergency response systems for various vulnerable coastlines around the world, especially for those neighbouring the Indian Ocean. The second deadliest tsunami prior to 2004 in South Asia occurred on 28 November 1945 (Heck 1947; Dominey-Howes et al. 2007; Heidarzadeh et al. 2007; Jaiswal et al. 2009; Hoffmann et al. 2013). It originated off the southern coast of Pakistan and was destructive in the Northern Arabian Sea and caused fatalities as far away as Mumbai (Berninghausen 1966; Quittmeyer & Jacob 1979; Ambraseys & Melville 1982; Heidarzadeh et al. 2008; Jaiswal et al. 2009). More than 4000 people were killed by both the earthquake and the tsunami (Ambraseys & Melville 1982). Several researchers have different estimates about the location of the earthquake epicentre. Heck (1947) reported the epicentre at 25.00° N and 61.50° E. According to Pendse (1948), the epicentre was at 24.20° N and 62.60° E, about 120 km away from Pasni. Ambraseys and Melville (1982) reported the epicentre at 25.02° N and 63.47° E. By recalculating the seismic parameters of the 1945 earthquake, Byrne et al. (1992) suggested that the epicentre was at 25.15° N and 63.48° E, which is used in the present study. The earthquake mainly affected the region between Karachi and the Persian border. In Karachi, ground motions lasted approximately 30 sec, stopping the clock in the Karachi Municipality Building and interrupting the communication cable link between Karachi and Muscat (Oman). According to Pendse (1948), the tsunami that was generated reached a height of 12–15 m in Pasni and Ormara on the Makran coast and caused great damage to the entire coastal region of Pakistan. However, several researchers have estimated the tsunami height of about 5–7 m near Pasni (Page et al. 1979; Ambraseys & Melville 1982; Heidarzadeh et al. 2008b). The tsunami wave was observed at 8:15 am on Salsette Island, i.e. Mumbai, and reached a height of 2 m (Jaiswal et al. 2009; Newspaper archives, Mumbai).

Furthermore, tsunami waves reached a height of about 11–11.5 m in the Gulf of (Kutch) Kachchh, Gujarat (Pararas-Carayannis 2006; Jaiswal et al. 2009). The tsunami reached as far south as Mumbai, Bombay Harbour, Versova (Andheri), Haji Ali (Mahalaxmi), Juhu (Ville Parle) and Danda (Khar). There was no report on damage at Bombay Harbour. Five people died at Versova (Andheri, Mumbai), and six more at Haji Ali (Mahalaxmi, Mumbai); several fishing boats were torn off their moorings at Danda and Juhu.

Pendse (1948) has reports that the first wave reached Pasni on the Makran coast about half an hour after the earthquake, but greater damage was caused by a wave that came 3.25 h after the first. Ambraseys and Melville (1982) also state that the first wave did not cause much damage. They report that two destructive waves arrived 90–120 min after the first at Pasni and Ormara. Further, they report that high waves persisted for a long duration at Karachi. Pendse (1948) concurs and notes the times of a series of high waves that struck there for a period of ~3 h after the arrival of the initial wave. On the other hand, only one high wave was reported (Pendse 1948; Ambraseys & Melville 1982) from Mumbai.

The oldest known tsunami in the region may have been generated by a large-magnitude earthquake, which occurred in the Indus delta/Kachchh region in 326 BC.

Table 1. Historical tsunami that affected the western coast of India.

No.	Year	Longitude (°E)	Latitude (°N)	Moment magnitude	Tsunami source	Loss of life	Source
		Location					
1	326 BC	67.30	24.00		Earthquake		
2	1008	60.00[a] 52.3[b]	25.00[a] 27.7[b]	?	Earthquake	1000*	
3	1524	Gulf of Cambay			Earthquake		
4	1819	Rann of Kutch		7.8	Earthquake	>2000*	
5	1883	Krakatau			Volcanic		
6	1845	Rann of Kutch		7.0	Earthquake		
7	1945	63.00	24.50	8.1	Earthquake	4000*	
8	2007	101.36	−4.43	8.4	Earthquake		
9	2013	62.26	25.18	7.7	Earthquake Volcanic		

[a]Rastogi and Jaiswal (2006).
[b]Ambraseys and Melville (1982).
*Both by earthquake and tsunami: Ambraseys and Melville (1982), Bilham (1999), Byrne et al. (1992), Dominey-Howes et al. (2006), Heck (1947); Merewether (1852); Murty and Rafiq (1991), Murty et al. (1999); Okal et al. (2006), Pararas-Carayannis (2006); Pendse (1946), Rastogi and Jaiswal (2006), Quittmeyer and Jacob (1979), Walton (1864), National Oceanic and Atmospheric Administration (NOAA); United States Geological Survey (USGS); Jaiswal et al. (2011), Jaiswal et al. (2008).

Alexandra's fleet was washed away by the oldest tsunami of the Arabian Sea. Destructive earthquakes and tsunami have occurred in the North Arabian Sea throughout geologic history and in recent times. Many of these events have not been adequately documented. On the western side of India, the earthquakes of 1524 and 1819 in the Kachchh region probably generated destructive tsunamis (Jaiswal et al. 2009). Many researchers (Rastogi & Jaiswal 2006; Benjamin 2008; Heidarzadeh et al. 2008a) have studied historical tsunamis in the Arabian Sea that affected the western coast of India (table 1).

Although large earthquakes along the Makran subduction zone (MSZ) are infrequent, the potential for the generation of destructive tsunamis in the northern Arabian Sea cannot be ruled out. It is quite possible that historical tsunamis in this region have not been properly reported or documented. Such past tsunamis very likely affected southern Pakistan, India, Iran, Oman, and possibly other areas as well (Hoffmann et al. 2013). Smith et al. (2013) argued that the MSZ has the potential to generate earthquakes similar in magnitude to the 2004 Andaman–Sumatra subduction zone. Furthermore, tsunami hazard assessment studies along the MSZ revealed that the southern coast of Iran and Pakistan, as well as Muscat and Oman are the most vulnerable areas (Heidarzadeh et al. 2008b, 2009; Heidarzadeh & Kijko 2011). The seismotectonics of the MSZ, historical earthquakes in the region and the recent earthquake of 24 September 2013 are indicative of the active tectonic collision process that is taking place along the entire southern and south-eastern boundary of the Eurasian plate as it collides with the Indian plate and the adjacent microplates. In view of this, an attempt is made to generate tsunami risk zone maps for the Gujarat state, which hosts the longest coastline of about 1600 km in India. These maps will be useful to the tsunami emergency response system and in planning the protection measures due to tsunami.

2. Demand for tsunami risk 3D visualizations

Gujarat state has the longest coastline in India, and has massive capital and infrastructure investments in its coastal regions (Singh et al. 2008). With rapid developmental activities along the coastline of Gujarat, there is a need for preparing tsunami risk three-dimensional (3D) visualization database using geoinformation technology. The coast of Gujarat has experienced many disasters in the past (e.g. Singh et al. 2008). Some of the most devastating disasters that have struck the state in the last few decades include the Morbi floods of 1978; the Kandla (port) cyclone of 1998; the killer earthquake in Kutch, 26 January 2001; and the flash floods in South Gujarat in 2005 and Surat in 2006. The coast of Gujarat was also affected by tsunami in the past (Singh et al. 2008; Jaiswal et al. 2009; Patel et al. 2010, 2013). Visualization is the graphical presentation of information, with the goal of improving the viewer's understanding of the information contents. Comprehension of 3D visualized models is easier and more effective than 2D models. 3D visualization models are important tools to simulate disasters from different angles that help users to comprehend the situation in more detail and help decision-makers plan for appropriate rescue operations. 3D visualizations are tools for rescue operations during disasters, e.g., cyclone, tsunami, earthquake, flooding, fire, etc. (Patel et al. 2013).

Thus, 3D graphical representations significantly reduce the amount of cognitive effort and improve the effectiveness of the communication tool (Raper 1989; Zlatanova et al. 2005; Zlatanova 2006; Marincioni 2007; Kolbe et al. 2008). For example, city and building models created by CAD software and scanned into a -computer from real-world objects; convey information that is useful to the decision-making process in natural disaster risk assessment and management.

3. Methodology

3.1. Numerical modelling

In the present study, 17 tsunami forecast stations were selected for output of tsunami simulation along the coast of India, Pakistan, Oman and Iran. Most of the tsunami forecast stations were selected in such a way that sea depth is less than 10.0 m to better examine the tsunami effect (Onat & Yalciner 2012). The location of tsunami forecast points along the west coast of India including Pakistan, Iran and Oman is illustrated in figure 1 and given in table 2. Bathymetry and elevation data are the principal data-sets required for the model to capture the generation, propagation and inundation of the tsunami wave from the source to the land. The bathymetry database taken from General Bathymetric Chart of the Oceans (GEBCO) 30 sec is used for tsunami modelling and the topography data taken from SRTM 90 m resolution is used for preparation of the inundation map. The bounding coordinates selected are 55°–76° E longitudes and 10°–30° N latitudes. The rupture parameters are taken from Byrne et al. (1992), which were used to model the source of the 1945 earthquake in this study (table 3). The initial wave amplitude (elevation and depression) for the source is computed using Okada (1985) method. The water elevation in the source is about 3 m, and the depression is about 1 m.

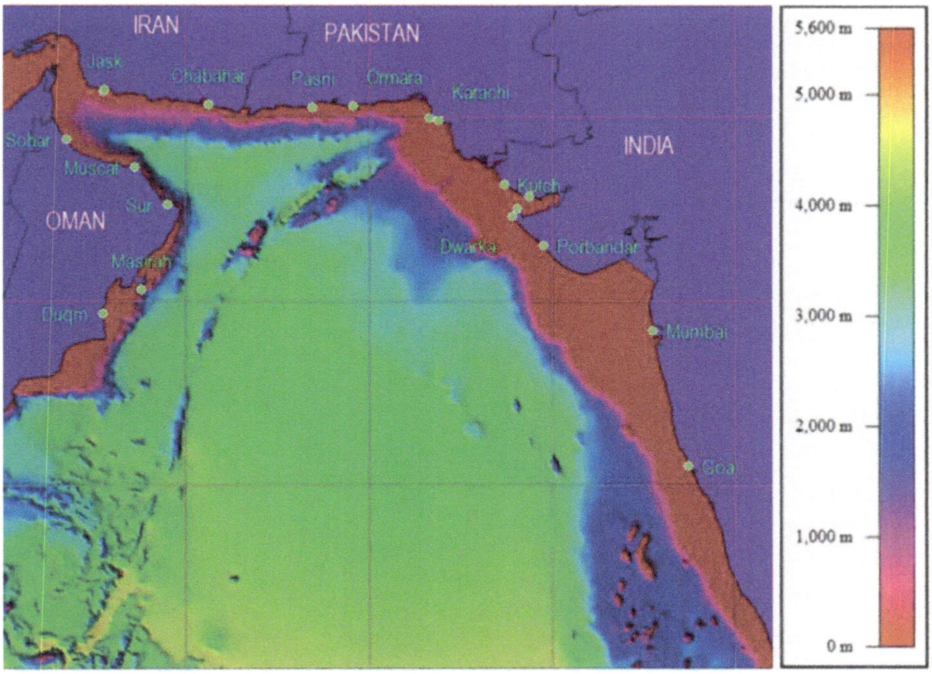

Figure 1. Tsunami forecast points indicated by circles, where the time series for tsunami run-up were calculated and presented in figure 7.

Furthermore, tsunami simulation basically aims to calculate the tsunami heights and its arrival times in space and time. The tsunami is assumed as a shallow water wave, where wavelength is much larger than the depth of the sea floor. The governing equations in tsunami numerical modelling are non-linear forms of shallow water equations with a friction term. The formulas are solved in Cartesian coordinate system (Imamura et al. 2006). Based upon these approximations and neglecting the

Table 2. Locations and depth of forecast points.

Name of gauge point	Depth of gauge point (m)	Latitude (°N)	Longitude (°E)	Name of gauge point	Depth of gauge point (m)	Latitude (°N)	Longitude (°E)
Kutch, India	0.84	23.1158	68.7153	Muscat, Oman	37.52	23.5919	58.61
Okha, India	0.17	22.4826	69.0647	Sohar, Oman	1.98	24.3535	56.7543
Dwarka, India	5.99	22.2528	68.9445	Pasni, Pakistan	2.96	25.2022	63.4809
Porbandar, India	0.24	21.457	69.7793	Karachi, Pakistan	1.33	24.8351	66.9055
Mumbai, India	0.71	19.1829	72.7854	Karachi, Pakistan	0.6	24.9143	66.6777
Goa, India	0.47	15.5088	73.7597	Ormara, Pakistan	2.24	25.2489	64.5757
Duqm, Oman	1.58	19.6358	57.7365	Jask, Iran	1.1	25.6574	57.7906
Masirah, Oman	2.66	20.2889	58.7791	Chabahar, Iran	1.5	25.2713	60.6373
Sur, Oman	3.74	22.57	59.53				

Table 3. The rupture parameter of 1945 Makran earthquake provided by Byrne et al. (1992).

Epicentre of earthquake		Fault length (km)	Fault width (km)	Strike angle (°)	Rake angle (°)	Dip angle (°)	Slip magnitude (m)	Focal depth (km)
Latitude	Longitude							
25.15° N	63.48° E	200	100	246	90	7	7	15

vertical acceleration, the equations of mass conservation and momentum in the 3D problem are given as follows:

Continuity equation:

$$\frac{\partial \eta}{\partial t}+\frac{\partial M}{\partial x}+\frac{\partial N}{\partial y}=0 \tag{1}$$

Momentum equation:

$$\frac{\partial M}{\partial t}+\frac{\partial}{\partial x}\left(\frac{M^2}{D}\right)+\frac{\partial}{\partial y}\left(\frac{MN}{D}\right)+gD\frac{\partial \eta}{\partial x}+\frac{gn^2}{D^{7/3}}M\sqrt{M^2+N^2}=0 \tag{2}$$

$$\frac{\partial N}{\partial t}+\frac{\partial}{\partial x}\left(\frac{MN}{D}\right)+\frac{\partial}{\partial y}\left(\frac{N^2}{D}\right)+gD\frac{\partial \eta}{\partial y}+\frac{gn^2}{D^{7/3}}N\sqrt{M^2+N^2}=0 \tag{3}$$

$$M=u(h+\eta)=uD \tag{4}$$

$$N=v(h+\eta)=vD \tag{5}$$

where x and y are the horizontal axes; z is the vertical axis; t is the time; h is the still water depth; η is the vertical displacement of water surface above the still water surface; u, v and w are the water particle velocities in x, y and z directions, respectively; g is the gravitational acceleration and g is the gravitational acceleration and D is the total water depth given by $h+\eta$.

In this study, the NAMI DANCE was used for simulation and efficient visualization of tsunamis, and understanding and investigation of tsunami generation and propagation mechanisms. This code is developed by Profs. Andrey Zaytsev, Ahmet Yalciner, Anton Chernov, Efim Pelinovsky and Andrey Kurkin in collaboration with Ocean Engineering Research Center, Middle East Technical University, Turkey and Institute of Applied Physics, Russian Academy of Science, Russia, especially for tsunami modelling (Yalciner et al. 2006b). It has advantages to simulate, animate, and visualize the generation, propagation, coastal amplification and inundation of tsunami in a given arbitrary shaped bathymetry under the input wave and current conditions (Yalciner et al. 2006a).

3.2. GIS/CAD-based 3D modelling

3.2.1. Raster layer operation. The first step of raster layer operation is geo-referencing of satellite images. Satellite images collected from Google Earth orthophotos were used for geo-referencing. Nine pairs of images (RGB, jpeg format) from Google Earth cover the study area. There was an average of 12 ground control points

(GCPs) collected for each image registration. The World Geodetic System 84 (WGS84) geographic projection was selected for image-to-image registration with Digital Globe image. After creating the geo-referenced images, all images are mosaicked to generate a whole geo-referenced image. The next step of raster layer operation is the digital elevation model (DEM) generation. A DEM is defined as a file or a database containing points over contiguous areas (Manual of photogrammetry 2004). The needed field height data for DEM generation can be obtained from point, line or polygonal vector height maps or stereo satellite/vector images. The SRTM data were used to create DEM. The geo-referenced image and DEM results constitute the basis of vector layer operations. This study uses SRTM data to generate the tsunami risk zone map.

3.2.3. Vector layer operation. The purpose of 3D GIS/CAD modelling was the thematic visualization of the disaster vulnerability of Okha generated from a 3D

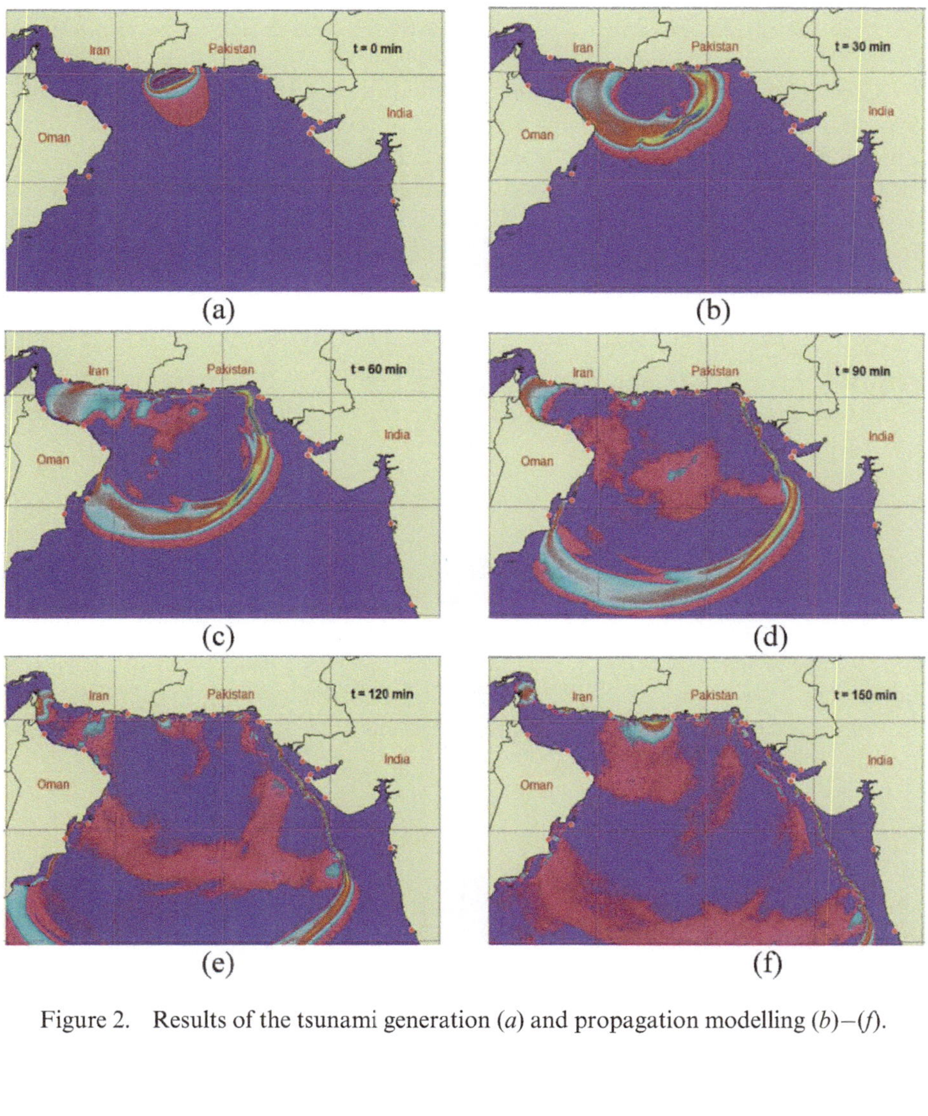

Figure 2. Results of the tsunami generation (*a*) and propagation modelling (*b*)–(*f*).

model. The geo-referenced image of the Okha coast was used for 2D building footprint in AutoCAD Map 3D. AutoCAD Map 3D provides access to geographic information system (GIS) and mapping data to support planning, design and data management. The building footprint layer was generated according to height value governed by the number of floors to create 3D building shapes.

4. Results and discussion

Tsunami snapshots show that the 1945 Makran event affected all the neighbouring countries including Iran, Oman, Pakistan and India (figure 2). The results of initial tsunami generation based on the fault parameters given by Byrne et al. (1992) are shown in figure 2 (*a*). Tsunami snapshots (figure 2(*b*)–(*f*)) show the estimated wave propagation at $t = 30$, 60, 90, 120 and 150 min after the tsunamigenic earthquake, respectively. Along the southern coast of Pakistan, the tsunami wave reaches Pasni in about 5–15 min, Ormara in about 60 min and Karachi in about 110 min. While along the southern coast of Iran, the tsunami wave reaches Chabahar in about 30–35 min and Jask in about 70–75 min. After the earthquake, the tsunami wave reaches the coast of Oman, namely at Muscat, in about 40 min, Sur in about 30–40 min, Masirah in about 60–70 min, Sohar in about 80 min and Duqm in about 130 min. Furthermore, the tsunami wave reaches the western coast of India along the Gulf of Kachchh in about 240 min, Okha in about 185 min, Dwarka in about 150 min, Porbandar in about 155 min, Mumbai in about 300 min and Goa in about 215 min. It is also observed that the distance from epicentre to Mumbai is less than Goa, but the arrival time of the first tsunami wave at the Mumbai is more than Goa. It could be due to the fact that Mumbai offshore is shallower that Goa and also due to the directivity of tsunami wave propagation. It is well known that most of the tsunami's energy travels perpendicular to the strike of the fault which is due to directivity (Ben-Menahem & Rosenman 1972; Singh et al. 2012). Due to this effect, most of the tsunami energy propagates in the direction shown in figure 3. The tsunami

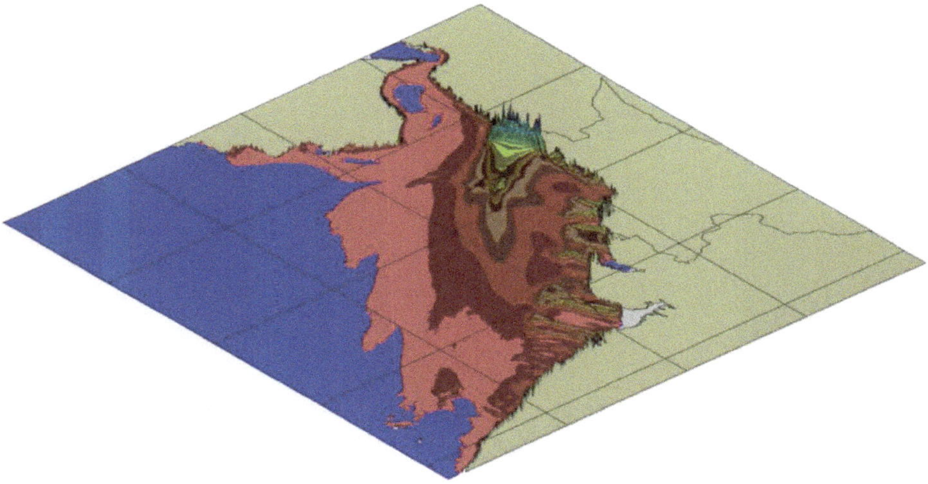

Figure 3. Direction of tsunami's energy propagation.

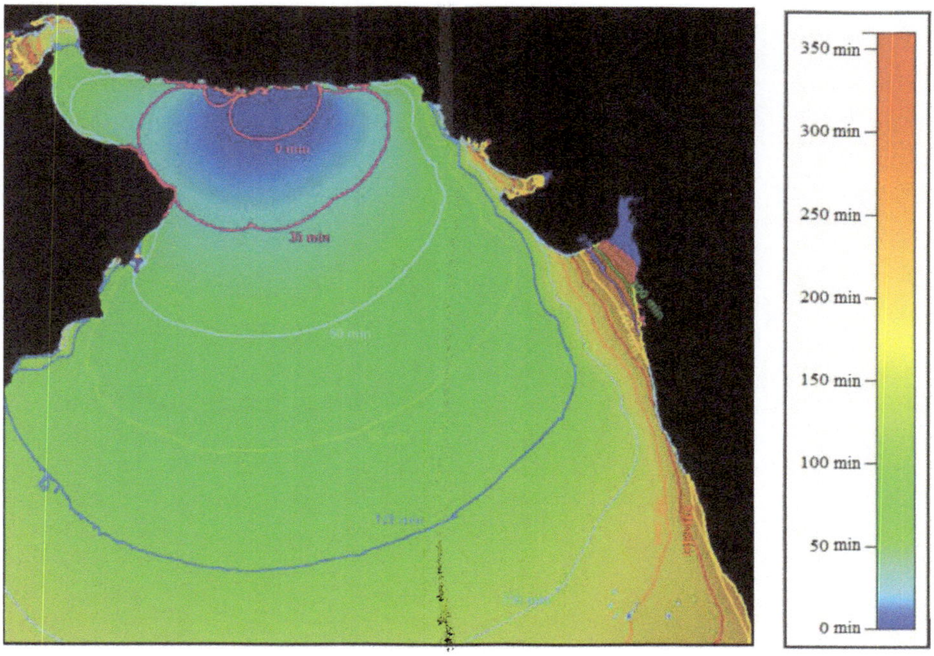

Figure 4. Tsunami travel time contour map.

travel time map is shown in figure 4. It is clear from the map that closely spaced contours along the western part of India may be due to shallower bathymetry features or any natural barrier. It is also noticed from figure 5 that the ocean depth suddenly changes from deeper (>2000 m) to shallower depth (<225 m) along the western part

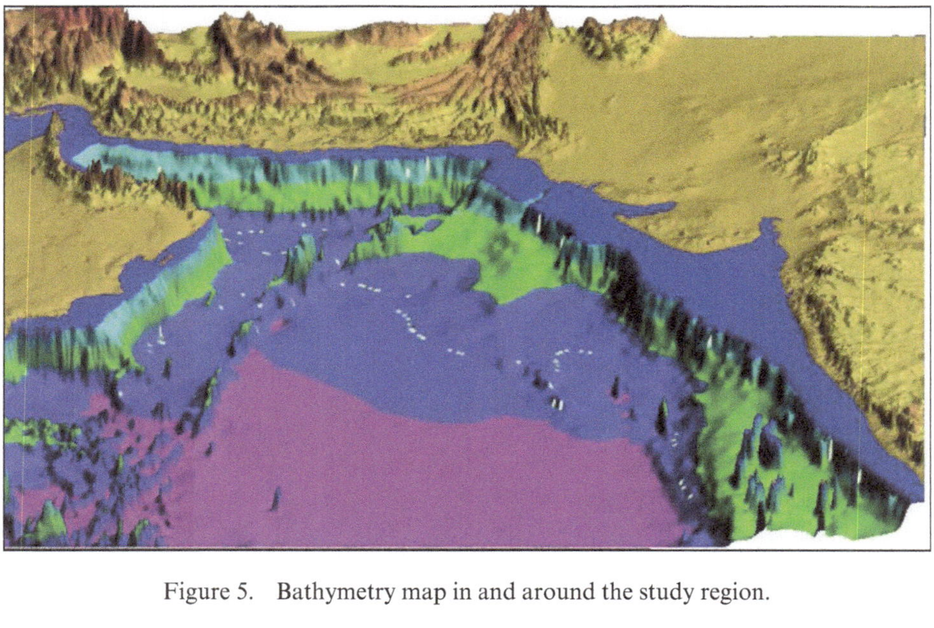

Figure 5. Bathymetry map in and around the study region.

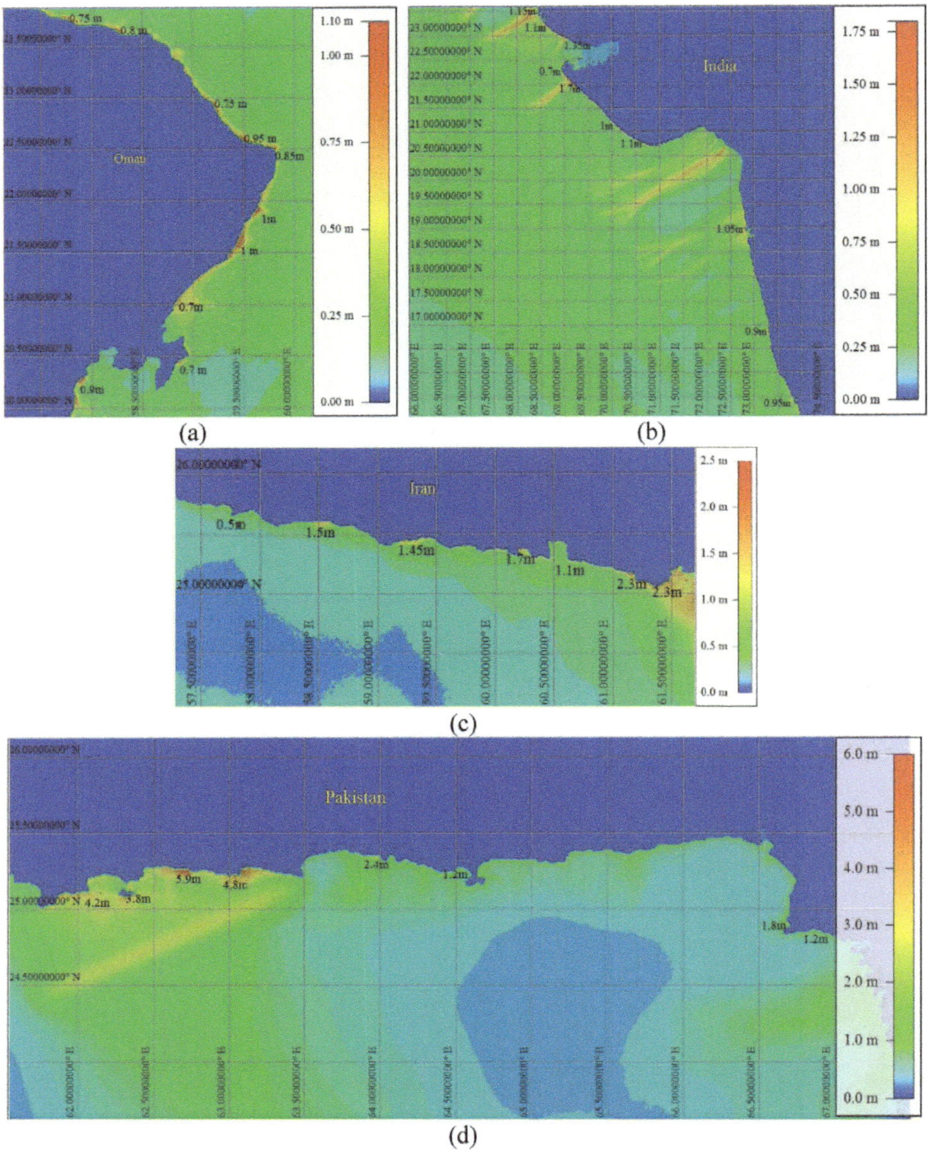

Figure 6. Maximum calculated tsunami run-up for a tsunami simulation of 360 min.

of India. Several researchers reported that the delay in the tsunami arrival from MSZ was probably due to a submarine landslide generated by the earthquake (Bilham et al. 2007; Neetu 2011).

Figure 6 shows the maximum calculated tsunami run-ups along the neighbouring countries including Iran, Oman, Pakistan and India for a tsunami simulation of 360 min. The simulated results show that the maximum tsunami height is about 5–6 m near the southern coast of Pakistan (figure 6 (*d*)), which is corroborated with the previous researchers in the same region (Page et al. 1979; Ambraseys &

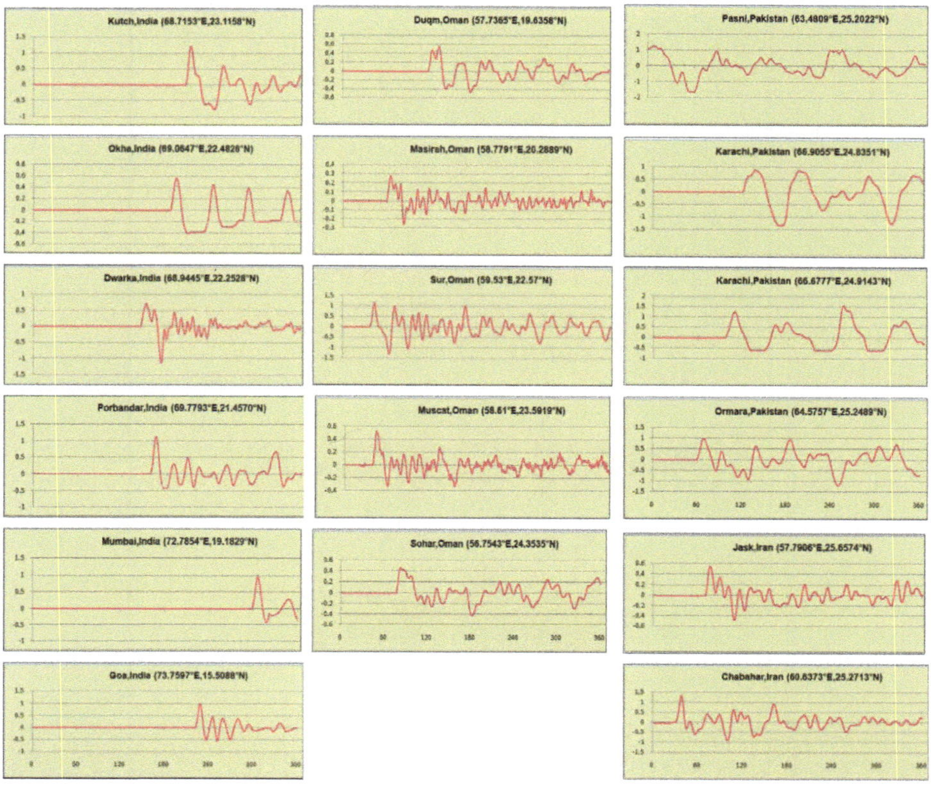

Figure 7. Time series of the calculated tsunami run-up at several locations along the coast of Iran, Oman, Pakistan and India.

Melville 1982; Heidarzadeh et al. 2008b). The maximum calculated tsunami run-ups were about 0.7–1.1 m along the coast of Oman, 0.7–1.35 m along the western coast of India, 0.5–2.3 m along the southern coast of Iran and 1.2–5.8 m along the southern coast of Pakistan, as shown in figure 6 (*a*)–(*d*), respectively. The tsunami run-up along the southern coast of Pakistan is far larger than that along the other coasts and may be due to the directivity of the tsunami. The time series for tsunami forecast points along Pakistan, India, Iran and Oman are presented in figure 7. The tsunami run-up and its arrival time at various tide gauge stations are listed in table 4.

It is believed that the digital topographical data are very important in detecting tsunami-prone area. The SRTM data are used to provide digital elevation information. Based on the processed SRTM data in GIS/CAD, all low-lying coastal areas potentially at risk of tsunami flooding have been identified and are shown in figure 8. The classification of tsunami risk zone is based on elevation vulnerability followed by Sinaga et al. (2011). However, for high-resolution mapping of tsunami risk zone along the coastal region, very-high-resolution topographical data and satellite images are needed. In this study, we developed the methodology for creation of 3D infrastructure located in tsunami risk zones using easily available and low-cost Google

Table 4. The maximum tsunami run-up and its arrival time at various forecast points.

Name of gauge point	Depth of gauge point (m)	Latitude (°N)	Longitude (°E)	Time first wave (min)	Time max. wave (min)	+Amplitude (m)	−Amplitude (m)
Kutch, India	0.84	23.1158	68.7153	206.433	211.7	1.21	−0.76
Okha, India	0.17	22.4826	69.0647	187.467	192.3	0.56	−0.4
Dwarka, India	5.99	22.2528	68.9445	146.1	151.467	0.71	−1.13
Porbandar, India	0.24	21.4570	69.7793	159	164.467	1.13	−0.43
Mumbai, India	0.71	19.1829	72.7854	299.367	305.2	0.96	−0.43
Goa, India	0.47	15.5088	73.7597	222.967	227.767	1.01	−0.56
Duqm, Oman	1.58	19.6358	57.7365	117.5	130.167	0.56	−0.46
Masirah, Oman	2.66	20.2889	58.7791	61.333	63.9	0.28	−0.26
Sur, Oman	3.74	22.5700	59.5300	38.033	42.6	1.16	−1.3
Muscat, Oman	37.52	23.5919	58.6100	38.9	42.433	0.53	−0.33
Sohar, Oman	1.98	24.3535	56.7543	78.333	82.067	0.46	−0.42
Pasni, Pakistan	2.96	25.2022	63.4809	0	7.133	1.25	−1.65
Karachi, Pakistan	1.33	24.8351	66.9055	120.733	135.333	0.88	−1.33
Karachi, Pakistan	0.60	24.9143	66.6777	97.2	252.467	1.55	−0.6
Ormara, Pakistan	2.24	25.2489	64.5757	60.6	67.333	1.02	−1.19
Jask, Iran	1.10	25.6574	57.7906	71.333	75.167	0.54	−0.48
Chabahar, Iran	1.50	25.2713	60.6373	32.567	38.233	1.33	−0.92

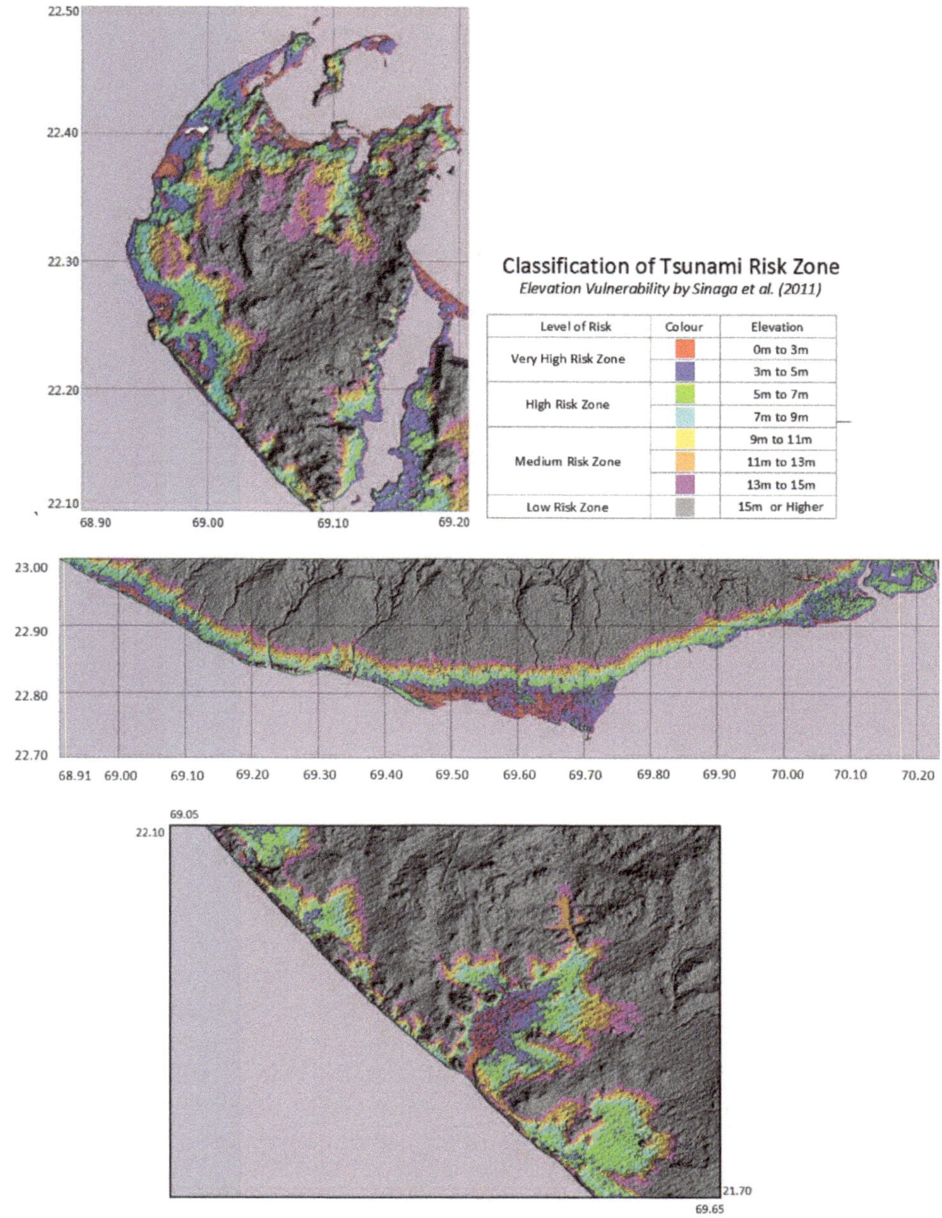

Figure 8. Classification of tsunami risk zone.

earth images and SRTM data in AutoCAD Map 3D software. The 3D tsunami risk model of Okha at different viewing angles is presented in figure 9 (*a*)–(*c*). A red, blue or green colour scheme was used to indicate the respective susceptibility to tsunami risk as shown in figure 9. It shows structures that are classified as very high risk, high risk and medium risk based on tsunami run-up height.

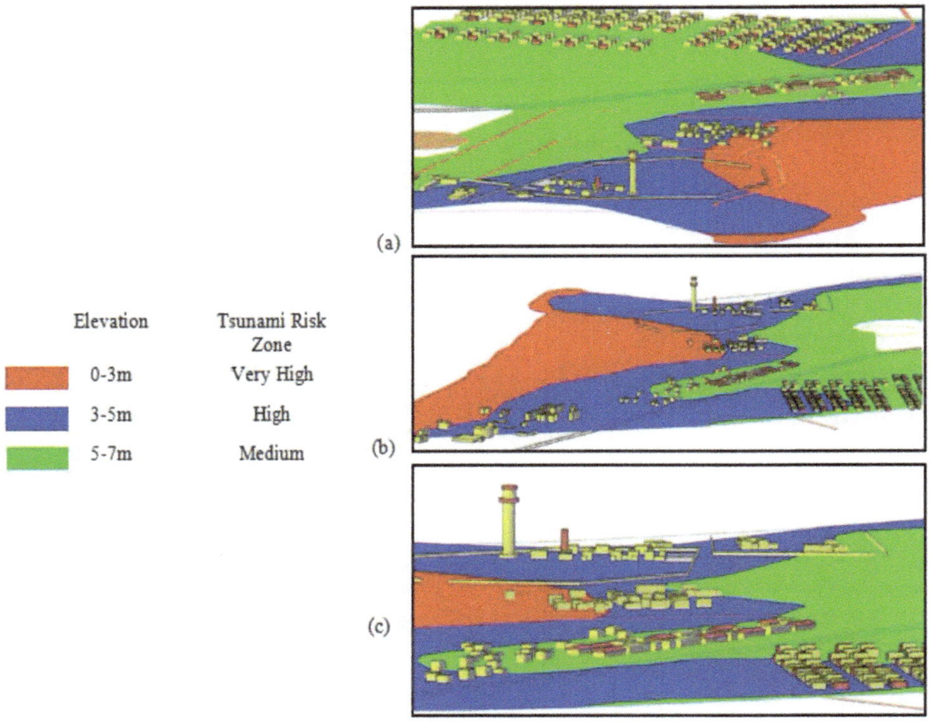

Figure 9. Visualization of 3D tsunami risk model of Okha with different viewing angles. To view this figure in colour, please see the online version of the journal.

5. Conclusions

The 1945 Makran tsunamigenic earthquake is modelled using rupture parameters suggested by Byrne et al. (1992). In most cases, the coastal regions which are far from the source have smaller tsunami height and longer tsunami travel times compared with the coastal regions near the source that have higher tsunami heights and shorter tsunami travel times. As a part of a tsunami emergency response system, the 3D coastal maps should be produced for countries in the vicinity of the MSZ, namely Pakistan, India, Iran and Oman. The lessons learnt from the December 2004 tsunami could be used for future planning. Ports, jetties, estuarine areas, river deltas and population in and around the coast of Pakistan, India, Iran and Oman could be protected with proper methods of mitigation and disaster management. In the future, scientists/researchers need to focus on 3D visualization and animation of tsunami risk. The study was performed to show the advantages of 3D GIS/CAD models and satellite images in tsunami risk assessment of the Okha coast, Gujarat. The main aim of the 3D Okha model is to visualize each building's tsunami risk level which improves decision-maker's understanding of the disaster level. Merging of SRTM elevation data with satellite images is suitable for tsunami risk zone classification. Combining the advanced computer-aided modelling, GIS-based modelling, marine parameter measurements by ocean bottom seismometers and satellite, installations of tide gauges and tsunami detection systems and also using conventional and traditional knowledge, it is possible to develop a suitable tsunami disaster management plan.

Acknowledgements

The authors thank Profs Andrey Zaytsev, Ahmet Yalciner, Anton Chernov, Efim Pelinovsky and Andrey Kurkin for providing NAMI-DANCE software and for their valuable assistance in tsunami numerical modelling of this study. Profs Nobuo Shuto, Costas Synolakis, Emile Okal and Fumihiko Imamura are acknowledged for invaluable endless collaboration. The VMP is grateful to Dr B. K. Rastogi, Director General, Institute of Seismological Research (ISR), for permission to use ISR library and other resource materials. A. P. Singh is thankful to Director General, ISR, for permission and encouragement to conduct such studies for the benefit of science and society.

References

Ambraseys NN, Melville CP. 1982. A history of Persian earthquakes. Cambridge (NY): Cambridge University Press; p. 219.

Benjamin J. 2008. Tsunamis of the Arabian Peninsula: a guide of historic events. Sci Tsunami Hazard. 27(1):31–46.

Ben-Menahem A, Rosenman M. 1972, Amplitude patterns of tsunami waves from submarine earthquakes. J Geophys Res. 77:3097–3128.

Berninghausen WH. 1966. Tsunamis and seismic seiches reported from regions adjacent to the Indian Ocean. Bull Seismol Soc Am. 56:69–74.

Bilham R. 1999. Slip parameters for the Rann of Kachchh, India, 16 June 1819 earthquake quantified from contemporary accounts. In: Stewart IS, Vita-Finzi C, editors. Coastal tectonics. London (UK): Geological Society (Special Publication); p. 295–318.

Bilham R, Lodi S, Hough S, Bukhary S, Khan AM, Rafeeqi SFA. 2007. Seismic hazard in Karachi, Pakistan: uncertain past, uncertain future. Seismol Res Lett. 78:601–613.

Byrne DE, Sykes LR, Davis DM. 1992. Great thrust earthquakes and a seismic slip along the plate boundary of the Makran subduction zone. J Geophys Res. 97: 449–478.

Dominey-Howes D, Cummins P, Burbidge D. 2007. Historic records of teletsunami in the Indian Ocean and insights from numerical modelling. Nat Hazard. 42:1–17.

Geist E, Titov V, Synolakis C. 2006. Tsunami: wave of change. Sci Am. 294:56–63.

Heck NH. 1947. List of seismic sea wave. B Seismol Soc Am. 37:269–284.

Heidarzadeh M, Kijko A. 2011. A probabilistic tsunami hazard assessment for the Makran subduction zone at the northwestern Indian Ocean. Nat Hazard. 56:577–593.

Heidarzadeh M, Pirooz MD, Zaker N, Mokhtari M. 2007. Evaluating the potential for tsunami generation in southern Iran. Int J Civil Eng. 5:312–329.

Heidarzadeh M, Pirooz MD, Zaker NH, Yalciner AC, Mokhtari M, Esmaeily A. 2008a. Historical tsunami in the Makran subduction zone off the southern coasts of Iran and Pakistan and results of numerical modeling. Ocean Eng. 35:774–786.

Heidarzadeh M, Pirooz M, Zaker NH, Synolakis C. 2008b. Evaluating tsunami hazard in the northwestern Indian Ocean. Pure Appl Geophys. 165:2045–2058.

Heidarzadeh M, Pirooz M, Zaker NH, Yalciner AC. 2009. Preliminary estimation of the tsunami hazards associated with the Makran subduction zone at the northwestern Indian Ocean. Nat Hazard. 48:229–243.

HoffmannG, RupprechterM, Al Balushi N, Grützner C, Reicherter K. 2013. The impact of the 1945 Makran tsunami along the coastlines of the Arabian Sea (northern Indian Ocean) – a review. Z Geomorphol. 57:257–277.

Imamura F, Yalciner AC, Ozyurt G. 2006. Tsunami modeling manual. Manual (TUNAMI-N2 model), April. p. 58.

Jaiswal RK, Rastogi BK, Singh AP. 2008. Past tsunamis in the Arabian Sea and future possibilities. Indian Miner. 61–62:75–82.

Jaiswal RK, Singh AP, Rastogi BK. 2009. Simulation of the Arabian Sea tsunami propagation generated due to 1945 Makran earthquake and its effect on western parts of Gujarat (India). Nat Hazard. 48:245–258.

Jaiswal RK, Singh AP, Rastogi BK, Murty TS. 2011. Aftershock sequences of two great Sumatran earthquakes of 2004 and 2005 and simulation of the minor tsunami generated on September 12, 2007 in the Indian Ocean and its effect. Nat Hazard. 57 (1):7–26.

Kolbe TH, Gröger G, Plümer L. 2008. CityGM - 3D city models and their potential for emergency response. In: Zlatanova S, Li J, editiors. Geospatial information technology for emergency response. London: Taylor & Francis; p. 257–274.

Manual of Photogrammetry. 5th ed. 2004. American Society for Photogrammetry and Remote Sensing ASPRS, Falls Church, Virginia; p. xvi + 1151.

Marincioni F. 2007. Information technologies and the sharing of disaster knowledge: the critical role of professional culture, disasters. Disasters. USA: Blackwell Publishing; p. 459–476.

Murty TS, Bapat A, Prasad V. 1999. Tsunamis on the coastlines of India. Sci Tsunami Hazard. 17:167–172.

Murty T, Rafiq M. 1991. A tentative list of tsunamis in the marginal seas of the north Indian Ocean. Nat Hazard. 4:81—83.

Neetu S, Suresh I, Shankar R, Nagarajan B, Sharma R, Shenoi SSC, Unnikrishnan AS, Sundar D. 2011. Trapped waves of the 27 November 1945 Makran tsunami: observations and numerical modeling. Nat Hazard. 59:1609–1618.

Okada Y. 1985. Surface deformation due to shear and tensile faults in a half-space. Bull Seismol Soc Am. 5:1135–1154.

Okal EA, Fritz HM, Raad PE, Synolakis C, Al-Sheibli Y, Al-Saifi M. 2006. Oman field survey after the December 2004 Indian Ocean Tsunami. Earthquake Spectra. 22:S203–S218.

Okal EA, Synolakis CE. 2008. Far-field tsunami hazard from mega-thrust earthquakes in the Indian Ocean. Geophys J Int. 172:995–1015.

Onat Y, Yalciner AC. 2012. Tsunami analysis for southern Aegean Sea. In: Proceedings of twenty-second (2012) international offshore and polar engineering conference; 2012 June 17–22; Rhodes (Greece). p. 236–241.

Page WD, Alt JN, Cluff LS, Plafker G. 1979. Evidence for the recurrence of large magnitude earthquakes along the Makran Coast of Iran and Pakistan. Tectonophysics. 52:533–547.

Pararas-Carayannis G. 2006. The potential of tsunami generation along the Makran subduction zone in the northern Arabian Sea – case study: the earthquake of November 28, 1945. Sci Tsunami Hazard. 24:358–384.

Patel VM, Dholakia MB, Singh AP. 2013. Tsunami risk 3D visualizations of Okha coast, Gujarat (India). Int J Eng Sci Innov Technol. 2:130–138.

Patel VM, Patel HS, Singh AP. 2010. Tsunami propagation in Arabian Sea and its effect on Dwarka city of Gujarat, India. Int J Adv Struct Eng. 2:163–174.

Pendse CG. 1946. The Mekran earthquake of the 28th November 1945. India Meteorol Depart Sci Notes. 10(125):141–145.

Pendse CG. 1948. The Mekran earthquake of the 28th November 1945. Scientific Notes Indian Met. Deptt. 10:141–145.

Pendse CG. 1948. The Mekran earthquake of the 28th November 1945. Sci Notes. X.

Quittmeyer RC, Jacob KH. 1979. Historical and modem seismicity of Pakistan, Afghanistan, northwestern India, and southeastern Iran. Bull Seismol Soc Am. 69:773–823.

Raper JF. 1989. The 3D geoscientific mapping and modeling system: a conceptual design, In: Raper JF, editor. Three-dimensional applications in geographical information systems. London: Taylor and Francis; p. 11–20.

Rastogi BK, Jaiswal R. 2006. A catalog of tsunamis in the Indian Ocean. Sci Tsunami Hazard. 25:128–143.

Sinaga TPT, Nugroho A, Lee Y-W, Suh Y. 2011. GIS mapping of tsunami vulnerability: case study of the Jembrana Regency in Bali, Indonesia. KSCE J Civil Eng. 15:537–543.

Singh AP, Bhonde U, Rastogi BK, Jaiswal RK. 2008. Possible inundation map of coastal areas of Gujarat with a tsunamigenic earthquake. Indian Minerals. 61(3–4) & 62 (1–4), p. 59–64.

Singh AP, Murty TS, Rastogi BK, Yadav RBS. 2012. Earthquake generated tsunami in the Indian Ocean and probable vulnerability assessment for the east coast of India. Mar Geod. 35:49–65.

Smith GL, McNeill LC, Wang K, He J, Henstock TJ. 2013. Thermal structure and mega thrust seismogenic potential of the Makran subduction zone. Geophys Res Lett. 40(8): 1528–1533.

Titov VV, Rabinovich AB, Mofjeld HO, Thomson R, Gonzálezl F. 2005. Global reach of the 26 December 2004 Sumatra tsunami. Science. 309:2045–2048.

Walton HI. 1864. Transactions of the Bombay Geographical Society from January 1863 to December 1864. Byculla: Education Society's Press.

Yalciner AC, Karakus H, Kuran U. 2006a. Modeling of tsunamis in the eastern Mediterranean and comparison with Caribbean. In Mercado A, Liu PL, editors. Caribbean tsunami hazard. World Scientific; p. 326–340.

Yalciner AC, Pelinovsky E, Zaytsev A, Chernov A, KurkinA, Ozer C, Karakus H. 2006b. NAMI DANCE Manual. Ankara, Turkey: METU, Civil Engineering Department, Ocean Engineering Research Center. Available from: http://namidance.ce.metu.edu.tr

Zlatanova S, Fabbri A, Li J. 2005. Geo-information for disaster management: large scale 3D data needed by urban areas. GIM Int. 19:10–13.

Zlatanova S, van Oosterom P, Verbree E. 2006. Geoinformation supports management of urban disasters. Open House Int. 31:62–79.

9

Mapping a burned forest area from Landsat TM data by multiple methods

W. CHEN*†, K. MORIYA†, T. SAKAI†, L. KOYAMA† and C.X. CAO‡
†Department of Social Informatics, Biosphere Informatics Laboratory, Graduate School of Informatics, Kyoto University, Kyoto 606-8501, Japan
‡State Key Laboratory of Remote Sensing Science, Institute of Remote Sensing and Digital Earth, Chinese Academy of Sciences, Beijing 100101, P. R. China

Forest fire is one of the dominant disturbances in boreal forests. It is the primary process responsible for organizing the physical and biological attributes of the boreal biome, shaping landscape diversity and influencing biogeochemical cycles. The Greater Hinggan Mountain of China is rich in forest resources while suffers from a high incidence of forest fires simultaneously. In this study, focusing on the most serious forest fire in the history of P. R. China which occurred in this region, we made use of two Landsat-5 TM (Thematic Mapper) images, and proposed to map the overall burned area and burned forest area by multiple methods. During the mapping, the fire perimeter, as well as rivers, roads and urban areas were first extracted and masked visually, and then four indices of Normalized Difference Vegetation Index, Enhanced Vegetation Index, Vegetation Fractional Cover and Disturbance Index were calculated. For each index, the optimal threshold for separating burned from unburned forest area was determined using their histograms. For comparison, threshold segmentation using single-band reflectance was performed, in addition to a Maximum Likelihood Classifier (MLC) based supervised classification of all features and forest area alone; their accuracies were also evaluated and analysed. Among all the methods compared here, mapping by EVI threshold segmentation proved to be optimal by the comparisons of overall accuracy (99.78%) and the kappa coefficient (0.9946). Finally, the calculated burned area and burned forest area were compared with the values from official statistics. Compared with the classical methods used to report official statistics on burned areas, the remote sensing-based mapping is more objective and efficient, less labour- and time-consuming, and more repeatable.

1. Introduction

Forests are multi-functional and multi-value ecosystems, which are widely distributed across land surfaces and have a complex composition, comprising one of the richest species resources on the land. Although they occupy less than 14% of the Earth's surface, forests and savannahs together account for more than 40% of the total solar energy captured each year by green plants, as well as containing the largest concentrations of organic material compared to all other global ecosystems (Gibbs et al. 2007; David et al. 2008; Chen et al. 2010). 'Forest' is a comprehensive

*Corresponding author. Email: chenwei@bre.soc.i.kyoto-u.ac.jp

description that has the greatest biodiversity among all terrestrial ecosystems (Krishnaswamy et al. 2009). For example, tropical rainforests, which cover only 7% of the total land area, contain 50%–70% of all global terrestrial species (Martinuzzi et al. 2009). Forests contain the most abundant biological resources and thus play an irreplaceable role in maintaining ecological balance on Earth.

The detection of forest disturbances is a key process in land ecosystem research, especially in fields related with global carbon cycles and climate change (Cao et al. 2011; Schroeder et al. 2006, 2012). Forest disturbance and recovery has been proposed as a primary mechanism for transferring carbon between the land surface and atmosphere, and thus plays an important role in both regional and global carbon budgets, as well as in long-term forest monitoring processes (Healey et al. 2005; Masek et al. 2008; Sönmez et al. 2011). Forest fire is one of the dominant disturbances, especially in boreal forests. It is the primary process responsible for organizing the physical and biological attributes of a boreal biome over most of its range, as well as shaping landscape diversity and influencing energy flows and biogeochemical cycles (Soja et al. 2007; Wotton et al. 2010; Stroppiana et al. 2012). Therefore, it is crucial to quantify the distribution and frequency of forest fire disturbances, which helps to understand the changes in the continental environment and climate. It is also useful for identifying spatial and temporal trends in forest management (Dyukarev et al. 2011; Pflugmacher et al. 2012; Schroeder et al. 2012).

The Greater Hinggan Mountain area, which is located in the north-east of China, is the northernmost and largest modern state-owned forest area within China, with a dense virgin forest. This area is rich in forest resources while also suffers from a high incidence of forest fires (Chen et al. 2011). As such, the annual burned forest area within this region ranks the first in all of China, making it the most serious forest fire hazard area in the country. Taking records in the 20 years from 1987 to 2006 as examples (Tian et al. 2011), 1059 fires occurred in this period, with a total burned area of 2.81×10^6 ha, including 1.36×10^6 ha of forest area. Tian and colleagues (2011) further indicated that, on average, each fire burned an area of 2660 ha, 1324 ha of which was ever covered by forest. There was an average of 53 fires each year, resulting in a burned area of 1.40×10^5 ha, including 6.77×10^4 ha of burned forest. The burned areas in 1987 and 2003, which experienced the most serious fires, account for 47.3% and 27.2% of the total burned area within this period, respectively.

Among all the fires in this region, one of the most noteworthy occurred on 6 May 1987 (abbreviated to the '5.6 Fire' hereinafter). On that day, the four forestry bureaus of Xilinji, Tuqiang, Amuer and Tahe were on fire at almost the same time, causing the most serious forest fire in the history of P. R. China. This fire continued to burn for 28 days and caused a great loss of life and property, ranking top among all the fire-caused losses in China (Zhao et al. 1994). Two indicators of the consequences of this fire were the total burned area, of 1.7×10^6 ha, and the burned forest area, of 1.01×10^6 ha. These data were acquired through the classical methods used to report official statistics on burned areas around 27 years ago, which are subjective, costly and time-consuming, but inefficient. Further, it is difficult to achieve repeatable and real-time dynamic monitoring in this way. In addition, ground surveys are not always standardized, and are completely lacking in many remote areas that are difficult to access. Thus, there is a great need for remotely sensed estimates of burned forest area over a variety of landscapes, in order to obtain accurate information for forest monitoring, assessment and management (David et al. 2008; Bergen et al. 2009).

At regional and greater scales, the only feasible and effective means of continuously and regularly monitoring the state of a forest and the dynamics is remote sensing. Satellite remote sensing offers an affordable and effective tool for forest change monitoring over large areas (Cohen & Goward 2004; Chen & Cao 2012; Pleniou & Koutsias 2013). The application of optical remote sensing technology is the most common and traditional approach to extracting forest characteristics and monitoring forest change. Optical imageries, particularly the extensive Landsat records, are well suited to the detection of forest disturbances (Masek et al. 2008; Kennedy et al. 2012; Townshend et al. 2012).

Using remote sensing data, various algorithms of supervised and unsupervised classification have been adopted to detect forest disturbances resulting from fires, logging and other causes (Comber et al. 2012; Margono et al. 2012). For serious stand-replacing forest disturbances, a much simpler method of threshold segmentation has also performed well, and been proven to be effective in the mapping of burned forest areas. Thus, this study employs threshold segmentation based on four indices to accurately extract burned area information from Landsat TM (Thematic Mapper) data. For comparison, threshold segmentation using single-band reflectance was conducted, as well as supervised classification of all features and forest areas alone; the accuracies of all these methods were evaluated and compared. Finally, the optimal method for burned forest area mapping was selected, and the estimated areas were compared with official statistical values.

2. Materials and methods

2.1. Study area

The Greater Hinggan Mountain area, which spans the northern part of Heilongjiang Province and the Inner Mongolia Autonomous Region of China, is the watershed of the Mongolian Plateau and the flat Songliao Plain. Its geographic coordinates range from 50°10′ to 53°33′ N in latitude, and from 121°12′ to 127°00′ E in longitude (figure 1). The region has a total length of over 1200 km, and a width of 200–300 km, as well as an average altitude of 600 m.

This region is an important climatic zone, having a typical cold temperate continental monsoon climate with warm summers and cold winters. The annual average temperature is −2.8 °C, with a minimum temperature of −52.3 °C, and an average annual precipitation of 746 mm. The Greater Hinggan Mountain area is also China's largest state-owned modern forest area, with a total ground area of 8.46×10^6 ha and a forest-covered area of 6.46×10^6 ha. Thus, the forest coverage is around 76.4%, providing a total stand volume of up to 5.01×10^8 m^3, which accounts for around 7.8% of the total national volume. The area contains more than 400 species of wild animals and 1000 varieties of wild plants, and is, therefore, a significant collection of biological resources. It is a mixed forest area dominated by the coniferous species, Mongolian pine (*Pinus sylvestris* L.) and Larch (*Larix gmelini* R.), and the deciduous species, Birch (*Betula platyphylla* S.) and Aspen (*Populus davidiana* D.). This region encompasses the location of the serious '5.6 Fire', making it a key focus for studies of forest fire prevention and post-fire monitoring (Sun et al. 2011).

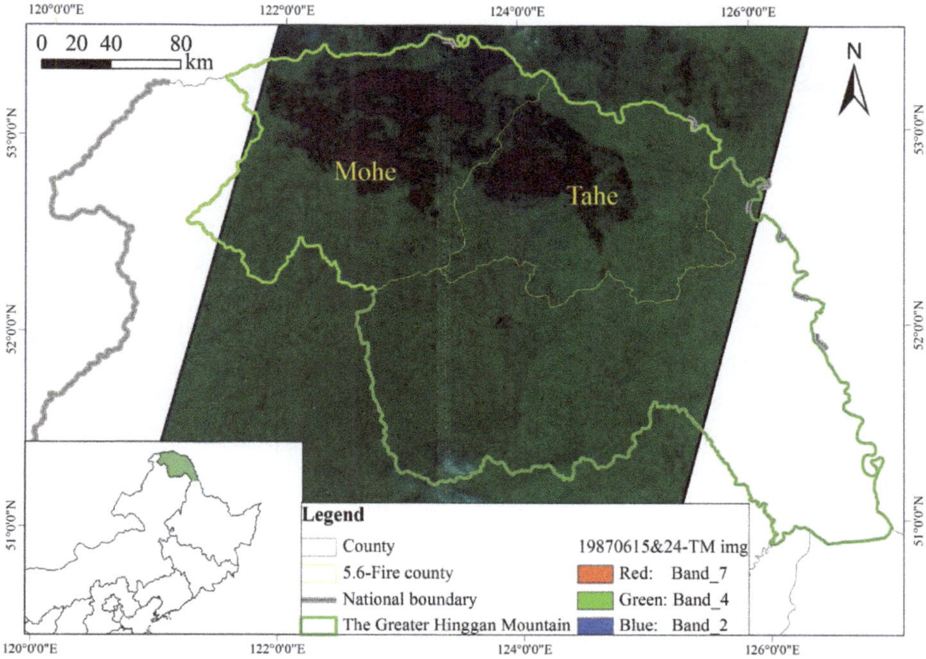

Figure 1. Location of the study area and the Landsat TM data used.

2.2. *Data and pre-processing*

Considering the '5.6 Fire' burned from 6 May to 2 June 1987, we have selected two TM images that were acquired just after the fire was extinguished (path 121 row 23: June 24; path 122 row 23: June 15). They were provided by the USGS-EROS (Earth Resources Observation and Science Centre).

For pre-processing of the data, the two scenes were first processed by geometric correction, terrain correction, radiometric calibration and atmospheric correction (based on the moderate resolution atmospheric transmission model-4). Following these, surface reflectance images could be acquired. The scene of path 121 row 23 was then normalized to the scene of path 122 row 23 in order to minimize the impact of potential sun-sensor-view angle effects (Schroeder et al. 2006). Finally, the two scenes were combined, using georeferenced mosaicking (figure 1). Since there were some small burned patches located in Russia about which we had no knowledge of the field survey and statistics, to reduce their impacts, we extracted the fire perimeter within the Greater Hinggan Mountain area of China using visual interpretation by the Geographic Information System software, and then calculated the dimensions of the burned area.

Additionally, some rivers, roads and developed areas were within the burned area while some were not, so in order to minimize their impacts on the mapping of the burned forest area, we extracted the rivers, roads and buildings by visual interpretation, and masked them out. In this way, only the forest area remained, facilitating the mapping of the burned forest area. The extracted fire perimeter, as well as rivers, roads and buildings, is shown in figure 2.

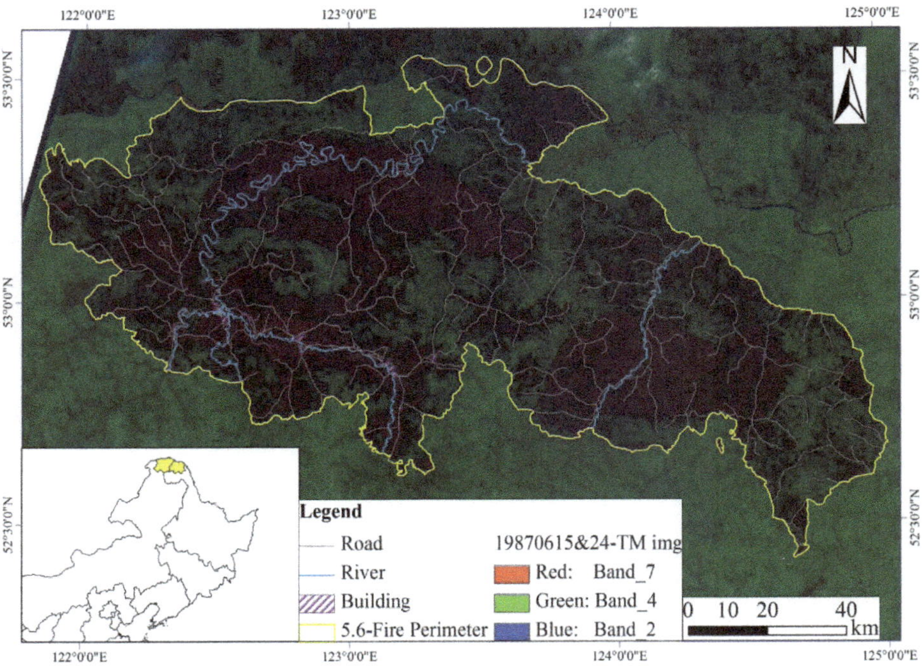

Figure 2. The fire perimeter and rivers, roads, building area extracted by visual interpretation.

2.3. Calculation of the four indices

Vegetation Indices (VIs) are combinations of single bands within a remote sensing image, and are commonly used for a simple, effective and well-referenced characterization of ground vegetation status. It is intended for geographical mapping of the relative amounts of different vegetation components, which can then be interpreted in terms of ecosystem conditions (Carlson & Ripley 1997; Koutsias et al. 2009; Sharma et al. 2013). Here, we selected two VIs, the Normalized Difference Vegetation Index (NDVI) and the Enhanced Vegetation Index (EVI). Additionally, we calculated the Vegetation Fractional Cover (VFC) and Disturbance Index (DI) for comparison.

2.3.1. NDVI. NDVI is one of the oldest, most well-known and most frequently used VIs. The combination of its normalized difference formulation and use of the highest absorption and reflectance regions of chlorophyll make it robust over a wide range of conditions (Rouse et al. 1974; Tucker 1979). It can, however, saturate in dense vegetation conditions, when the Leaf Area Index (LAI) becomes high. It is defined by the following equation:

$$\mathrm{NDVI} = \frac{(\rho_{\mathrm{NIR}} - \rho_{\mathrm{Red}})}{(\rho_{\mathrm{NIR}} + \rho_{\mathrm{Red}})} \tag{1}$$

where ρ_{NIR} and ρ_{Red} indicate the reflectance of the near-infrared and red bands, respectively. The calculated NDVI is shown in figure 3(*a*).

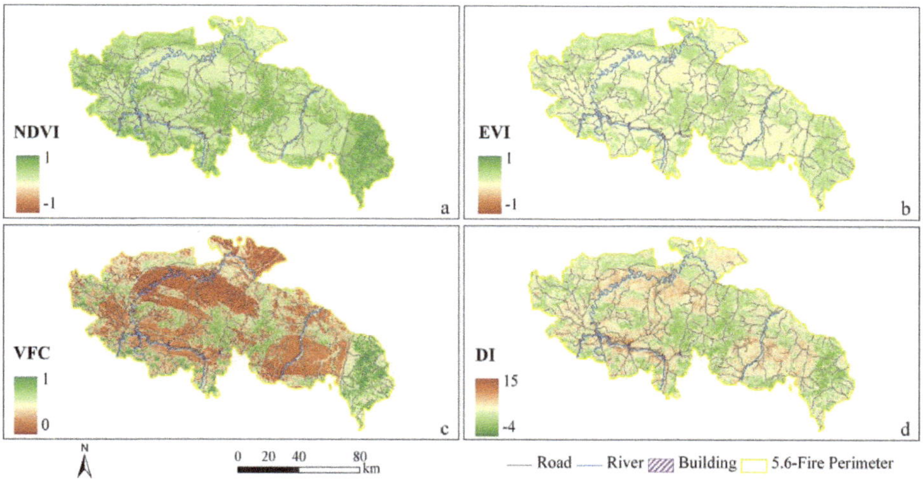

Figure 3. Maps of the burned area of (*a*) NDVI, (*b*) EVI, (*c*) VFC and (*d*) DI.

2.3.2. EVI. EVI was developed to improve the NDVI by optimizing the vegetation signal in high-LAI regions, by using the blue reflectance to correct for soil background signals and reduce atmospheric influences (Huete et al. 1997). It is, therefore, more useful in the high-LAI regions. EVI is defined by the following equation:

$$\mathrm{EVI} = \frac{2.5(\rho_{\mathrm{NIR}} - \rho_{\mathrm{Red}})}{\rho_{\mathrm{NIR}} + c_1\,\rho_{\mathrm{Red}} - c_2\,\rho_{\mathrm{Blue}} + L} \tag{2}$$

where ρ_{NIR} and ρ_{Red} have the same meaning as in equation (1), and ρ_{Blue} indicates the reflectance of the blue band. We set $c_1 = 6.0$, $c_2 = 7.5$ and $L = 1$, following those values used in the Moderate-resolution Imaging Spectroradiometer (MODIS) product. The acquired EVI is shown in figure 3(*b*).

2.3.3. VFC. VFC, which defines the amount of vegetation on the land surface, plays an important role in the fields of resource surveys and environment management, as well as climate change monitoring (Jiang et al. 2008). Many methods have been developed to estimate VFC, among which, the linear spectral mixture analysis (Collins et al. 2001; Fernandez-Manso et al. 2012) has been widely favoured. In actual applications, we often use VIs in place of reflectance of a single band to take full advantage of multi-band information (Cao et al. 2011). NDVI was selected in this case, and the calculation of VFC was performed using the following equation:

$$f_{\mathrm{cover}} = \frac{\mathrm{NDVI} - \mathrm{NDVI}_{\mathrm{soil}}}{\mathrm{NDVI}_{\mathrm{veg}} - \mathrm{NDVI}_{\mathrm{soil}}} \tag{3}$$

where f_c is the VFC, and $\mathrm{NDVI}_{\mathrm{veg}}$ and $\mathrm{NDVI}_{\mathrm{soil}}$ are the NDVI values of pure forest and pure soil pixels in the study area, respectively. The acquired VFC is shown in figure 3(*c*).

2.3.4. DI. DI is derived from the Tasseled Cap transformation (Crist & Cicone 1984; Healey et al. 2005; Chen et al. 2012). It is a linear combination of the three Tasseled Cap indices: brightness (*B*), greenness (*G*) and wetness (*W*). However, as there are variations in the acquisition date between the images used, the detection model should be designed to be relatively insensitive to phenology and BRDF (bidirectional reflectance distribution function) variation. Thus, spectral normalization steps should be taken, which use within-image statistics to normalize radiometric change. The normalized equations are as follows:

$$\begin{aligned} B_n &= (B - B_\mu)/B_\sigma \\ G_n &= (G - G_\mu)/G_\sigma \\ W_n &= (W - W_\mu)/W_\sigma \end{aligned} \tag{4}$$

where B_μ, G_μ and W_μ represent the mean Tasseled Cap brightness, greenness and wetness of the dense forest for a particular scene, respectively; B_σ, G_σ, W_σ are the corresponding standard deviations, and hence B_n, G_n and W_n indicate the normalized brightness, greenness and wetness, respectively. Here, the 'dense forest of a particular scene' that was used for normalization was selected from the corresponding NDVI image based on the criterion $\text{NDVI} \geq \text{NDVI}_{95\%}$.

Once the three component indices are normalized, they can be combined linearly to acquire the DI (Healey et al. 2005) as follows:

$$\text{DI} = B_n - (G_n + W_n) \tag{5}$$

Thus, the DI is able to record the spectral distance of a given pixel from the dense forest 'centroid' for that scene. The DI transformation is based on the observation that disturbed stands of forest typically have a higher Tasseled Cap brightness value and lower greenness and wetness values than undisturbed forest areas. A disturbed forest area that shows a high positive B_n and a low negative G_n and W_n will give high DI values. Conversely, an undisturbed forest area should present low DI values. The estimated DI image is shown in figure 3(*d*).

2.4. Determination of thresholds for the four indices

In order to determine the thresholds in each index that will permit separation of burned forest from unburned forest, we selected samples in both the burned and unburned forest area, by a stratified random sampling method, and then plotted their distribution histograms from the NDVI, EVI, VFC and DI images. The results are shown in figures 4(*a*), (*b*), (*c*) and (*d*), respectively.

From these figures, we found that the histograms of the burned forest area and unburned forest area were clearly separated from one another, making the threshold segmentation feasible. For accurate determination of the thresholds, we focused on the overlap area in each histogram, between the burned and unburned forest areas, and selected the optimal threshold with which the highest accuracy of burned forest mapping could be achieved.

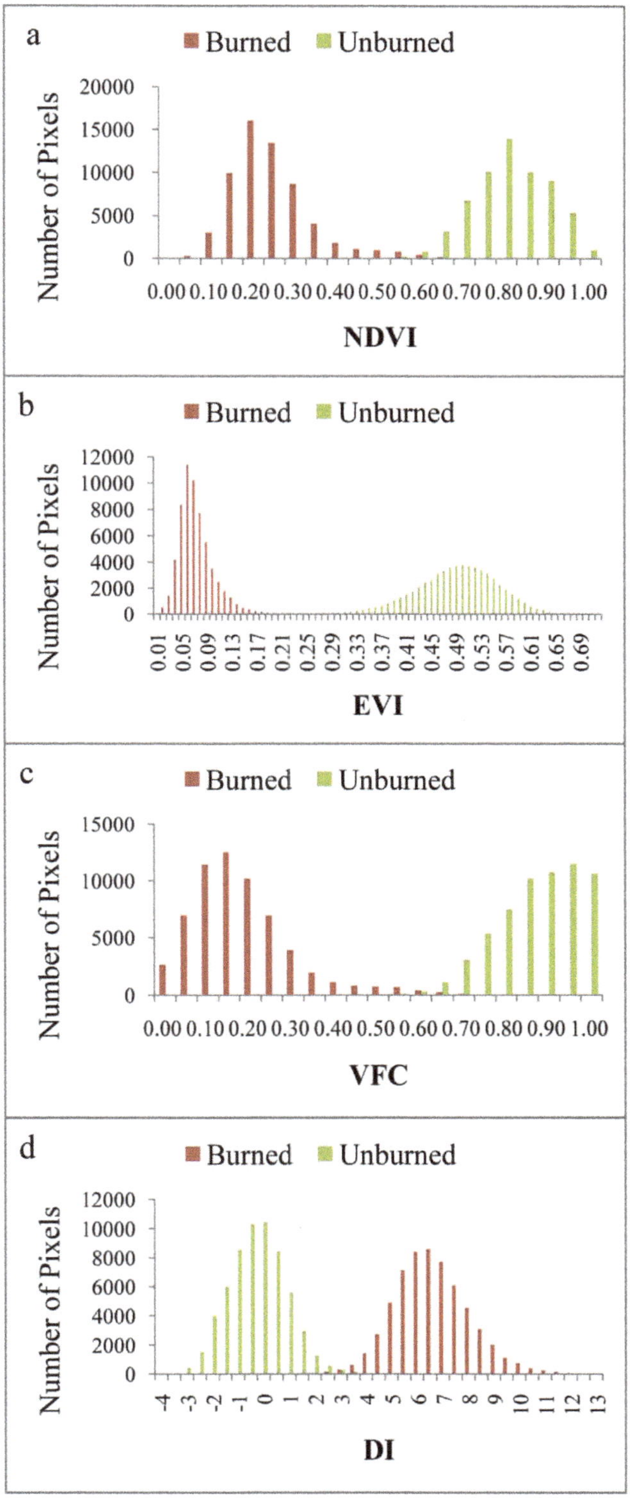

Figure 4. Histograms of forest area of (*a*) NDVI, (*b*) EVI, (*c*) VFC and (*d*) DI.

2.5. Mapping by single-band threshold segmentation

The NDVI, EVI, VFC and DI indices, which incorporate information from different bands, gave different accuracies in their mapping of the burned forest area. As such, what may we expect if only the surface reflectance of single-band is used for mapping? For comparison, we performed a burned forest area mapping based on threshold segmentation of single-band reflectance.

The threshold for each band should be determined. Here, the same samples used above were employed for threshold determination. The histograms of the six TM multispectral bands were plotted. As above, we focused on the overlap area in each histogram and selected the optimal threshold with which the highest accuracy of burned forest area mapping could be achieved.

2.6. Supervised classification

As methods of supervised and unsupervised classification are currently widely used to detect forest disturbances resulting from fires, logging and other causes (Castellana et al. 2007; Comber et al. 2012; Margono et al. 2012), for further comparison, we also conducted an mapping of burned forest area using a supervised classification. Here, the Maximum Likelihood Classifier (MLC) (Kavzoglu & Reis 2008) was used to classify the TM surface reflectance images. In fact, we also trained some other classifiers, such as Parallelepiped, Minimum Distance and Spectral Angle Mapper; however, their classification accuracies did not show significant improvement, and thus they have not been included here.

First, classification of all features was performed. The features we selected in training samples were separated into five classes: roads, rivers, buildings, burned forest and unburned forest. Additionally, as the mapping by threshold segmentation with the various indices was achieved using only the forest area, from which the rivers, roads and buildings had been masked; supervised classification of the forest area alone was also performed for a more targeted comparison. Thus, only two classes, burned forest and unburned forest, were included in the training samples. Finally, these two types of classification were carried out using six different band combinations: bands 2 and 5; bands 3 and 4; bands 4 and 7; bands 1, 3 and 4; bands 1, 2, 3 and 5; and all 6 bands.

3. Results

3.1. Mapping using the four indices

From figures 5(*a*), (*b*), (*c*) and (*d*), we concluded that the optimal thresholds for segmentation using NDVI, EVI, VFC and DI were 0.56, 0.24, 0.58 and 2.7, respectively. Based on the acquired thresholds for NDVI, EVI, VFC and DI, we mapped the burned forest area using each index, respectively. To fully verify and evaluate the mapping accuracy, we selected another group of test samples consisting of 2,023,605 burned forest pixels and 500,325 unburned forest pixels by visual interpretation following the stratified random sampling criterion. Using the index of Moran's I, we examined the spatial autocorrelation of these burned and unburned forest pixels, respectively. The results indicated that the interpreted burned and unburned forest pixels were both randomly distributed and can be used for the accuracy assessment.

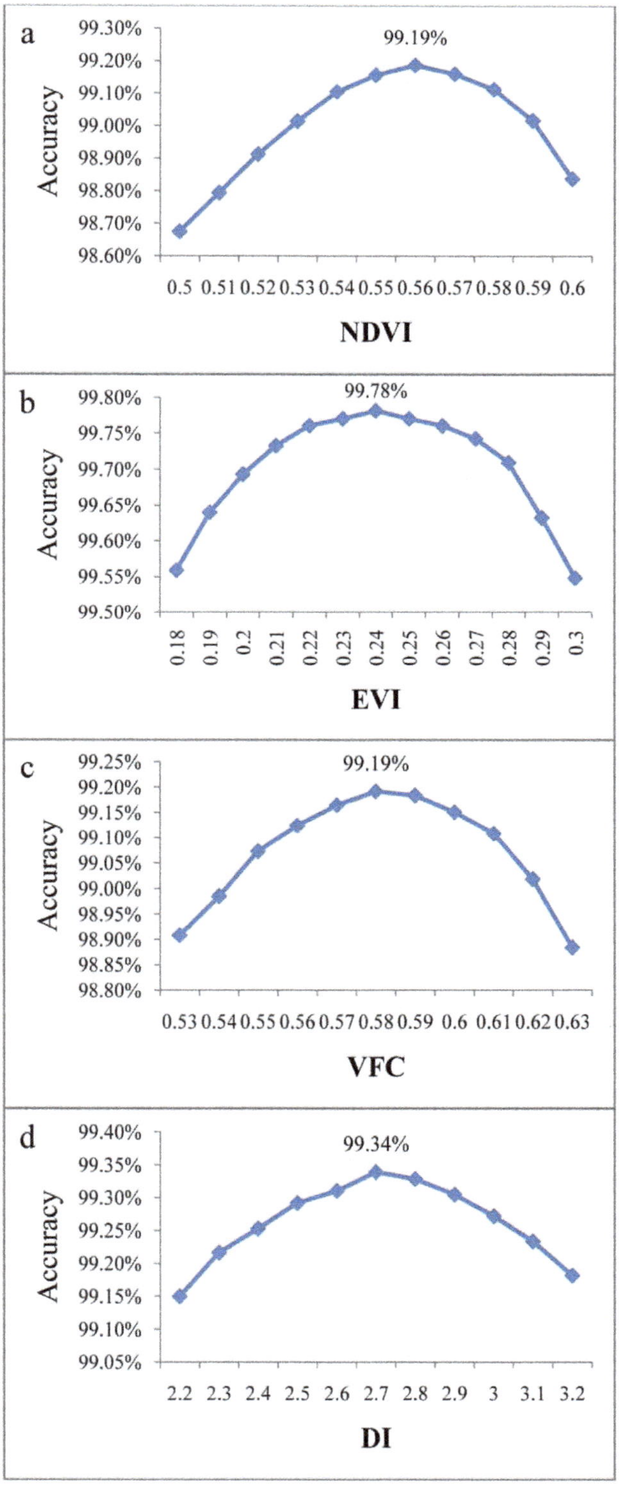

Figure 5. Determination of optimal thresholds by focusing on the overlap area of histograms of (*a*) NDVI, (*b*) EVI, (*c*) VFC and (*d*) DI.

Table 1. The accuracies of burned forest area mapping using: (a) four indices and (b) six TM bands.

Indices/variables	Overall accuracy	Kappa coefficient
(a) Four indices		
NDVI	99.07%	0.9712
EVI	99.83%	0.9946
VFC	99.06%	0.9716
DI	99.44%	0.9824
(b) Six TM bands		
Band 1	70.18%	0.2098
Band 2	53.61%	0.1411
Band 3	80.62%	0.4703
Band 4	99.72%	0.9914
Band 5	46.03%	0.1253
Band 7	99.16%	0.9738

The assessment results based on the whole collection of these pixels are shown in table 1(a). The reference samples used to estimate the accuracy were created using photo interpretation of the same Landsat image; therefore, usually high accuracies were justified.

Using the test samples, the overall mapping accuracies of the NDVI, EVI, VFC and DI indices were found to be 99.07%, 99.83%, 99.06% and 99.44%, with corresponding kappa coefficients of 0.9712, 0.9946, 0.9716 and 0.9824, respectively. By comparison, we found that the mapping accuracies using the four indices differed, and that their relative merits followed EVI > DI > NDVI ≈ VFC. To explore the causes of this difference, we plotted the spectral signatures of the Landsat TM multispectral bands for the burned and unburned forest area (figure 6).

In figure 6, we found that the differences in spectral characteristics of bands 1, 3, 4 and 7 were significant, with band 4 most prominent. In contrast, the differences in bands 2 and 5 were not significant. From the algorithms of the four indices (equations (1)–(5)), we knew that NDVI incorporates information from bands 3 and 4, and thus reflected the significant difference between the burned and unburned forest area in the two bands. EVI, which utilized the information from bands 1, 3 and 4, had an improved performance over NDVI. Since VFC was acquired from NDVI, their accuracies were approximately equal. DI, derived from the Tasseled Cap transformation which made use of all TM multispectral bands (bands 1–5 and 7), also performed well in the mapping. However, as it incorporated the indistinct information from bands 2 and 5, the accuracy of the DI mapping was less than that of EVI, but higher than those of NDVI and VFC.

3.2. Mapping using the six TM bands

We also mapped the burned forest area using the acquired optimal thresholds for bands 1–5 and 7 which were 0.04, 0.04, 0.04, 0.15, 0.16 and 0.11, respectively. The

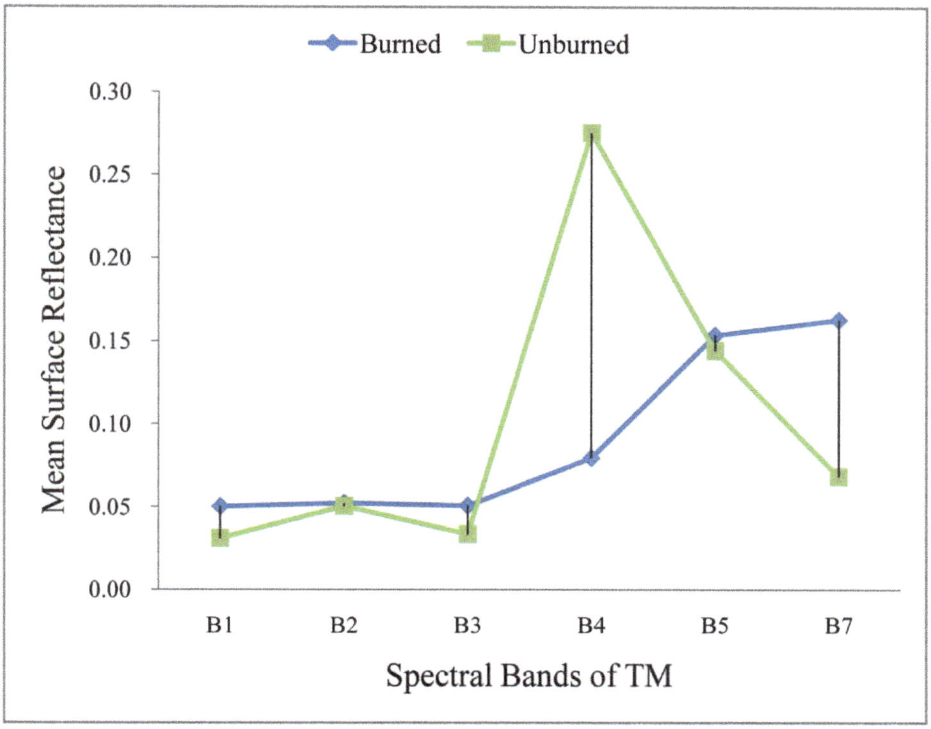

Figure 6. The spectral signatures of Landsat TM multispectral bands of forest area.

same test samples were used for accuracy evaluation. From table 1(b), we found that the overall mapping accuracies based on bands 1–5 and 7 were 70.18%, 53.61%, 80.62%, 99.72%, 46.03% and 99.16%, with corresponding kappa coefficients of 0.2098, 0.1411, 0.4703, 0.9914, 0.1253 and 0.9738, respectively. From comparison among the bands, we concluded that the mapping based on bands 2 and 5 had the lowest accuracy, followed by those of bands 1 and 3, while the mapping using bands 4 and 7 performed much better. This could also be explained by figure 6, in which the spectral difference between burned and unburned forest areas was least in bands 2 and 5, followed by that in bands 1 and 3, while the difference in bands 4 and 7 was the most significant.

3.3. *Mapping using supervised classification*

With the results of the two types of classification (all features and only forest area), the same test samples as used above were taken as ground truth references for error analysis and accuracy evaluation. The overall accuracies and kappa coefficients were calculated as the measure of classification accuracy (table 2).

From table 2, we found that the classification accuracy based on the combination of bands 2 and 5 was the lowest, and was significantly lower than the accuracies from other combinations. This result can be explained by the little spectral difference between the burned and unburned forest areas in bands 2 and 5, which is shown in figure 6. As there were significant differences in the spectral characteristics of burned

Table 2. Mapping accuracies of (a) all features classification (five classes) and (b) only forest classification (two classes) using six different band combinations.

Band combinations	Overall accuracy	Kappa coefficient
(a) All features classification (five classes)		
Bands 2 and 5	47.95%	0.1864
Bands 3 and 4	91.54%	0.7996
Bands 4 and 7	93.86%	0.8500
Bands 1, 3 and 4	93.06%	0.8310
Bands 1, 2, 3 and 5	90.40%	0.7711
All six bands	95.18%	0.8800
(b) Only forest classification (two classes)		
Bands 2 and 5	52.78%	0.1599
Bands 3 and 4	99.49%	0.9841
Bands 4 and 7	99.73%	0.9915
Bands 1, 3 and 4	99.51%	0.9847
Bands 1, 2, 3 and 5	95.42%	0.8629
All six bands	99.71%	0.9910

and unburned forest in bands 1, 3, 4 and 7, the classifications based on the combinations of bands 3 and 4, bands 4 and 7, as well as bands 1, 3 and 4 were able to achieve more desirable results. Among them, the accuracy based on bands 1, 3 and 4 was higher than that of bands 3 and 4, but still lower than that of bands 4 and 7. The combination of bands 1, 2, 3 and 5 gave a lower classification accuracy than the above three combinations. It is probably because of the exclusion of band 4 from this combination, and thus the exclusion of the most significant information.

3.4. Area calculation and comparison with official statistics

Among all the mapping methods studied here, including the six single-band reflectance threshold segmentation, four indices threshold segmentation and two types of supervised classification, the mapping by EVI threshold segmentation achieved the highest accuracy (overall accuracy of 99.83% and kappa coefficient of 0.9946). As such, the burned forest area estimated using the EVI was determined to be optimal and taken as the final output. The burned and unburned forest areas given by the EVI threshold segmentation, in addition to rivers, roads and building areas input by visual interpretation, are shown in figure 7.

Based on the mapping result, we calculated the burned area, using the fire perimeter from visual interpretation, and the burned forest area given by the EVI threshold segmentation, and compared them with values from the official statistics (figure 8). The comparison revealed a discrepancy between the remote sensing extracted values, of 1.298 million and 0.862 million ha, respectively, and the official statistical values, of 1.7 million and 1.01 million ha, respectively. The official statistical values came from the measurement and estimation of area within the range delineated on the topographic map and forest form map. The delineation was achieved mainly by visual inspection in the field survey with the help of corresponding land use map.

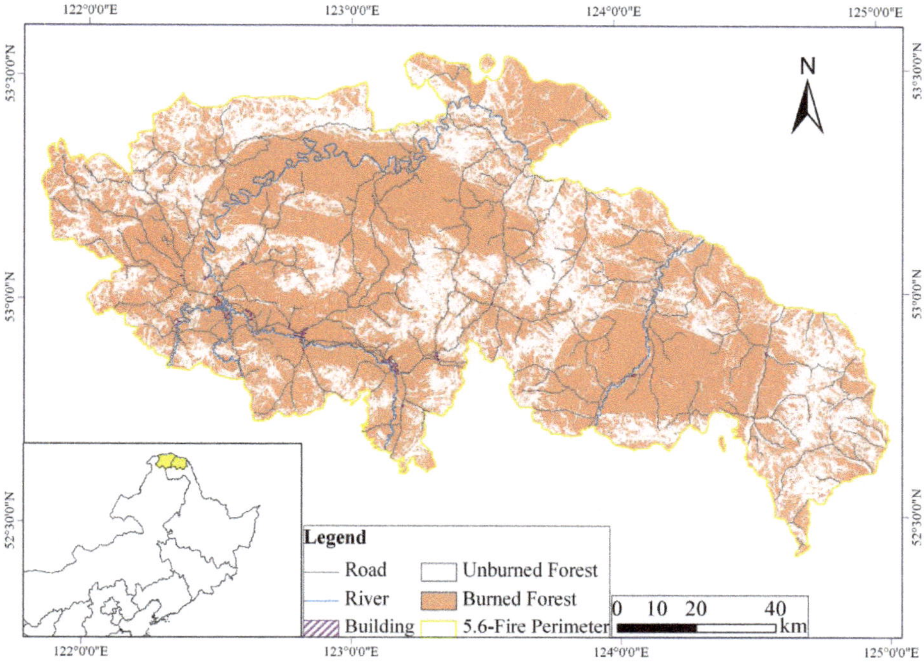

Figure 7. The burned/unburned forest area by EVI threshold segmentation as well as the rivers, roads and building area by visual interpretation.

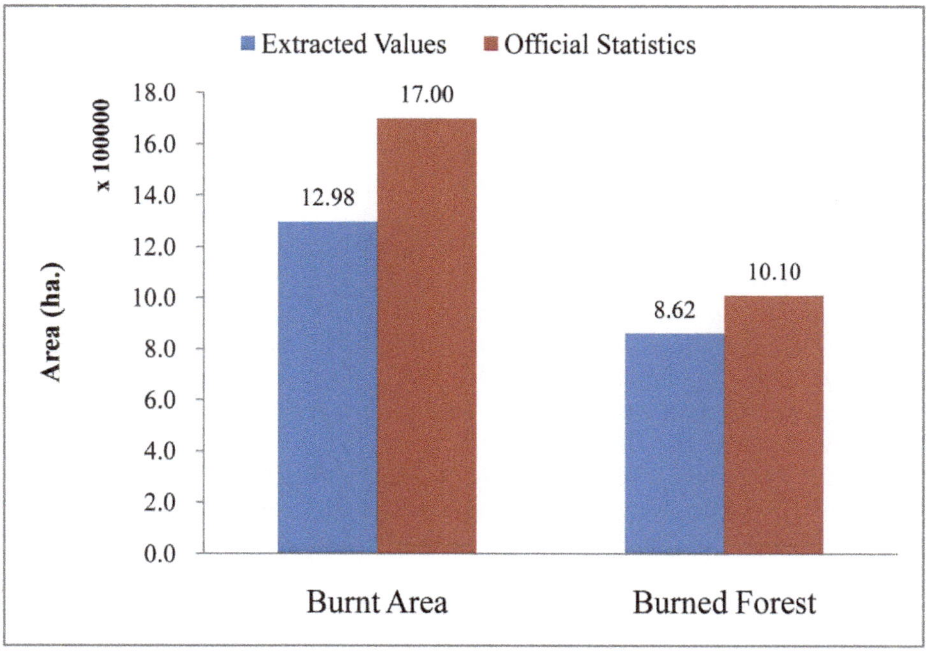

Figure 8. The comparison of burned area and burned forest area between remote sensing extracted values and official statistics.

Compared with human experience in the field survey and range delineation of burned area, the mapping and calculation of burned area using remote sensing data appeared to be more objective, making the results more credible. In addition, due to the advantages of large-scale, multi-temporal and high-precision of remote sensing technology, the Landsat-based mapping was repeatable, less labour-intensive and more efficient, and proved to be an excellent choice of burned forest area mapping.

4. Discussion

The burned area and burned forest areas of the '5.6 Fire' were mapped from remote sensing images in this study. In this area, the discrepancy between the remote sensing extracted and official statistical values was found for the first time. This finding allows a more objective and accurate understanding of the losses caused by that fire and thus provides crucial information for the local forest monitoring and management.

Remote sensing approaches offer a feasible and effective tool for the detection of forest disturbances in regional and even global scales (Healey et al. 2005; Masek et al. 2008; Rykhus & Lu 2011; Stroppiana et al. 2012). In previous studies, a variety of spectral indices (such as DI by Healey et al. 2005, NDVI and EVI by Stroppiana et al. 2012 and NDVI, DVI, RVI, SAVI, etc. by Veraverbeke et al. 2012) have been utilized and compared, and are being proven to achieve high accuracies. However, discussion on the causes of the variations in accuracies under certain circumstances still remains to be further explored.

In this study, we calculated four indices of NDVI, EVI, VFC and DI for the burned forest area mapping. For comparison, the mapping using six TM multi-spectral bands was also performed individually by the same approach of threshold segmentation. The high accuracies achieved by the bands 4 (NIR) and 7 (SWIR) were consistent with the findings in some previous studies (Koutsias & Karteris 2000; Pleniou & Koutsias 2013). However, here we found that the reflectance of band 5 was apparently not sensitive to forest fire disturbance. Actually, unlike Landsat TM7, in Landsat TM5, the spectral pattern of burned surfaces does not follow a unique pattern. This is consistent with the findings of the separability analysis by Maingi (2005) who concluded that information recorded in Landsat ETM+7 was more informative than in Landsat ETM+5 for the separation of old and new fire scars except for shaded deciduous forests. Other studies carried out in the middle-infrared area also indicate a varied behaviour of burned areas (Pereira et al. 1997; Arino et al. 2001). In particular, the comparison between the VIs and the single spectral bands showed the accuracy of the mapping based on band 4 was even higher than that using NDVI, VFC and DI, and quite close to that by EVI. In addition, the mapping accuracy based on band 7 was also higher than that using NDVI and VFC although the magnitude of the difference was not significant. These results indicated that the accuracy of burned forest area mapping depends not upon the number of bands used, but on the proportion of significant information contained within the mapping application. This provides more consideration for the selection of indices or bands used for burned forest area mapping in the Greater Hinggan Mountain area.

The MLC-based supervised classification was also conducted to detect forest fire disturbance using different band combinations. Here, we considered two groups of training samples. Except the samples consisting of all features, another group only included two classes of burned forest and unburned forest. The comparative study gave the result that the forest-only classification had higher accuracies than the all features classification. Especially, the classification accuracy from the combination of bands 4 and 7 was even higher than that from all six bands combination. This demonstrates that the classification accuracy is not only determined by the amount of significant information contained, but also influenced by any insignificant information included. The combination of bands 4 and 7 was demonstrated to be most suitable for the classification in this case, when only considering the forest area.

The above analyses indicate that simple indices or methods can also produce better performance than complex calculations under some specific conditions. It is probably due to the blurring or loss of some original significant characteristics in the reorganization of more complex models or indices. Thus, the complexity in the models is better to be avoided if simpler but also effective approaches exist, especially in actual applications.

The Landsat records are particularly useful for forest change monitoring, as they have unprecedented historical coverage, spanning 40 years of observations, as well as the necessary spatial and spectral resolutions, of up to 30 m pixels and seven bands, respectively. This permits the capture of most natural and managed disturbance events at regional and global scales (Wilson & Sader 2002; Tucker et al. 2004; Koutsias et al. 2009; Kennedy et al. 2012). In future studies, the monitoring of post-fire forest recovery can also be conducted based on these images, as well as with a combination of data from other sensors, including Quickbird, MODIS and SAR (Synthetic Aperture Radar), etc.

Acknowledgements

The authors are grateful to USGS-EROS (Earth Resources Observation and Science Center) for providing the study with the Landsat TM data. We express our sincere thanks to the anonymous referee for the detailed and useful comments and suggestions.

Funding

The research in this paper was supported by the Japanese Government MEXT Scholarship [Grant No. 113378].

References

Arino O, Piccolini I, Kasischke E, Siegert F, Chuvieco E, Martin P, Li Z, Fraser R, Eva H, Stroppiana D, et al. 2001. Methods of mapping surfaces burned in vegetation fires. In: Ahern FJ, Goldammer JG, Justice CO, editors. Global and regional vegetation fire monitoring from space: planning a coordinated international effort. The Hague: SPB Academic Publishing; p. 227–255.

Bergen KM, Goetz SJ, Dubayah RO, Henebry GM, Hunsaker CT, Imhoff ML, Nelson RF, Parker GG, Radeloff VC. 2009. Remote sensing of vegetation 3-D structure for biodiversity and habitat: review and implications for lidar and radar spaceborne missions. J Geophys Res Biogeosci. 114:G00E06. DOI:10.1029/2008JG000883.

Cao CX, Chen W, Li GH, Jia HC, Ji W, Xu M, Gao MX, Ni XL, Zhao J, Zheng S, et al. 2011. The retrieval of shrub fractional cover based on a geometric-optical model in combination with linear spectral mixture analysis. Can J Remote Sens. 37:348–358.

Carlson TN, Ripley DA. 1997. On the relation between NDVI, fractional vegetation cover, and leaf area index. Remote Sens Environ. 62:241–252.

Castellana L, D'Addabbo A, Pasquariello G. 2007. A composed supervised/unsupervised approach to improve change detection from remote sensing. Pattern Recognit Lett. 28:405–413.

Chen CX, Tang P, Bian Z. 2012. Tasseled cap transformation for HJ-1A/B charge coupled device images. J Appl Remote Sens. 6:063575.

Chen HW, Hu YM, Chang Y, Bu RC, Li YH, Liu M. 2011. Simulating impact of larch caterpillar (*Dendrolimus superans*) on fire regime and forest landscape in Da Hinggan Mountains, Northeast China. Chin Geogr Sci. 21:575–586.

Chen W, Cao CX. 2012. Topographic correction-based retrieval of leaf area index in mountain areas. J Mt Sci. 9:166–174.

Chen W, Cao CX, He QS, Guo HD, Zhang H, Li RQ, Zheng S, Xu M, Gao MX, Zhao J, et al. 2010, Quantitative estimation of the shrub canopy LAI from atmosphere-corrected HJ-1 CCD data in Mu Us Sandland. Sci China Earth Sci. 53:26–33.

Cohen WB, Goward SN. 2004. Landsat's role in ecological applications of remote sensing. Bioscience. 54:535–545.

Collins EF, Roberts DA, Borel CC. 2001. Spectral mixture analysis of simulated thermal infrared spectrometry data: an initial temperature estimate bounded TESSMA search approach. IEEE Trans Geosci Remote Sens. 39:1435–1446.

Comber A, Fisher P, Brunsdon C, Khmag A. 2012. Spatial analysis of remote sensing image classification accuracy. Remote Sens Environ. 127:237–246.

Crist EP, Cicone RC. 1984. A physically-based transformation of Thematic Mapper data-the TM Tasseled Cap. IEEE Trans Geosci Remote Sens. 22:256–263.

David AP, Ram O, Stephen CH. 2008. Forest ecosystems. 2nd ed. Baltimore (MD): The Johns Hopkins University Press; p. 3–5.

Dyukarev EA, Pologova NN, Golovatskaya EA, Dyukarev AG. 2011. Forest cover disturbances in the South Taiga of West Siberia. Environ Res Lett. 6:035203. DOI:10.1088/1748-9326/6/3/035203.

Fernandez-Manso A, Quintano C, Roberts D. 2012. Evaluation of potential of multiple endmember spectral mixture analysis (MESMA) for surface coal mining affected area mapping in different world forest ecosystems. Remote Sens Environ. 127:181–193.

Gibbs HK, Brown S, Niles JO, Foley JA. 2007. Monitoring and estimating tropical forest carbon stocks: making REDD a reality. Environ Res Lett. 2:045023. DOI:10.1088/1748-9326/2/4/045023.

Healey SP, Cohen WB, Yang ZQ, Krankina ON. 2005. Comparison of Tasseled Cap-based Landsat data structures for use in forest disturbance detection. Remote Sens Environ. 97:301–310.

Huete AR, Liu HQ, Batchily K, van Leeuwen W. 1997. A comparison of vegetation indices global set of TM images for EOS-MODIS. Remote Sens Environ. 59:440–451.

Jiang XG, Wang D, Tang LL, Hu J, Xi XH. 2008. Analyzing the vegetation cover variation of China from AVHRR-NDVI data. Int J Remote Sens. 29:5301–5311.

Kavzoglu T, Reis S. 2008. Performance analysis of maximum likelihood and artificial neural network classifiers for training sets with mixed pixels. Gisci Remote Sens. 45:330–342.

Kennedy RE, Yang ZQ, Cohen WB, Pfaff E, Braaten J, Nelson P. 2012. Spatial and temporal patterns of forest disturbance and regrowth within the area of the Northwest Forest Plan. Remote Sens Environ. 122:117–133.

Koutsias N, Karteris M. 2000. Burned area mapping using logistic regression modeling of a single post-fire Landsat-5 Thematic Mapper image. Int J Remote Sens. 21:673–687.

Koutsias N, Mallinis G, Karteris M. 2009. A forward/backward principal component analysis of Landsat-7 ETM + data to enhance the spectral signal of burnt surfaces. ISPRS J Photogramm Remote Sens. 64:37–46.

Krishnaswamy J, Bawa KS, Ganeshaiah KN, Kiran MC. 2009. Quantifying and mapping biodiversity and ecosystem services: utility of a multi-season NDVI based Mahalanobis distance surrogate. Remote Sens Environ. 113:857–867.

Maingi JK. 2005. Mapping fire scars in a mixed-oak forest in Eastern Kentucky, USA, using Landsat ETM + data. Geocarto Int. 20:51–63.
Margono BA, Turubanova S, Zhuravleva I, Potapov P, Tyukavina A, Baccini A, Goetz S, Hansen MC. 2012. Mapping and monitoring deforestation and forest degradation in Sumatra (Indonesia) using Landsat time series data sets from 1990 to 2010. Environ Res Lett. 7:034010. DOI:10.1088/1748-9326/7/3/034010.
Martinuzzi S, Vierling LA, Gould WA, Falkowski MJ, Evans JS, Hudak AT, Vierling KT. 2009. Mapping snags and understory shrubs for a LiDAR-based assessment of wildlife habitat suitability. Remote Sens Environ. 113:2533–2546.
Masek JG, Huang CQ, Wolfe R, Cohen W, Hall F, Kutler J, Nelson P. 2008. North American forest disturbance mapped from a decadal Landsat record. Remote Sens Environ. 112:2914–2926.
Pereira JMC, Chuvieco E, Beaudoin A, Desbois N. 1997. Remote sensing of burned areas: a review. In: Chuvieco E, editor. A review of remote sensing methods for the study of large wildland fires. Alcala de Henares: Universidad de Alcala; p. 127–183.
Pflugmacher D, Cohen WB, Kennedy RE. 2012. Using Landsat-derived disturbance history (1972-2010) to predict current forest structure. Remote Sens Environ. 122:146–165.
Pleniou M, Koutsias N. 2013. Sensitivity of spectral reflectance values to different burn and vegetation ratios: a multi-scale approach applied in a fire affected area. ISPRS J Photogramm Remote Sens. 79:199–210.
Rouse JW, Haas RH, Schell JA, Deering DW. 1974. Monitoring vegetation systems in the great plains with ERTS. In: Freden SC, Mercanti EP, Becker MA, editors, Proceedings of Third Earth Resources Technology Satellite-1 Symposium-Volume I: Technical Presentations; 1974 Jan; Washington, D.C.: NASA Special Publication.
Rykhus R, Lu Z. 2011. Monitoring a boreal wildfire using multi-temporal Radarsat-1 intensity and coherence images. Geomatics, Nat Hazard Risk. 2:15–32.
Schroeder TA, Cohen WB, Song C, Canty MJ, Yang Z. 2006. Radiometric correction of multi-temporal Landsat data for characterization of early successional forest patterns in western Oregon. Remote Sens Environ. 103:16–26.
Schroeder TA, Wulder MA, Healey SP, Moisen GG. 2012. Detecting post-fire salvage logging from Landsat change maps and national fire survey data. Remote Sens Environ. 122:166–174.
Sharma RC, Kajiwara K, Honda Y. 2013. Estimation of forest canopy structural parameters using kernel-driven bi-directional reflectance model based multi-angular vegetation indices. ISPRS J Photogramm Remote Sens. 78:50–57.
Soja AJ, Tchebakova NM, French NHF, Flannigan MD, Shugart HH, Stocks BJ, Sukhinin AI, Parfenova EI, Chapin FS, Stackhouse PW. 2007. Climate-induced boreal forest change: predictions versus current observations. Glob Planet Change. 56:274–296.
Sönmez İ, Erdi E, Tekeli AE, Demir F, Arslan M. 2011. Foogle: fire monitoring tool for EUMETSAT's active fire product over Turkey using Google Earth. Geomatics, Nat Hazard Risk. 2:1–13.
Stroppiana D, Bordogna G, Carrara P, Boschetti M, Boschetti L, Brivio PA. 2012. A method for extracting burned areas from Landsat TM /ETM plus images by soft aggregation of multiple Spectral Indices and a region growing algorithm. ISPRS J Photogramm Remote Sens. 69:88–102.
Sun L, Hu HQ, Guo QX, Lv XS. 2011. Estimating carbon emissions from forest fires during 1980 to 1999 in Daxing'an Mountain, China. Afr J Biotechnol. 10:8046–8053.
Tian XR, McRae DJ, Jin JZ, Shu LF, Zhao FJ, Wang MY. 2011. Wildfires and the Canadian Forest Fire Weather Index system for the Daxing'anling region of China. Int J Wildland Fire. 20:963–973.
Townshend JR, Masek JG, Huang CQ, Vermote EF, Gao F, Channan S, Sexton JO, Feng M, Narasimhan R, Kim D, et al. 2012. Global characterization and monitoring of forest cover using Landsat data: opportunities and challenges. Int J Digit Earth. 5:373–397.

Tucker CJ. 1979. Red and photographic infrared linear combinations for monitoring vegetation. Remote Sens Environ. 8:127–150.

Tucker CJ, Grant DM, Dykstra JD. 2004. NASA's global orthorectified Landsat data set. Photogramm Eng Remote Sens. 70:313–322.

Veraverbeke S, Gitas I, Katagis T, Polychronaki A, Somers B, Goossens R. 2012. Assessing post-fire vegetation recovery using red-near infrared vegetation indices: accounting for background and vegetation variability. ISPRS J Photogramm Remote Sens. 68:28–39.

Wilson EH, Sader SA. 2002. Detection of forest harvest type using multiple dates of Landsat TM imagery. Remote Sens Environ. 80:385–396.

Wotton BM, Nock CA, Flannigan MD. 2010. Forest fire occurrence and climate change in Canada. Int J Wildland Fire. 19:253–271.

Zhao KY, Zhang WF, Zhou YW. 1994. The impact of Da xing'an ling forest fires on environment and its countermeasures. Beijing: Science Press; p. 17–24.

10

Determination of stability of epimetamorphic rock slope using Minimax Probability Machine

MANOJ KUMAR†, PIJUSH SAMUI*‡ and AJAY KUMAR NAITHANI§
†National Institute of Rock Mechanics, Kolar Gold Fields-563 117, Karnataka, India
‡Centre for Disaster Mitigation and Management, VIT University, Vellore-632014, India
§Department of Engineering Geology, National Institute of Rock Mechanics, P.O.- Champion Reefs, Kolar Gold Fields–563117, Karnataka, India

The article employs Minimax Probability Machine (MPM) for the prediction of the stability status of epimetamorphic rock slope. The MPM gives a worst-case bound on the probability of misclassification of future data points. Bulk density (d), height (H), inclination (β), cohesion (c) and internal friction angle (ϕ) have been used as input of the MPM. This study uses the MPM as a classification technique. Two models {Linear Minimax Probability Machine (LMPM) and Kernelized Minimax Probability Machine (KMPM)} have been developed. The generalization capability of the developed models has been checked by a case study. The experimental results demonstrate that MPM-based approaches are promising tools for the prediction of the stability status of epimetamorphic rock slope.

1. Introduction

The prediction of natural hazards is a challenging task due to uncertainty. Researchers use Artificial Neural Network (ANN) for prediction of different natural hazards (Lee & Liu 2000; Huang et al. 2003; Lui et al. 2006; Mishra et al. 2007; Pradhan & Lee 2007; Lee 2008; Adeli & Panakkat 2009). Although ANN gives reasonable performance, it has various limitations such as black box approach, arriving at local minima, low generalization capability, overfitting, etc. (Park & Rilett 1999; Kecman 2001). Researchers successfully adopted Adaptive Neuro Fuzzy Inference System (ANFIS) for modelling different natural hazards (Konstantanaras et al. 2004; Bacanli et al. 2009; Sezer et al. 2011). However, ANFIS does not give any probabilistic output. Support Vector Machine (SVM) has been successfully adopted to model different natural hazards (Goh & Goh 2007; Samui 2012). The SVM suffers the following limitations:

- SVM has high computational complexity due to quadratic programming (Vapnik 1998).
- It has three tuning parameters {capacity factor(C), error insensitive zone (ε) and kernel parameter (Samui 2008)}. The determination of the design value of these tuning parameters is a difficult task.

*Corresponding author. Email: pijushsamui@vit.ac.in

This article examines the capability of Minimax Probability Machine (MPM) for the determination of the stability of epimetamorphic rock slope. The failure of epimetamorphic rock slope is quite common due to the presence of thicker weathered layer and extremely broken rock mass (Chen et al. 2011). Researchers use different techniques for slope stability analysis (Fellenius 1936; Bishop 1955; Bishop & Morgenstern 1960; Morgenstern & Price 1965; Michalowski 1994, 1995, 2002; Sah et al. 1994; Griffiths & Lane 1999; Yang et al. 2004; Kumar & Samui 2006; Gao 2009). Fellenius (1936) used the method of slices for assessing the stability of slopes. It was seen that this method generally provides a conservative estimate of the factor of safety. Bishop (1955) used the method of slices in obtaining the stability of slopes. The results obtained from this method compare very closely with the more rigorous approaches such as the finite element method. Morgenstern and Price (1965) attempted to satisfy all the equations of statical equilibrium in obtaining the solution of the stability problem using the method of slices. They assumed a different distribution of inter-slice forces so as to obtain the solution. It should be mentioned that the previously available methods of slices, namely, Fellenius (1936) and Bishop (1955) do not satisfy all the conditions of statical equilibrium. Using the upper bound limit analysis, Michalowski (1995) presented a stability analysis of slopes based on a translational mechanism failure mechanism. Michalowski (2002) also used the upper bound theorem of limit analysis in order to obtain the stability numbers for homogeneous slopes in the presence of pore water

Table 1. Summary of the available methods for slope stability.

Reference	Brief methodology
Fellenius (1936)	Fellenius assumed that the resultant of inter-slice forces acts in a direction parallel to the base of each slice.
Taylor (1948)	Taylor (1948) used the friction circle method to obtain the stability numbers (N_s) for homogeneous soil slopes.
Bishop (1955)	Bishop (1955) used the method of slices in obtaining stability of slopes.
Janbu (1957)	This method also used the method of slices to solve the problem. Janbu solved the problem by assuming the point of application of the inter-slices forces.
Morgenstern and Price (1965)	Morgenstern and Price (1965) attempted to satisfy all the equations of statical equilibrium in obtaining the solution of the stability problem using the method of slices.
Chen (1975)	Chen (1975) used the upper bound theorem of the limit analysis to obtain the critical heights for homogenous soil slopes.
Michalowski (1994)	Michalowski (1994) also used the upper bound theorem of limit analysis in order to obtain the stability numbers for homogenous soil slopes.
Michalowski (1995)	Using the upper bound limit analysis, Michalowski (1995) presented a stability analysis of slopes based on a translational mechanism failure mechanism.
Michalowski (2002)	Michalowski (2002) also used the upper bound theorem of limit analysis in order to obtain the stability numbers for homogeneous slopes in the presence of pore water pressures as well as pseudo-static horizontal earthquake body forces.

Table 2. Statistical parameters of the data-set.

Input variable	Mean	Standard deviation	Skewness	Kurtosis
d (kN/m^3)	23.88	2.60	−0.19	1.33
H (m)	46.45	16.09	0.50	3.82
β (°)	32.67	7.60	0.17	3.90
C (kPa)	33.69	8.77	−1.35	4.70
ϕ (°)	29.94	5.85	−0.26	1.74

pressures as well as pseudo-static horizontal earthquake body forces. The available methods have their own limitations. Table 1 shows the summary of the available methods for the determination of slope stability. This study uses the database collected by Chen et al. (2011). The data-set contains information about bulk density (d), height (H), inclination (β), cohesion (c) and internal friction angle (ϕ) and the status of slope (the status of slope means whether the slope was stable or had failed). Table 2 shows the statistical parameters of the data-set. MPM is a discriminant classifier and it is derived from probabilistic framework (Lanckriet et al. 2002a, 2002b). It has been successfully used for solving different problems (Hoi & Lyu 2004; Zhou et al. 2011; Zhou et al. 2013). The developed MPM has been applied for a case study.

Details of MPM

This section will give a small description about the MPM model. More details are given by Lanckriet et al. (2002a, b). The MPM assumes positive definite covariance matrices for each of the two classes. Assume two random vectors x and y represent two classes of data points. The mean and covariance of x is $\bar{x}$ and $\sum_x x$, respectively. The mean and covariance of y is $\bar{y}$ and $\sum_y y$, respectively, where $x, y, \bar{x}, \bar{y} \in R^n$ and $\sum_x x, \sum_y y \in R^{n \times n}$.

In this study, $x = [d, H, c, \beta, \phi]$ and $y = [d, H, c, \beta, \phi]$.

The MPM uses the following optimal hyperplane that separates the data into two classes

$$a^T z = b \qquad (a, z \in R^n, a \neq 0, b \in R). \tag{1}$$

Lanckriet et al. (2002a) give the following mathematical formulation:

$$\max_{\alpha, b, a \neq 0} \alpha.$$

Subjected to

$$\begin{aligned} \inf P_r\{a^T x \geq b\} \geq \alpha \\ \inf P_r\{a^T y \leq b\} \geq \alpha. \end{aligned} \tag{2}$$

The above optimization problem is solved by the Lagrangian Multiplier. After applying the Lagrangian Multiplier, the optimization problem takes the following form:

$$\max_{\kappa,a} \kappa.$$

Subjected to

$$\begin{aligned} -b + a^T x &\geq \kappa\sqrt{a^T \textstyle\sum_x a} \\ b - a^T y &\geq \kappa\sqrt{a^T \textstyle\sum_y a}. \end{aligned} \tag{3}$$

The above optimization problem turns into (after eliminating κ) the following expression

$$\min_a \sqrt{a^T \sum_y a} + \lambda\sqrt{a^T \sum_x a}.$$

Subjected to

$$a^T(\bar{x} - \bar{y}) = 1. \tag{4}$$

The above MPM method has been adopted to predict the stability of epimetamorphic rock slope. To develop the MPM, 41 data-sets (see table 3) have been used as training data-sets. The remaining 12 data-sets (see table 4) have been used as testing data-sets. Testing data-set has been used to verify the developed MPM. The training and testing data-sets are the same as used by Chen et al. 2011. The data-sets are normalized between 0 and 1. Radial basis function has been used as kernel function. A value of -1 is assigned to the failed rock slope while a value of $+1$ is assigned to the stable rock slope so as to make this a two-class classification problem. The MPM program has been developed by using MATLAB.

Results and discussion

Linear MPM (LMPM) and Kernelized MPM (KMPM) have been tried to get the best performance. The performance of training data-set is expressed in percentage and is determined as the ratio of the number of data predicted accurately by MPM to the total number of data in the training set. The performance of testing data-set is expressed in percentage and is determined as the ratio of the number of data predicted accurately by MPM to the total number of data in the testing set. For training data-set, the developed LMPM correctly classified 39 data-sets. So, the training performance is 95.12%. The developed LMPM gives 91.67% testing performance. The performance of training and testing has been shown in table 3 and 4, respectively. Table 5 shows the value of α. The value of α is greater than the training as well as testing performance. So, the validity of α has been checked for the LMPM model. For KMPM, the design value of width (σ) of radial basis function has been determined by the trial and error approach. Figure 1 shows the effect of σ on the training performance (%). It is clear from figure 1 that the developed KMPM gives the best performance at $\sigma = 0.1$. The developed KMPM gives training performance = 100%

Table 3. Performance of training data-set.

d (kN/m^3)	H (m)	β (°)	C (kPa)	ϕ (°)	Actual class	Predicted class by LMPM	Predicted class by KMPM
20	10	10	8	20	−1	−1	−1
27.3	30	30	37.3	31	1	1	1
20.6	35	25	26.31	22	−1	−1	−1
21.6	50	40	6.5	19	−1	−1	−1
22.4	35	28	28.9	24	−1	−1	−1
23.2	33	30	31.2	23	−1	1	−1
26.8	26	30	37.5	32	1	1	1
27.4	42	25	38.1	31	1	1	1
21.8	50	50	32.7	27	−1	−1	−1
21.8	60	35	27.6	25	−1	−1	−1
26.5	21	30	35.4	32	1	1	1
26.5	39	35	36.1	31	1	1	1
27	69	30	35.8	32	1	1	1
27	22	25	38.4	33	1	1	1
21.4	52	50	28.8	20	−1	−1	−1
26	55	38	42.4	37	1	1	1
26	30	25	39.4	36	1	1	1
25.6	26	25	38.8	36	1	1	1
20	53	45	30.3	25	−1	−1	−1
25.8	50	30	34.7	33	1	1	1
21.8	99	35	28.8	26	−1	−1	−1
21.8	60	30	31.2	25	−1	−1	−1
24	51	30	41.5	36	1	1	1
24	50	35	40.8	35	1	1	1
20.6	70	35	27.8	27	−1	−1	−1
20.6	55	35	32.4	26	−1	−1	−1
25.8	40	27	38.2	33	1	1	1
25.8	45	25	39.4	33	1	1	1
21.1	31	40	33.5	28	−1	−1	−1
21.1	75	30	34.2	26	−1	−1	−1
26.6	52	25	42.4	37	1	1	1
26.6	42	35	44.1	38	1	1	1
26.6	60	35	40.7	35	1	1	1
25.8	40	30	41.2	35	1	−1	1
25.8	33	30	43.3	37	1	1	1
21.7	60	45	32	27	−1	−1	−1
20.6	65	40	28.5	27	−1	−1	−1
21.5	70	40	29.8	26	−1	−1	−1
26.5	36	34	42.9	38	1	1	1
20.8	45	30	15.6	20	−1	−1	−1
20.8	40	30	14.8	21	−1	−1	−1

and testing performance = 100%. Tables 2 and 3 show the performance of KMPM model. For KMPM, the value of α is given by table 4. The value of α is equal to the performance of KMPM. Therefore, the developed KMPM validates the value of α. The developed LMPM and KMPM have been applied on the Wangjiazhai slope in Kaili–Sansui highway. The details of the slope are as follows (Chen et al. 2011): $d = 19.8$ kN/m^3, $H = 98$ m, $\beta = 26°$, $c = 8.6$ kPa and $\phi = 17.8°$. The actual

Table 4. Performance of testing data-set.

d (kN/m^3)	H (m)	β (°)	C (kPa)	ϕ (°)	Actual class	Predicted class by LMPM	Predicted class by KMPM
19.6	58	40	29.6	23	−1	−1	−1
25.4	35	20	33	33	1	1	1
22.4	50	50	29.3	26	−1	−1	−1
26.2	30	35	41.5	36	1	1	1
26.2	36	23	42.3	36	1	−1	1
25.6	32	30	39.8	36	1	1	1
25.6	60	35	36.8	34	1	1	1
26.2	37	30	42.8	37	1	1	1
26.2	68	35	43.8	38	1	1	1
20.6	42	30	32.4	26	−1	−1	−1
26.5	54	42	41.8	36	1	1	1
20.8	53	30	15.4	21	−1	−1	−1

Table 5. Performance of the LMPM and KMPM.

Models	Training performance (%)	Testing performance (%)	α
LMPM	95.12	91.67	96%
KMPM	100	100	100%

condition of the Wangjiazhai slope is failure. The status of the Wangjiazhai slope has been checked by the developed LMPM and KMPM. The output of LMPM and KMPM is −1. So, the output from LMPM and KMPM was matched with the actual condition.

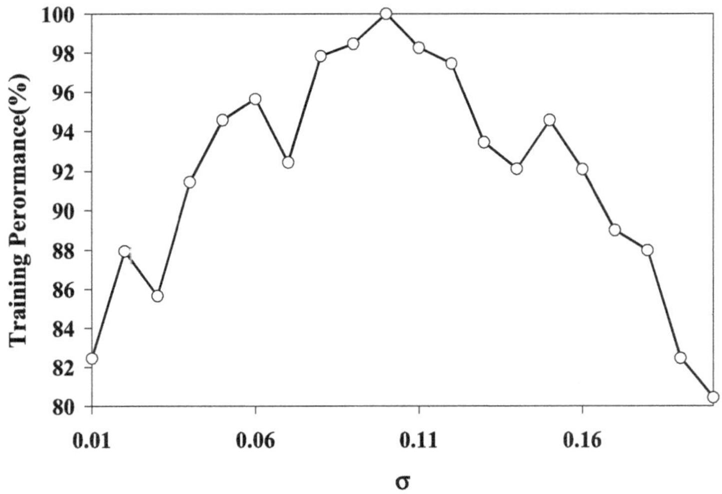

Figure 1. The effect of σ on training performance (%).

Conclusion

This article has proposed models based on the MPM for the prediction of the stability of epimetamorphic rock slope. Fifty-three data-sets have been utilized to construct the MPM models. Two models (LMPM and KMPM) have been developed. The prediction of MPM matches well the actual result. The main advantage of the developed MPM model is that it provides low bound on classification accuracy. The performance of the KMPM is better than the LMPM model. The developed models (LMPM and KMPM) give accurate prediction for a case study. So, it shows good generalization capability. The developed MPM can be used to model different problems in natural hazard.

References

Adeli H, Panakkat A. 2009. A probabilistic neural network for earthquake magnitude prediction. Neural Netw. 22:1018–24.

Bacanli UG, Firat M, Dikbas F. 2009. Adaptive neuro-fuzzy inference system for drought forecasting. Stochastic Environ Res Risk Assess. 23(8):1143–1154.

Bishop AW. 1955. The use of slip circle in the stability of slopes. Geotechnical. 5(1):7–17.

Bishop AW, Morgenstern NR. 1960. Stability coefficients for earth slopes. Geotechnical. 10(4):129–150.

Chen Ch F, Xiao Zh Y, Zhang GB. 2011. Stability assessment model for epimetamorphic rock slopes based on adaptive neuro-fuzzy inference system Electronic. J Geotech Eng Bund A. 93–107.

Chen WF. 1975. Limit analysis and soil plasticity. Amsterdam: Elsevier.

Fellenius W. 1936. Calculation of stability of earth dams. Transactions of the Second Congress on Large Dams. Washington, 4–445.

Gao W. 2009. Analysis of stability of rock slope based on ant colony clustering algorithm. Rock Soil Mechanics. 30(11):3476–3480.

Goh ATC, Goh SH. 2007. Support vector machines: their use in geotechnical engineering as illustrated using seismic liquefaction data. Comput Geotechnics. 34:410–21.

Griffiths DV, Lane PA. 1999. Slope stability analysis by finite elements. Geotechnical. 49(3):387–403.

Hoi CH, Lyu MR. 2004. Robust face recognition using minimax probability machine. Technical Report, Department of Computer Science and Engineering: The Chinese University of Hong Kong, Hong Kong.

Huang CS, Hung SL, Wen CM, Tu TT. 2003. A neural network approach for structural identification and diagnosis of a building from seismic response data. Earthquake Eng Struct Dyn. 32:187–206.

Janbu N. 1957. Earth pressures and bearing capacity calculation by generalized procedure by slices. Proceedings, 4th International Conference on Soil Mechanics, London. 2:207–212.

Kecman V. 2001. Learning and soft computing: support vector machines, neural networks, fuzzy logic models. London: The MIT press).

Konstantanaras A, Varley MR, Vallianatos F, Collins G, Holifield P. 2004. A neuro-fuzzy approach to the reliable recognition of electric earthquake precursors. Nat Hazards Earth Sci. 4:641–646.

Kumar J, Samui P. 2006. Determination for layered soil slopes using the upper bound limit analysis. Geotechnical Geol Eng. 24(6):1803–1819.

Lanckriet GRG, Ghaoui LE, Bhattacharyya C, Jordan MI. 2002a. Mini-max probability machine. . In: Dietterich TG, Becker S, Ghahramani Z., editors. Advances in neural information processing systems. 14; p. 801–807.

Lanckriet GRG, Ghaoui LE, Bhattacharyya C, Jordan MI. 2002b. A robust minimax approach to classification. J Machine Learn Res. 3:555–582.
Lee RST, Liu JNK. 2000. Tropical cyclone identification and tracking system using integrated neural oscillatory leastic graph matching and hybrid RBF network track mining techniques. IEEE Trans Neural Netw. 11:680–689.
Lee TL. 2008. Back-propagation neural network for the prediction of the short-term storm surge in Taichung harbor, Taiwan. Eng Appl Artif Intell. 21:63–72.
Lui Y, Guo HC, Zou R, Wang, LJ. 2006. Neural network modelling for regional hazard assessment of debris flow in Lake Qionghai Watershed, China. Environ Geol. 49:968–976.
Michalowski RL. 1994. Limit analysis of slopes subjected to pore pressure. In: Srirwardane, Zaman, editors. Proceedings of the Conference on Comp. Methods and Advances in Geomech; Rotterdam, the Netherlands: Balkema.
Michalowski RL. 1995. Slope stability analysis: a kinematical approach. Geotechnical. 45(2):283–293.
Michalowski RL. 2002. Stability charts for uniform slopes. J Geotech Geoenviron Eng ASCE. 128(4):351–355.
Mishra AK, Desai VR, Singh, VP. 2007. Drought forecasting using a hybrid stochastic and neural network model. J Hydrol Eng. 12(6):626–638.
Morgenstern NR, Price VE. 1965. The analysis of the stability of general slips surfaces. Geotechnical. 15(1):79–93.
Park D, Rilett LR. 1999. Forecasting freeway link ravel times with a multi-layer feed forward neural network. Comput Aided Civil Infrastructure Eng. 14:358–367.
Pradhan B, Lee S. 2007. Utilization of optical remote sensing data and GIS tools for regional landslide hazard analysis by using an artificial neural network model. Earth Sci Front. 14(6):143–152.
Sah NK, Sheorey PR, Upadhyaya, LN. 1994. Maximum likelihood estimation of slope stability. Intl J Rock Mech Min Sci Geomech. 31(1):47–53.
Samui P. 2008. Support vector machine applied to settlement of shallow foundations on cohesionless soils. Comput Geotechnics. 419–427.
Samui P. 2012. Support vector classifier analysis of slope. Geomatics Nat Hazards Risk. 1–12.
Sezer EA, Pradhan B, Gökceoğlu C. 2011. Manifestation of an adaptive neurofuzzy model on landslide susceptibility mapping: Klang valley, Malaysia. Expert Syst Appl. 38(7):8208–8219.
Taylor DW. 1948. Fundamental of soil mechanics. New York: John Wiley.
Vapnik VN. 1998. Statistical learning theory. New York: Wiley.
Yang CX, Tham LG, Feng XT, Wang YJ, Lee PKK. 2004. Two-stepped evolutionary algorithm and its application to stability analysis of slopes. J Comput Civ Eng. 18(2):145–153.
Zhou Y, Huan J, Hao H, Li D. 2013. Video text detection via robust minimax probability machine. Int J Digital Content Technol Appl. 7(1):679–686.
Zhou Y, Wang R, Xia K. 2011. Nonlinear prediction of fast fading channel based on minimax probability machine. Proceedings of the 2011 6th IEEE Conference on Industrial Electronics and Applications, ICIEA 2011. No. 5975626, p. 451–454.

11

Assessment of glacial lake development and prospects of outburst susceptibility: Chamlang South Glacier, eastern Nepal Himalaya

DAMODAR LAMSAL*†¶, TAKANOBU SAWAGAKI‡, TEIJI WATANABE‡ and ALTON C. BYERS§

†Division of Earth System Science, Graduate School of Environmental Science, Hokkaido University, Sapporo 060-0810, Hokkaido, Japan

‡Section of Integrated Environmental Science, Faculty of Environmental Earth Science, Hokkaido University, Sapporo 060-0810, Hokkaido, Japan

§The Mountain Institute, 108 LA, Elkins, WV 26241, USA

Chamlang South Tsho has been identified as one of the six high-priority glacial lakes in terms of glacial lake outburst flood (GLOF) danger in Nepal Himalaya, despite the fact that no detailed investigations of the lake had been hitherto undertaken. We conducted detailed mapping of the lake and its surroundings along with field surveys in October 2009 to determine the developmental history of Chamlang South Tsho and to assess its potential for GLOF. The lake expanded rapidly between 1964 (0.04 km^2) and 2000 (0.86 km^2) and has been stable ever since. Future lake expansion is improbable as its sides are confined by relatively stable landforms. The lake is 87-m deep with a water volume of approximately $34.9-35.6 \times 10^6$ m^3. Hanging glaciers on the steep surrounding mountain slopes and prominent seepage water in the terminal moraine dam could be potential triggers for a future outburst flood. Additionally, the debris-covered dead-ice dam, which is higher than the lake water level, is narrow and low; therefore, it could be overtopped easily by surge waves. Furthermore, the pronounced difference in elevation between the lake and the base of the terminal moraine dam makes the lake susceptible for a large flood.

1. Introduction

As a result of glacial retreat, numerous new glacial lakes have formed in past decades, and existing lakes have grown in many glacierized mountainous regions of the world, such as in the Himalayas (Bajracharya & Mool 2009; Gardelle et al. 2011), Central Asia (Bolch et al. 2011; Mergili et al. 2013; Wang et al. 2013), the Canadian Cordillera (Clague & Evans 2000), the Andes (Carey 2005), and the European Alps (Huggel et al. 2003), probably in response to recent climate warming. Supraglacial ponds, a precursor of larger glacial lakes, are likely to form, in particular, at the termini of debris-covered glaciers with gentle surface gradients (Reynolds 2000), stagnated or low-velocity ice (Quincey et al. 2007), and high rates of ablation (Sakai & Fujita

*Corresponding author. Email: lamsal@nagoya-u.jp

¶Present address: Department of Earth and Environmental Sciences, Graduate School of Environmental Studies, Nagoya University, Furo-cho, Chikusa-Ku, Nagoya 464-8601, Aichi, Japan.

2010). In the eastern Himalayas (Nepal, Bhutan, and Tibet), debris-covered glaciers are ubiquitous and recession of many of these has led to the formation of supraglacial ponds (lakes) and eventually to the development of larger moraine-dammed glacial lakes (Watanabe et al. 1994, 2009; Richardson & Reynolds 2000a; Bolch et al. 2008a; Komori 2008; Fujita et al. 2009). Most often, these glacial lakes are located between the ice-front and unconsolidated moraine dams (Benn et al. 2000; Komori 2008; Watanabe et al. 2009; Nuimura et al. 2012).

The moraine dams are composed of loose and friable debris. As a result, they are susceptible to dam breaching when some triggering event such as an ice avalanche or rock fall into the lake produces surge waves. From the perspective of disaster risk management, better understanding of glacial lakes is required as these lakes sometimes produce destructive glacial lake outburst floods (GLOFs) (Ives 1986; Clague & Evans 2000; Richardson & Reynolds 2000b; Bolch et al. 2008b; Ives et al. 2010). Floods from glacial lake outbursts can travel downstream for tens of kilometres from the drained lake causing widespread damage to lives and property (Vuichard & Zimmermann 1987; Dwivedi et al. 2000; Narama et al. 2010). More than a score of GLOF events have been recorded in the Himalayas, and several have had catastrophic impacts (Richardson & Reynolds 2000a; Ives et al. 2010). At least 13 GLOFs have occurred in Nepal Himalaya since the 1960s. The Dig Tsho GLOF in 1985 claimed several lives and destroyed the Namche small hydropower plant and trails, bridges, and farmland over a distance of more than 60 km downstream (Vuichard & Zimmermann 1987; ICIMOD 2011), and this outburst in particular has been regarded as a pivotal event for the initiation of study on GLOF hazards in Nepal Himalaya. Similar catastrophic events have also occurred in the different regions of the world, including Central Asia (Mergili & Schneider 2011; Narama et al. 2010), the Canadian Cordillera (Clague & Evans 2000), the Andes (Emmer & Cochachin 2013), and the European Alps (Huggel et al. 2003). During the past decade, GLOFs have emerged as an urgent contemporary threat because of the risk presented to mountain communities (Ives 1986; Yamada 1998; Richardson & Reynolds 2000a; Clague & Evans 2000; Mergili & Schneider 2011; Emmer & Cochachin 2013).

Recently, remote sensing (RS) techniques, such as the interpretation of satellite imagery, have been increasingly used for the study of glaciers and glacial lakes, largely because of their widespread availability and the high cost of traditional, field-based methods. A number of useful parameters of glacial lakes and their surroundings (e.g., surface area, slope gradient) can be extracted with the modern geographic information system (GIS) and RS technology (Huggel et al. 2003). However, other parameters must still be obtained from ground surveys and observations (e.g., water volume, lake depth, presence of ice in a terminal moraine, presence of seepage, etc.). Moreover, field investigations are required to confirm data gathered through RS techniques in order to minimize the uncertainty associated with RS-based analyses and interpretations, particularly those pertaining to a lake's GLOF potential (Iwata et al. 2002; Fujita et al. 2008; Byers et al. 2013).

Twenty-one glacial lakes among 1466 lakes in Nepal Himalaya have been identified as being potentially dangerous (ICIMOD 2011), primarily by using remotely sensed data. Chamlang South Tsho (the word 'tsho' means glacial lake or lake) and six other lakes located in the remote Hongu valley in the Dudh Koshi basin were included in the list (Byers et al. 2013; Sawagaki et al. 2012). Chamlang South Tsho (figure 1) has been ranked as one of the six high-priority glacial lakes in terms of GLOF danger in Nepal Himalaya (ICIMOD 2011). This paper builds on the findings of Sawagaki et al. (2012) and Byers et al. (2013) in order to expand our

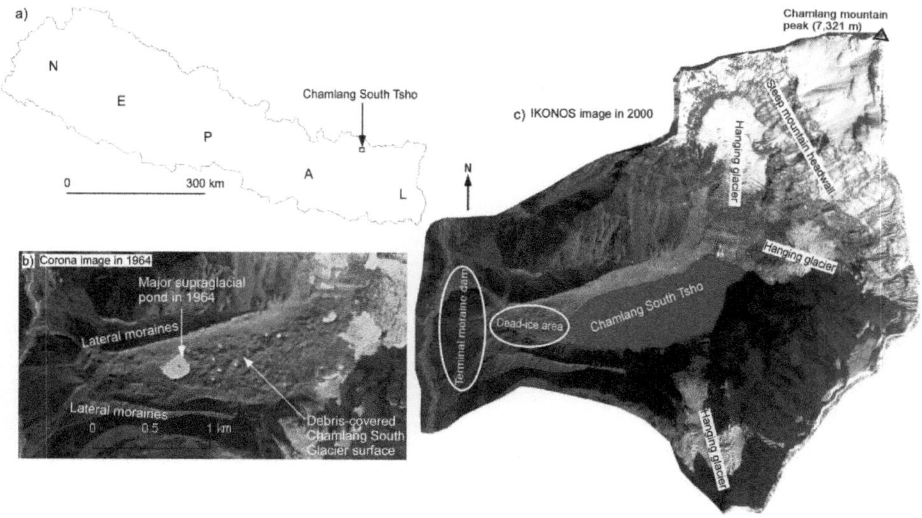

Figure 1. (*a*) Location of Chamlang South Glacier Lake in the map of Nepal, (*b*) closer views of the lake and its surroundings in a Corona image taken in 1964, and (*c*) IKONOS image taken in 2000. Note the substantial morphological change that occurred on the debris-covered glacier from 1964 to 2000.

understanding of the potential degree of danger posed by Chamlang South Tsho through the integration of field survey and RS data and methods, including photogrammetric analyses of stereo images.

2. Data and methods

Ground-based surveys, observational data, field photographs, and RS data were used in this study. Specifications of the RS data are summarized in table 1.

Table 1. Remote sensing and supplementary data utilized in the study.

Source/sensor	Date of acquisition	Spatial resolution (m)	Applications
Corona KH-4A (stereo-images)	1964-11-24	2.7	3D mapping and lake delineation
Corona Hexagon	1976-01-03	9.0	Lake delineation
NGS map	1988	1:50,000	GCP collection
Landsat TM5	1992-09-22	30.0	Lake delineation
Landsat TM5	2000-10-30	30.0	Lake delineation
IKONOS	2000-11-26	1.0	Detailed view
ALOS PRISM (stereo-images)	2006-12-04	2.5	3D topography, lake delineation, and GCP collection
ALOS AVNIR-2	2008-10-24	10.0	Reference image
ALOS AVNIR-2	2010-12-15	10.0	Lake delineation
ASTER DEM		29.3	Elevation data for orthorectification

2.1. Image processing to delineate glacial lakes

The study area is located in a high mountain region with extremely pronounced relief, such that uncorrected RS imagery will contain distortions. Elevation data are an essential tool for removing geometric distortions associated with relief. An ASTER digital elevation models (DEMs) was used as a source of elevation data to orthorectify all data-sets. Stereo Corona KH-4A (1964) and ALOS (Advanced Land Observing Satellite) PRISM (2006) data were orthorectified with the Leica Photogrammetric Suite (LPS). The remaining images were orthorectified with the image geometric correction module of ERDAS IMAGINE. To obtain better alignment among the data-sets, we performed image-to-image co-registration. The operation was carried out with the AutoSync, an add-on of ERDAS IMAGINE. The AutoSync generates a large number of tie points on the overlap portion of an input image and a reference image, thereby producing a mathematical model to tie the images together. An ALOS AVNIR-2 (2008) image, which was mostly free from cloud and seasonal snow, was set as the reference image with which remaining images were co-registered. Then, manual delineation of the glacial lakes was conducted.

2.2. Stereo-data processing to produce three-dimensional (3D) topographic maps and DEMs

Because of coarser resolutions, global DEMs such as SRTM (90 m) and ASTER (30 m) are problematic for detecting small landforms and local altitudinal variation typical on glacier surfaces and their moraines (Fujita et al. 2009). Optical photogrammetry can be a useful technique for producing finer-resolution topographic maps and DEMs with stereo-imagery data (Altmaier & Kany 2002), such as Corona KH and ALOS PRISM data (Lamsal et al. 2011; Sawagaki et al. 2012). We used declassified stereo Corona KH-4A (acquired in 1964) and ALOS PRISM with a rational polynomial coefficient (RPC) file (acquired in 2006). Corona KH-4A and ALOS PRISM imageries have spatial resolutions of 2.7 and 2.5 m, respectively. ALOS PRISM data with an RPC file contains all the necessary interior and exterior orientation information to process the stereo data to produce stereo models and DEMs. Furthermore, input of a few ground control points (GCPs) together with the RPC file enhances the accuracy of the stereo imagery processed by photogrammetric techniques when compared with image processing in which only RPC data are used. A hybrid-type map (containing both images of surface features and topographic information) produced by the National Geographic Society (NGS), Washington D. C., in 1988 was used to collect four GCPs to process ALOS PRISM stereo data with an RPC file. ALOS PRISM images were triangulated (root-mean-square error [RMSE] ~0.40 pixel) to create an ALOS stereo model and were subsequently used to produce topographic maps and DEMs. Unfortunately, Corona KH-4A data do not contain an RPC file and orientation information. As a result, Corona data require a larger number of GCPs for their processing. For Corona KH-4A stereo images, 20 GCPs were collected from the ALOS PRISM stereo model (described above), and an aerial triangulation was subsequently performed (RMSE ~1.25 pixel) to create topographic maps and DEM with the Corona stereo data, similar to those of the ALOS PRISM data mentioned previously. These data were processed with the aid of a Leica Photogrammetric Suite (LPS) workstation. Topographic maps (2-m contour interval) and DEMs (5 m $\times$ 5 m) were produced after rigorous editing of digital terrain models (DTMs) in a Leica Terrain Format to minimize errors pertaining to irregular micro-landforms, high relief, low contrast, and shadows on the images based on procedures recommended by Lamsal et al. (2011) and Sawagaki et al. (2012). Fine-resolution topographic maps and DEMs are crucial in the investigation of GLOF hazards and debris-covered glaciers. The image processing tasks of ALOS and Corona stereo data, including generation of DTMs and their editing were well explained by Lamsal et al. (2011) and Sawagaki et al. (2012). The relative accuracy of the topographic maps and DEMs from the Corona KH-4A and the ALOS PRISM stereo images in this study is believed to be $\sim\pm10$ m, similar to that obtained by Sawagaki et al. (2012).

2.3. Lake bathymetry

The depth of the lake was measured using a sonar-sounding machine (Eagle Fish Elite 500c) (figure 2(*a*)) with an accuracy of ±1 m. To calibrate accuracy, a depth measurement test was performed beforehand in a shallow supraglacial pond (figure 2(*b*)) on the same glacier (about 500 m down-glacier from the main lake) with both the machine and fishing line. It was found that the depth measurements from the two methods were nearly identical. The sonar and global positioning system (GPS) were mounted on the raft, and three longitudinal and nine traverse transects were conducted (figure 3(*a*)). Depth measurement points were simultaneously recorded with the GPS.

3. Development of Chamlang South Tsho

Figure 3(*b*) shows the bathymetric map of the lake with 10-m contour intervals. In October 2009, the lake was found to be 87 ± 1 m deep (i.e., deepest point), approximately 550-m wide (north–south), and approximately 1650-m long (east–west). The elevations of the lake water surface and bottom (deepest site) were 4940 and 4853 m, respectively, and the mean depth of the lake was calculated to be 40.2 ± 1 m (derived from a produced bathymetric DEM). The volume of water contained in the lake was determined to be between $\sim$34.9 and 35.6×10^6 m^3. The first value was calculated directly from the bathymetric DEM using a surface volume algorithm (volume below the 4940-m surface) of the 3D analyst extension of ArcGIS. The second value was obtained by summing the areas of the polygons between successive contours multiplied by their corresponding depths.

Lake surface areas and their expansion are weak indicators of GLOF hazard (Watanabe et al. 2009). In addition, larger glacial lakes do not necessarily produce larger flood volume during a GLOF event, as evidenced by the volumes experienced during actual GLOFs (cf. table 1 and Fujita et al. 2013). However, an understanding of a lake's potential for flooding requires an understanding of the characteristics of the glacier (e.g., down-glacier or up-glacier growth) and future growth prospects of the glacial lake because these affect water volume and alter dam stability. For example, down-glacier lake expansion obviously weakens the moraine dam by creating a narrower structure (Watanabe et al. 2009). Moreover, the downward shift of the lake shore can modify slope angles of moraine dam topography in relation to the lake surface, leading to larger potential flood volumes (PFVs) as witnessed at Nagma Pokhari, Nepal (Fujita et al. 2013). Consequently, it might result in an increased level of GLOF hazard and the risk associated with it.

Figure 2. (*a*) Boating on Chamlang South Glacier Lake to measure lake bathymetry and (*b*) test site on a supraglacial pond about 500-m downstream of the lake.

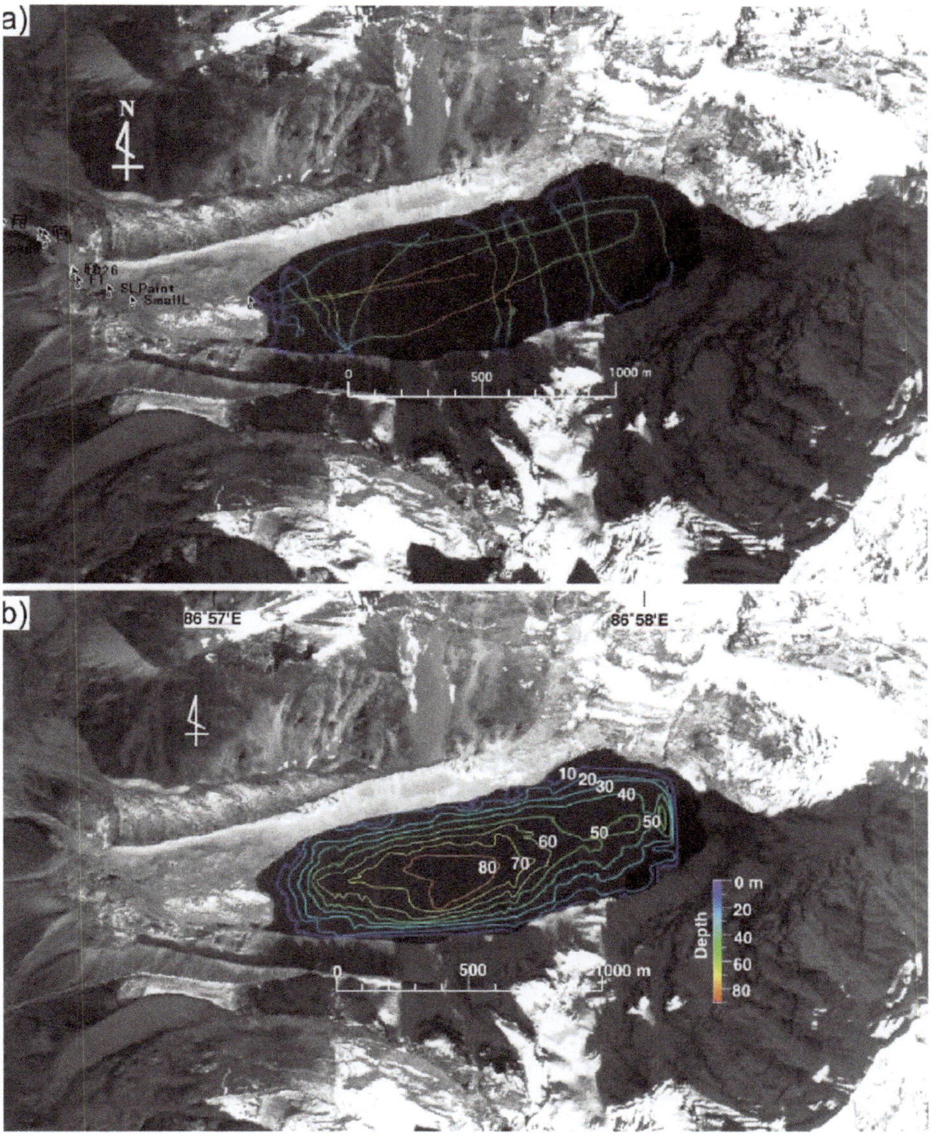

Figure 3. (*a*) Three longitudinal and nine transects routes and (*b*) bathymetric map of the lake with 10-m contour intervals. The base map is orthorectified ALOS PRISM imagery acquired in 2006.

Topographic maps produced in 1964 (figure 4(*a*)) and 2006 (figure 4(*b*)) and the lake bathymetry data obtained in 2009 (figure 3(*b*)) enabled us to demonstrate that substantial changes in the morphology and the elevation of the debris-covered glacier surface have occurred since 1964 and that evolution of supraglacial ponds into a formidable-size glacial lake has resulted. There were several supraglacial ponds on the glacier surface in 1964, the largest of which was located at an elevation of 4952 m (figure 4(*a*)). The glacial surface upward from the pond had a relatively steep gradient, while the gradient down-glacier from the supraglacial pond was gentle (figure 4(*a*)). Glacial surface lowering (or glacial melting) around the supra-glacial

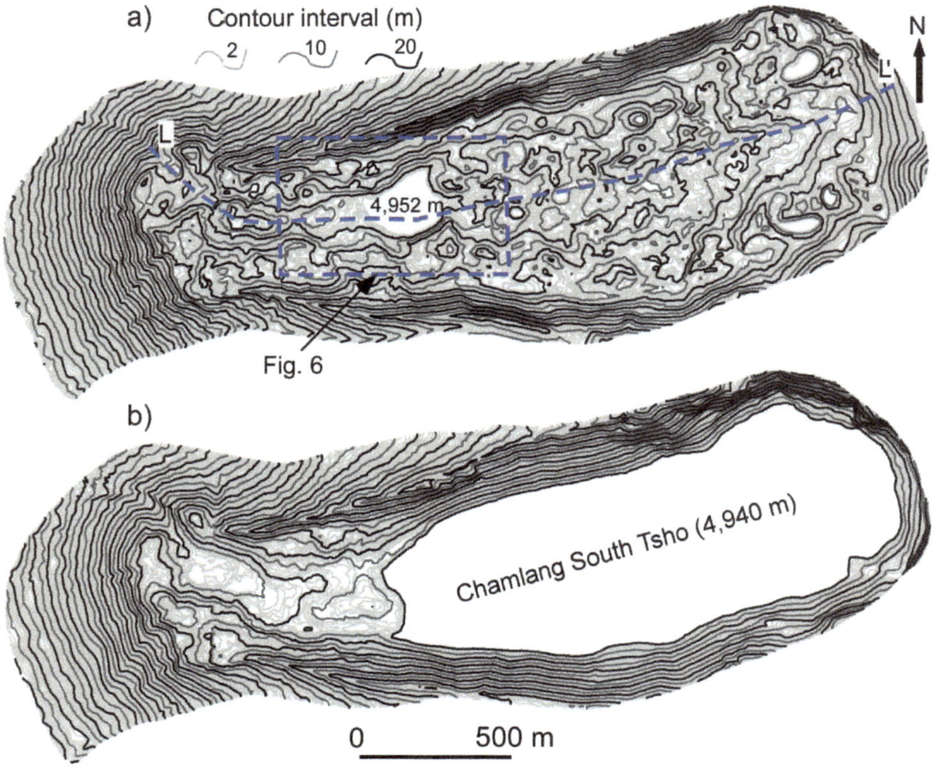

Figure 4. Detailed topographic maps around Chamlang South Glacier/Lake in (a) 1964 and (b) 2006. Contour intervals: 1 m (faint lines), 10 m (lighter bold lines), and 20 m (bold lines). The dotted line and rectangle in (*a*) represent the sites of the profile shown in figure 5 and the area of figure 6, respectively.

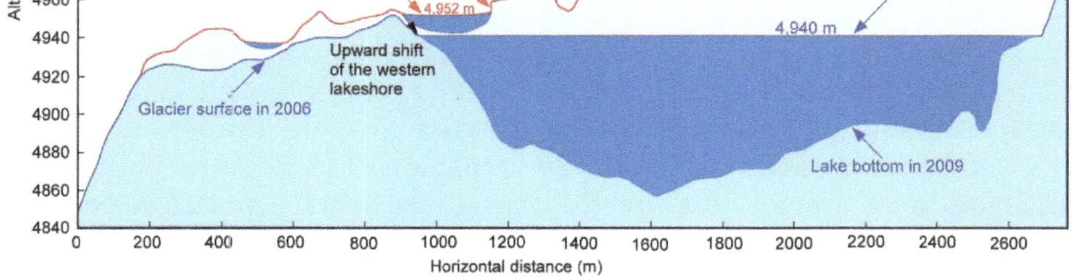

Figure 5. Longitudinal profile along L–L′ showing glacier surface topographies in 1964 and 2006 and the lake bathymetry in 2009. For location, see figure 4 (*a*).

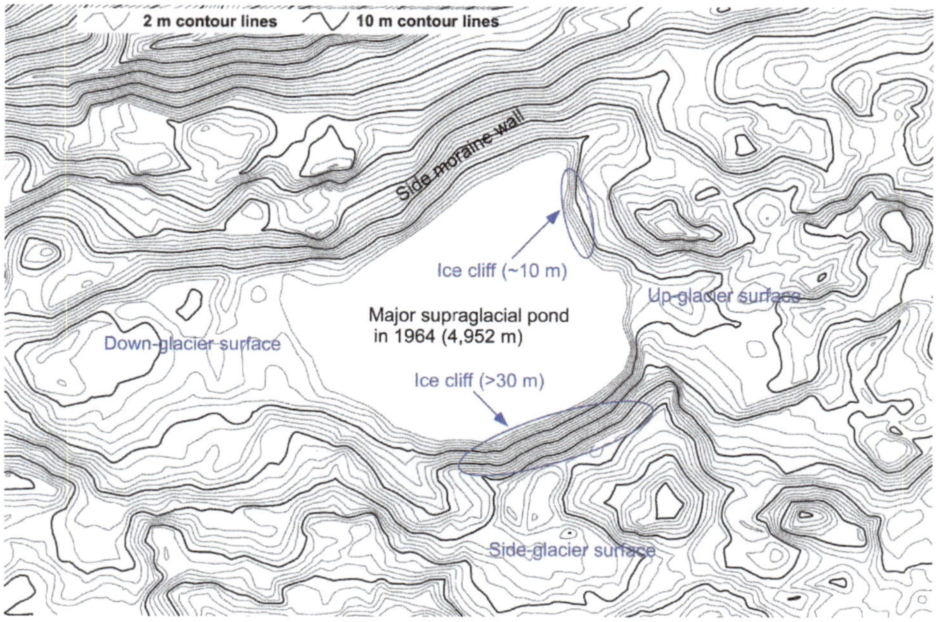

Figure 6. Detailed view of the major supraglacial pond and its surrounding topography in 1964. Note the prominent ice cliffs on the left side and up-shore of the pond and the gentle and ice-cliff free down-glacier area. The location is indicated by the dotted rectangular box in figure 4(*a*).

pond was smaller, while it was progressively higher in the up-glacier area (figure 5). The longitudinal profile (L–L′) of the glacier (figure 5) clearly shows a gentle glacier slope and surface at the lower part of the glacier. Such a condition favours the formation of supraglacial ponds on debris-covered glaciers (Reynolds 2000). The presence of a steeper up-glacier surface (figure 4(*a*)) and ice cliffs on the southern and eastern sides of the supraglacial pond in 1964 (figures 5 and 6) suggests ice-calving into the pond that resulted in faster up-glacial-lake expansion during the years after 1964. Ice-calving into the lake could have been the key mechanism for the faster lake growth in the up-glacial area after 1964; otherwise, downward lake expansion of Chamlang South Tsho would also have been large.

The surface areas of Chamlang South Tsho in six different years between 1964 and 2010 were 0.04, 0.13, 0.64, 0.86, 0.87, and 0.87 km^2 in 1964, 1976, 1992, 2000, 2006, and 2010, respectively (figure 7). The lake expanded at a relatively slow rate from 1964 to 1976, i.e., by 0.007 km^2/year (0.085 km^2 in 12 years). However, the lake grew rapidly from 1976 to 2000 at 0.031 km^2/year (0.74 km^2 in 24 years), even faster than the growth of Imja Tsho (by 0.51 km^2 during the same period in 24-year period), a rapidly expanding glacial lake in Nepal Himalaya (Yamada 1998; Watanabe et al. 2009). Lake growth virtually stopped sometime between 1992 and 2000 (figure 7). The reason for the steady state of Chamlang South Tsho in the up-glacial area is that the feeder glaciers had already retreated to the higher mountain slopes. Similarly, the lake has been confined on its two sides by lateral moraines. As a result, Chamlang South Tsho has no possibility of further upward or lateral expansion. Downward lake expansion was examined by using overlays of the lake outlines from 1964 to 2010, and no downward expansion was found (figure 7). Instead, tens of metres of upward shift of the western lakeshore were noticed (figures 5 and 7), and this shift

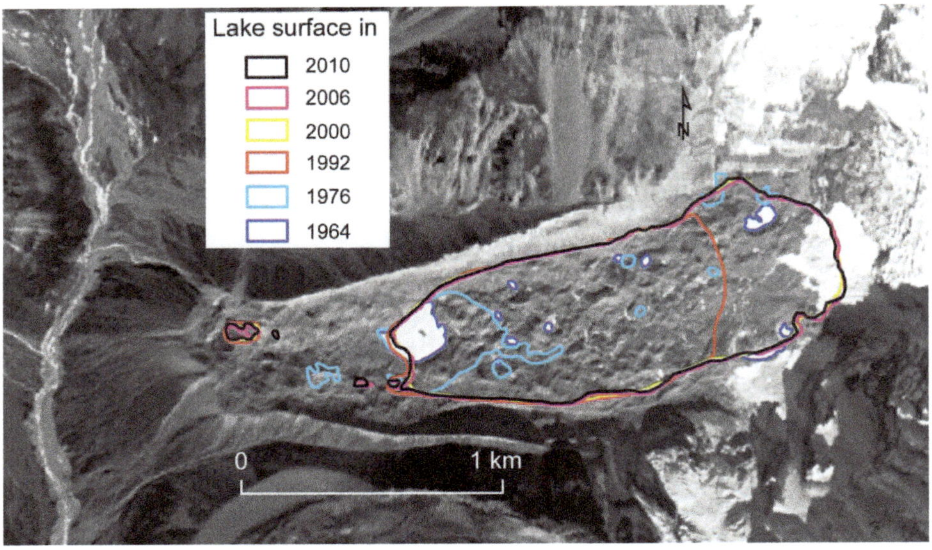

Figure 7. Growth of surface area of the lake from 1964 to 2010. The lake expanded rapidly to 2000 but subsequently became stagnant. The background image is a Corona KH-4A taken in 1964.

may have occurred as a result of lake water surface lowering along the inclined lakebed sloping eastwards (see figure 5).

A small lake area discrepancy during the years from 2000 to 2010 is believed to have occurred by delineation errors resulting from the diverse spatial resolution of the RS data (see table 1). For instance, we did not find significant differences in surface areas of the lake over the years; however, the lake boundaries did not exactly overlap at a few locations (figure 7). Obviously, lake boundaries derived from finer resolution satellite data (e.g., ALOS PRISM, 2.5 m) can illustrate small landform features, such as sharp notches, that coarser images cannot.

A spillway channel does not exist on the dead-ice surface to drain water from Chamlang South Tsho. There is a slightly raised, narrow dead-ice area (maximum height of 4–18 m and approximately 200-m wide) covered by thick debris, which effectively dams the lake water. The area between the western lakeshore and the thick blue line (same elevation as lake water level 4940 m) passing through the dead-ice surface in figure 8 represents the dead-ice dam. The dead-ice surface, including a large portion of the lake dam, slopes downward (figure 8). As a result, meltwater in most of the dead-ice area flows out readily and does not contribute to lake expansion. Even if dead-ice melting does not occur and the lake water level rises, the water level cannot rise more than 4 m from the present level. This suggests that the surface area and the water volume of the lake will not grow significantly in the future.

4. Susceptibility of GLOF hazard from Chamlang South Tsho

Among the recorded GLOF events in the Himalayas, the North American Cordillera, and the Andes, slope movements such as ice avalanches, rockfalls, and landslides have been found to be the most frequent GLOF triggers (Clague & Evans 2000; Richardson & Reynolds 2000a; Emmer & Cochachin 2013). In the Himalayas, in particular, eight out of 14 known GLOF triggers were reported to have been ice

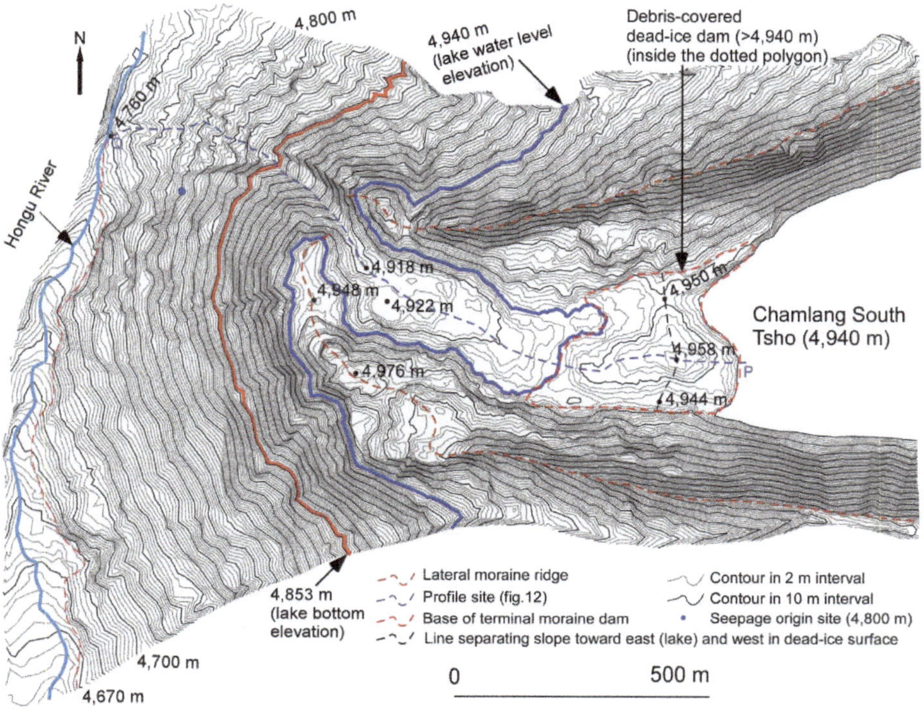

Figure 8. Detailed topographic map of the dead-ice area, lateral moraines, and terminal moraine complex. An orange bold contour line of 4940 m passing through the dead-ice area separates the areas higher (i.e., debris-covered dead-ice dam) and lower than the lake water surface. A red line with elevation of 4853 m passing through the terminal moraines represents the same elevation as the bottom of the lake. Note the altitudinal distributions and their differences among the various land features, such as the lake, lateral moraines, and the base of the terminal moraine dam along the Hongu River. The dotted line PQ represents the site of the profile shown in figure 12. The relative height difference between the lake water surface and the dead-ice dam is only about 4–18 m.

avalanches (Richardson & Reynolds 2000a), including those of the two most recent outbursts (Dig Tsho and Tam Pokhari in 1985 and 1998, respectively) that occurred in the Everest area of Nepal. Therefore, ice avalanches from hanging glaciers constitute one of the most important outburst triggering factors in the Himalayas. In consideration of this fact, hanging glaciers above glacial lakes need to be considered as a potential source of outburst triggers when evaluating GLOF hazards.

There are some prominent hanging glaciers above Chamlang South Tsho (figures 2(*a*) and 9(*a*)), which were readily observed during the 2009 field expedition and with fine-resolution IKONOS images. These hanging glaciers could dislodge and cascade into the lake producing surge waves. Three hanging glaciers from the eastern, north-eastern, and south-eastern mountain slopes (figures 9(*a*) and 10) are potential sites for ice avalanches, whereas the north-western slopes are potential sites for rockfalls or landslides into the lake. Well-developed crevasses and bergschrunds exist on these hanging glaciers (figure 9(*a*)), which further increase the possibility of ice detachment. The termini of the glaciers are >150 m above the lake, and there is no space in between to alleviate the impact of possible ice avalanches. Hanging glaciers reportedly triggered the 1998 Tam Pokhari GLOF in the adjoining Hinku valley and the

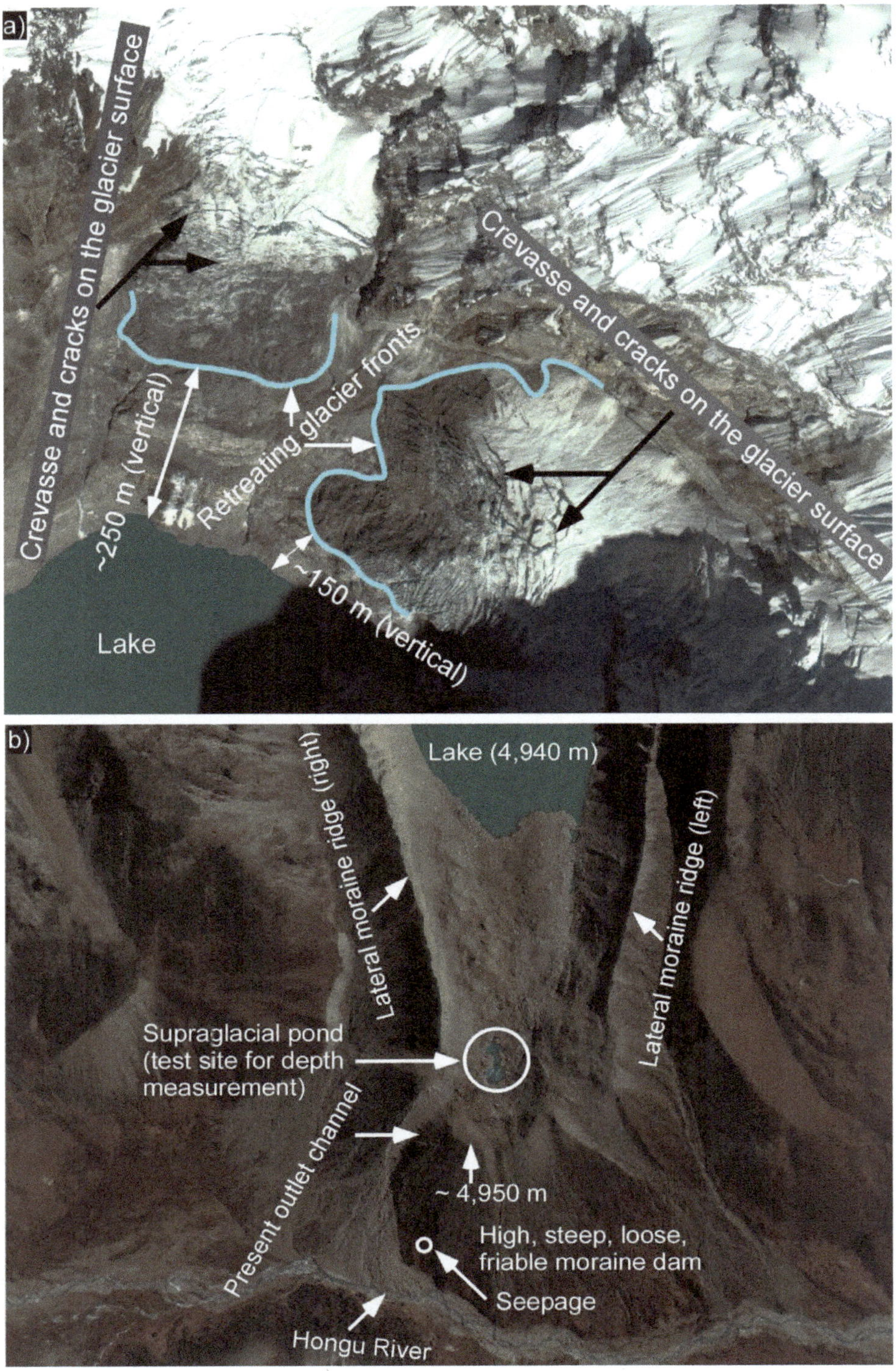

Figure 9. (*a*) Hanging glaciers with crevasses and cracks and (*b*) high, steep, and unconsolidated terminal moraine dam.

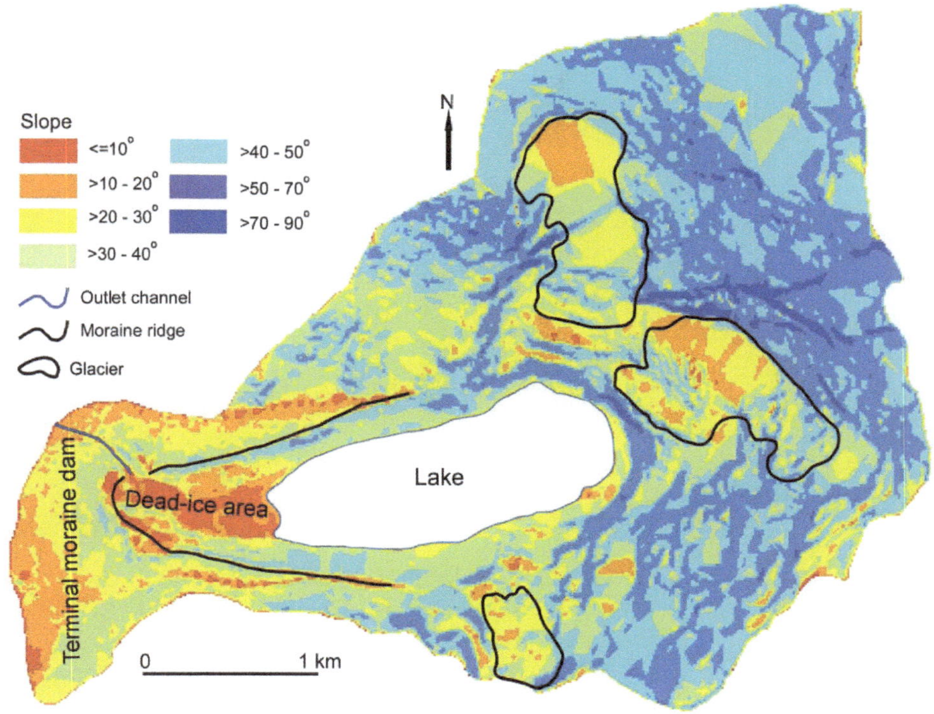

Figure 10. Slope angle of terrain surrounding of the lake. The steepness of hanging glaciers ranges from 20° to more than 50°, while the steepness of the terminal moraine dam is up to 50°.

1985 Dig Tsho GLOF in the Bhote Koshi valley, both located in the Everest region of Nepal (Vuichard and Zimmermann 1987; Dwivedi et al. 2000; ICIMOD 2011). Generally, the glaciers in this region are retreating (Bajracharya & Mool 2009). Chamlang South Tsho has also been receding rapidly upslope, and its surface gradients are becoming increasingly steep (figure 10). Consequently, an increase in the relative height between the lake and the hanging glaciers, concurrent with the change in the glacier's surface steepness, could increase both the possibility for and potential impact of an ice avalanche into the lake.

In addition, large surge waves, if generated, could easily overtop the debris-covered dead-ice dam in contact with the west lakeshore because it is only ~4–18 m higher than the lake water surface at 4940 m. The dimensions of the dead-ice dam are ~250-m wide and ~180–325-m long compared to the ~1650-m long and 500-m wide lake (figures 8). The dead-ice surface farther downward becomes increasingly lower than the lake water level. However, the existence of an approximately 750-m long by 250-m wide dead-ice area, surface roughness, a wide size range of debris from pebbles to large boulders, and uneven surface morphology (figure 2(*b*)), west of the lake might alleviate potential surge waves of relatively small magnitude. A wide dam area is likely to act as a buffer against surge waves from various triggering mechanisms (Byers et al. 2013).

In particular, the terminal moraine topography of Chamlang South Tsho's outer slope is steep (figure 9) and composed of unconsolidated moraine materials (figures 8 and 11(*a*)). Such a moraine dam is susceptible to dam failure when triggering events

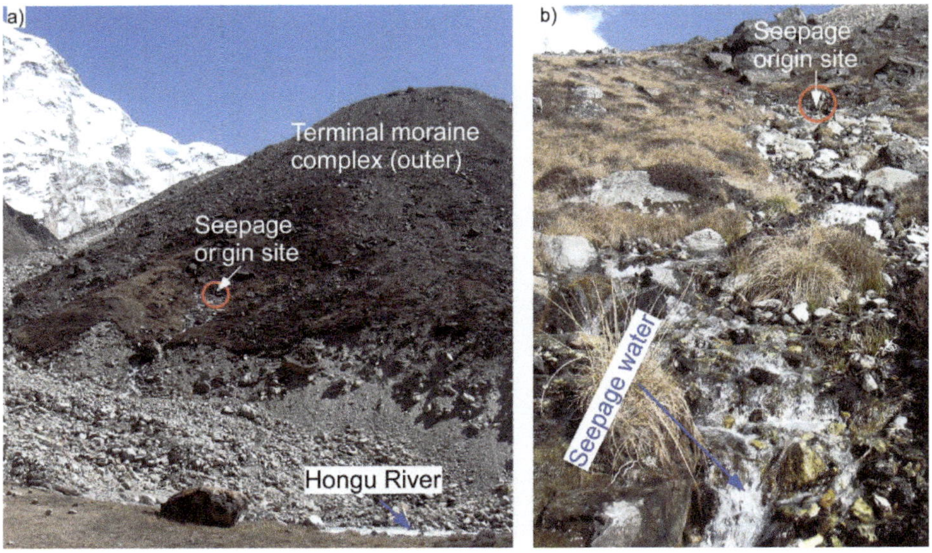

Figure 11. (*a*) Seepage site and outer slope of the terminal moraine dam and (*b*) close view of the seepage. Note the loose and unconsolidated moraine dam (*a*) and the amount of seepage water (*b*).

occur (Clague & Evans 2000; Richardson & Reynolds 2000b). In addition, a prominent seepage outlet (27° 45′ 20.65″ N, 86° 56′ 20.78″ E) exists in the lower belt of the terminal moraine dam well above the Hongu River, which flows at the base and far from the drainage channel of the lake (figures 8, 9(*b*), and 11(*a*) and (*b*)). It is unclear whether the conduit of this seepage is connected to the main lake or whether it begins from the dead-ice complex. In either case, the existence of such a seepage is regarded as a potential GLOF trigger (Clague & Evans 2000; Iwata et al. 2002) because the seepages weaken the ice-cored moraine dams, which may eventually collapse catastrophically (Richardson & Reynolds 2000b).

Figures 8 and 12 clearly show the surface topography from the western lakeshore to the base of the terminal moraines. There exists in this area a dead-ice surface substantially lower than the lateral and terminal moraine, as well as a well-developed drainage course through the terminal moraines. A drainage course through the dead-ice surface (here we call it a 'spillway channel') does not exist until a small supraglacial pond at 4918 m, as evidenced by our field observations and high-resolution RS data and maps (see figures 8 and 12). A drainage course outside the dead-ice area (here we call it an 'outlet channel') begins from a small supraglacial pond at 4922 m; however, the meltwater flows hidden beneath the debris surface for >100 m in the upper section (figure 12) and then re-emerges downstream. The outlet channel is noticeably incised compared to its side debris in the terminal moraine dam. Therefore, the outlet channel below the small supraglacial pond at 4918 m represents the lowest surface (indicated by the blue dotted line in figure 8) along which prospective surge waves would travel before reaching the Hongu River at the base if an outburst of Chamlang South Tsho were to occur.

The base level of the terminal moraine at the confluence of the Hongu River and the mouth of the present outlet channel (~4760 m) is located at a much lower elevation than the lake water level (4940 m) (figures 8 and 12). The base of the terminal

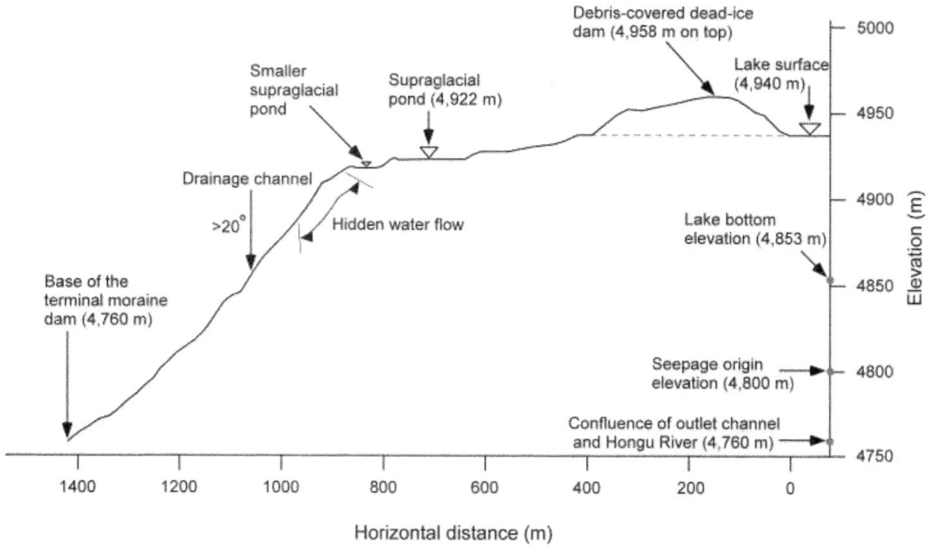

Figure 12. Graphical representation of the altitudinal distributions among various topographical and geomorphological units of the lake and its surroundings. See figure 8 for location of the profile.

moraine dam along the Hongu River downstream lies at an increasingly lower elevation (see figure 8). The lake has a maximum depth of 87 m, which means that the lake bottom has an elevation of 4853 m. It should be noted that the relative height between the lake water level and base level of the terminal moraine dam at Chamlang South Tsho (minimum ~180 m) is much larger than that at Tam Pokhari (~100 m) before its outburst in 1998 and that at widely known Imja Tsho (~20 m), which is regarded as one of the potentially dangerous glacial lakes in Nepal Himalaya (ICIMOD 2011). The outer slope of the terminal moraine attains a steepness of 50°, with the average being >25°, (figure 10), which is still steeper than that of Tam Pokhari, with an average of ~25°. The average slope of the drainage channel in the terminal moraine complex of Chamlang South Tsho is also >20° (figures 10 and 12).

Recently, Fujita et al. (2013) proposed an approach of steep lookdown area (SLA) determined by a 10° depression angle against a lake water level to its outer flank of terminal moraine to estimate the PFV of glacial lakes. Interestingly, several glacial lakes that have been considered potentially dangerous by previous studies (e.g., Mool et al. 2001; ICIMOD 2011) are found to contain insignificant PFVs (Fujita et al. 2013). Based on this 10° threshold, Fujita et al. (2013) estimated the PFV of West Chamlang to be 4.8×10^6 m^3. Unfortunately, Chamlang South Tsho is not contiguous to the terminal moraine because of the dead-ice body, which restricts the application of this method to Chamlang South Tsho. Moreover, we did not find distinguishable SLA on the terminal moraine topography of Chamlang South Tsho. At Tam Pokhari, where the topography of the terminal moraine dam before the outburst was comparable with that at Chamlang South Tsho, Osti and Egashira (2009) reported that the lake lowered by 50 m against an ~100-m high dam. At Chamlang South Tsho, the high altitudinal differences (minimum of ~180 m along the outlet channel) between the base of the loose and unconsolidated moraine dam and the glacial lake might suggest a larger flood volume from the lake in the event of dam breach because the moraine base is lower than even the lake bottom (figures 8 and 12).

It should be noted that smaller outburst floods do not necessarily correspond to lower risk, nor larger outburst floods necessarily pose higher risk. For instance, if downstream areas contain densely populated settlements and costly infrastructures, high fatalities and damage could be inflicted by even small outburst floods. Pertinent examples include the outburst floods of Dig Tsho in 1985 ($\sim 5.1 \times 10^6$ m^3) and Tam Pokhari in 1998 ($\sim 18.0 \times 10^6$ m^3), of which the former caused more devastation than latter. One reason why Dig Tsho swept away the nearly completed Namche hydroplant (costing approximately $3 million) is that it was located only about 12 km downstream from the lake. Vulnerability and risk assessments associated with Chamlang South Tsho that require detailed field surveying for tens of kilometres downstream have not yet been carried out and were also largely beyond the scope of this paper. As a result, detailed information on vulnerable people and properties in the downstream areas of Chamlang South Tsho and thus the actual risk associated with this lake, in particular, are not known well. Risk assessments associated with Imja Tsho were conducted by Khanal et al. (2009) along the neighbouring Dudh Koshi basin from Imja Tsho to the confluence of Dudh Koshi and Sun Koshi for about 120 km (figure 13). Of this region, about 50 km of lower river valley (from the Hongu River-Dudh Koshi confluence to the Dudh Koshi-Sun Koshi confluence) is common to both the Hongu River and Dudh Koshi basins (figure 13). The downstream areas for about 45 km along the Hongu River from Chamlang South Tsho to the confluence of the Hongu River and Dudh Koshi are almost uninhabited; however, many people from farther downstream commute and also install temporary huts along the river valleys during the pre- and post-monsoon seasons for grazing their sheep and goats. According to Khanal et al. (2009), there are several settlements, including Barkhubesi, Demiltar, Tuintar, Khatrigaon, Rabuwa, Tirpurabesi, Mahadevbesi, Khokselibesi, and Jayaramtar, along the river valley from the Hongu River-Dudh Koshi confluence to the Dudh Koshi-Sun Koshi confluence (figure 13). According to the study, a few hundred households and thousands of people, their property, livestock, crops, and agricultural land, along with several hotels, teashops, bridges, irrigation canals, and a feasibility-studied 300-MW hydropower plant (near Rabuwa) are located in flood-prone areas.

As mentioned earlier in the Introduction, this study expands on findings of Sawagaki et al. (2012) and Byers et al. (2013). Although Sawagaki et al. (2012) did not discuss GLOF hazards, the authors did investigate the surface lowering of Chamlang South Glacier and the development of a glacial lake on it over four decades and found substantial ice melting as high as 2.2 m/year that facilitated growth of the large lake. Byers et al. (2013) briefly discussed the GLOF hazard of the lake and inferred that the lake was 'not particularly dangerous' based primarily on the wide area between west lakeshore and the terminal moraine dam that might absorb displacement waves. This study integrated field surveys and a high-resolution photogrammetric data-set to present a more comprehensive analysis of the hazard from the lake by incorporating detailed topographical and geomorphological information around the lake (e.g., dead-ice dam, terminal moraine dam, and hanging glaciers) in relation to lake bathymetry, which had not been achieved previously. As discussed above, the existence of steep hanging glaciers above the lake, which could trigger an outburst, a huge volume of water in the lake, an easy-to-overtop dead-ice dam (low

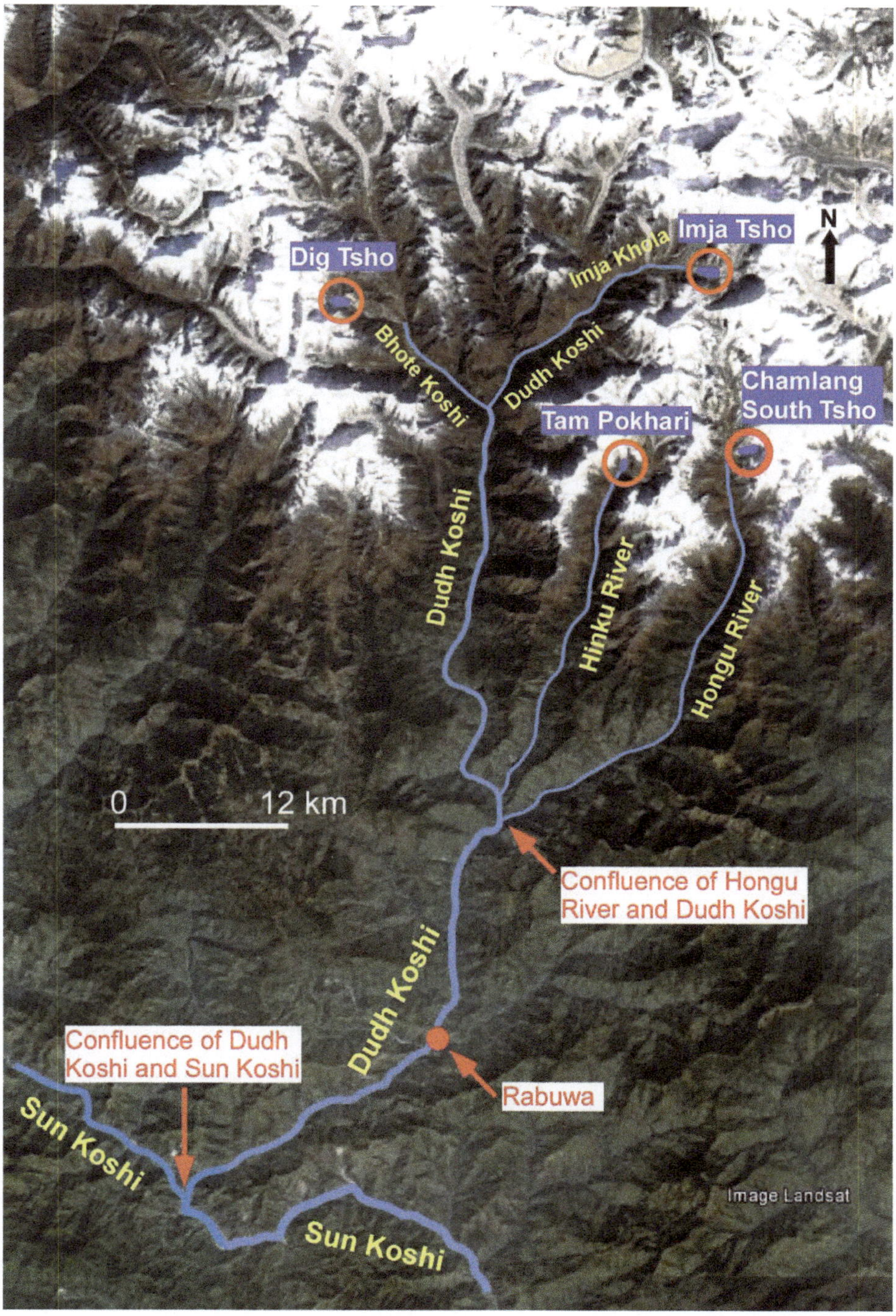

Figure 13. View from Chamlang South Tsho to areas about 100 km downstream, including Rabuwa, a settlement where a feasibility study for a 300-MW hydropower plant has been carried out. Two GLOF-occurred lakes (Dig Tsho and Tam Pokhari) and Imja Tsho, a repeatedly cited potentially dangerous glacial lake, along with river systems in the area are shown.

and narrow), an inclined dead-ice surface farther westward that would allow water to reach the terminal moraine area more quickly, and an unconsolidated terminal moraine dam (easy to erode), combined with a pronounced elevation difference between the lake (water surface and lake bottom) and base of the terminal moraine dam make the lake susceptible to an outburst flood. Based on our data-set, we suggest that the lake is potentially dangerous, and it should be monitored regularly.

In this study, we presented two pieces of crucial information essentially needed to assess GLOF hazards, a lake bathymetry, and a detailed topography around the lake (including the moraine dam). Despite this effort, our analysis of the GLOF hazard from Chamlang South Tsho is largely qualitative to semi-quantitative and does not adequately include a risk assessment. To address these issues further, we recommend future research for calculations pertaining to the volume of ice that may detach from the hanging glaciers (e.g., estimation of volume of rock avalanche as did by Hubbard et al. 2005), the size of potential displacement waves, estimation of the potential volume of water that may overtop the dam in the face of the wide dead-ice area, simulation of resulting floods or GLOF modelling (Byers et al. 2013, Bajracharya et al. 2007), and a detailed field survey to assess risk in the downstream areas, which should significantly improve understanding of both GLOF hazards and risk.

5. Conclusions

We carried out detailed surface area, bathymetric, and topographic mappings around Chamlang South Tsho to determine the lake's development processes and assess the prospect of GLOF from the lake. The volume of water contained in the lake in 2009 was determined to be $\sim 34.9-35.6 \times 10^6$ m^3. Surface areas of the lake in 1964, 1976, 1992, 2000, 2006, and 2010 were 0.04, 0.12, 0.64, 0.86, 0.87, and 0.87 km^2, respectively. Thus, the lake grew quickly from the 1960s to 2000 but has not been expanding since. Furthermore, the lake has little possibility of future expansion due to the confinement of its sides by either bedrock or lateral moraines and the relatively stable moraine dam. Hanging glaciers on the steep mountain slopes and prominent seepage water in the terminal moraine dam are considered potential triggers for an outburst flood. The existence of a low and narrow debris-covered dead-ice dam means that surge waves could overtop the dam and erode the terminal moraine dam, which is steep and unconsolidated along the present outlet channel. Furthermore, the lake can be considered potentially dangerous because of the pronounced height difference between the lake (water surface and bottom) and the base of the terminal moraine dam. Regular monitoring of Chamlang South Tsho is recommended.

Acknowledgements

We would like to thank Dr Martin Mergili and an anonymous reviewer for their valuable comments on the manuscript.

Funding

We like to thank the Global Centre of Excellence (GCOE) programme by the Japan Society for the Promotion of Science (JSPS) and the National Geographic Society Waitt Grant Program [grant number W67-09], Washington, DC for funding to carry out our research in the Hongu valley.

References

Altmaier A, Kany C. 2002. Digital surface model generation from CORONA satellite images. ISPRS J Photogramm Remote Sens. 56:221–235.

Bajracharya SR, Mool P. 2009. Glaciers, glacial lakes and glacial lake outburst floods in the Mount Everest region, Nepal. Ann Glaciol. 50:81–86.

Bajracharya B, Shrestha AB, Rajbhandari L. 2007. Glacial lake outburst floods in the Sagarmatha region. Mt Res Dev. 27:336–344.

Benn DI, Wiseman S, Warren CR. 2000. Rapid growth of a supraglacial lake, Zgozumpa glacier, Khumbu Himal, Nepal. Paper presented at: Symposium at Seattle 2000 on Debris-covered Glaciers. IAHS Publication No. 264, p. 177–185; Seattle, WA.

Bolch T, Buchroithner M, Pieczonka T, Kunert A. 2008a. Planimetric and volumetric glacier changes in the Khumbu Himal, Nepal, since 1962 using Corona, Landsat TM and ASTER data. J Glaciol. 54:592–600.

Bolch T, Buchroithner MF, Peters J, Baessler M, Bajracharya SR. 2008b. Identification of glacier motion and potentially dangerous glacier lakes at Mt. Everest area/Nepal using spaceborne imagery. Nat Hazards Earth Syst Sci. 8:1329–1340.

Bolch T, Peters J, Yegorov A, Pradhan B, Buchroithner M, Blagoveshchensky V. 2011. Identification of potentially dangerous glacial lakes in the northern Tien Shan. Nat Hazards. 59:1691–1714.

Byers AC, Mckinney DC, Somos-Valenzuela M, Watanabe T, Lamsal D. 2013. Glacial lakes of the Hinku and Hongu valleys, Makalu Barun National Park and Buffer Zone, Nepal. Nat Hazards. 69:115–139.

Carey M. 2005. Living and dying with glaciers: people's historical vulnerability to avalanches and outburst floods in Peru. Glob Planet Change. 47:122–134.

Clague JJ, Evans SG. 2000. A review of catastrophic drainage of moraine-dammed lakes in British Columbia. Quat Sci Rev. 19(17–18):1763–1783.

Dwivedi SK, Acharya MD, Simard R. 2000. The Tam Pokhari Glacier Lake outburst flood of 3 September 1998. J Nepal Geol Soc. 22:539–546.

Emmer A, Cochachin A. 2013. The causes and mechanisms of moraine-dammed lake failures in the Cordillera Blanca, North American Cordillera, and Himalayas. AUC Geogr. 48(2):5–15.

Fujita K, Sakai A, Nuimura T, Yamaguchi S, Sharma RR. 2009. Recent changes in Imja Glacial Lake and its damming moraine in the Nepal Himalaya revealed by in-situ surveys and multi-temporal ASTER imagery. Environ Res Lett. 4:045205.

Fujita K, Sakai A, Takenaka S, Nuimura T, Surazakov AB, Sawagaki T, Yamanokuchi T. 2013. Potential flood volume of Himalayan glacial lakes. Nat Hazards Earth Syst Sci. 13:1827–1839.

Fujita K, Suzuki R, Nuimura T, Sakai A. 2008. Performance of ASTER and SRTM DEMs, and their potential for assessing glacier lakes in the Lunana region, Bhutan Himalaya. J Glaciol. 54:220–228.

Gardelle J, Arnaud Y, Berthier E. 2011. Contrasted evolution of glacial lakes along the Hindu Kush Himalaya mountain range between 1990 and 2009. Glob Planet Change. 75:47–55.

Hubbard B, Heald A, Reynolds JM, Quincey D, Richardson SD, Luyo MZ, Portilla NS, Hambrey MJ. 2005. Impact of a rock avalanche on a moraine-dammed proglacial lake: Laguna Safuna Alta, Cordillera Blanca, Peru. Earth Surf Process Landforms. 30:1251–1264.

Huggel C, Kääb A, Haeberli W, Krummenacher B. 2003. Regional-scale GIS-models for assessment of hazards from glacier lake outbursts: evaluation and application in the Swiss Alps. Nat Hazards Earth Syst Sci. 3:647–662.

[ICIMOD] International Center for Integrated Mountain Development. 2011. Glacial lakes and glacial lake outburst floods in Nepal. Kathmandu: ICIMOD.

Ives JD. 1986. Glacial outburst floods and risk engineering in the Himalaya. *ICIMOD Occasional Paper 5*. Kathmandu: ICIMOD.

Ives JD, Shrestha RB, Mool PK. 2010. Formation of glacial lakes in the Hindu Kush-Himalayas and GLOF risk assessment. Kathmandu: ICIMOD.

Iwata S, Ageta Y, Naito N, Sakai A, Narama C, Karma. 2002. Glacial lakes and their outburst flood assessment in the Bhutan Himalaya. Glob Environ Res. 6:3–17.

Khanal NR, Koirala HL, Nepal P, Rai D, Khanal B, Sigdel S, Sharma AR. 2009. GLOF risk assessment of the Imja, Tsho Rolpa and Thulagi glacial lakes in Nepal. Report submitted to International Centre for Integrated Mountain Development (ICIMOD) (unpublished report).

Komori J. 2008. Recent expansions of glacial lakes in the Bhutan Himalayas. Quat Int. 184:177–186.

Lamsal D, Sawagaki T, Watanabe T. 2011. Digital terrain modelling using Corona and ALOS PRISM data to investigate the distal part of Imja Glacier, Khumbu Himal, Nepal. J Mt Sci. 8:390–402.

Mergili M, Müller JP, Schneider JF. 2013. Spatio-temporal development of high-mountain lakes in the headwaters of the Amu Darya River (Central Asia). Glob Planet Change. 107:13–24.

Mergili M, Schneider JF. 2011. Regional-scale analysis of lake outburst hazards in the southwestern Pamir, Tajikistan, based on remote sensing and GIS. Nat Hazards Earth Syst Sci. 11:1447–1462.

Mool PK, Bajracharya SR, Joshi SP. 2001. Inventory of glaciers, glacial lakes and glacial lakes outburst floods. Kathmandu: ICIMOD.

Narama C, Duishonakunov M, Kääb A, Daiyrov M, Abdrakhmatov K. 2010. The 24 July 2008 outburst flood at the western Zyndan glacier lake and recent regional changes in glacier lakes of the Teskey Ala-Too range, Tien Shan, Kyrgyzstan. Nat Hazards Earth Syst Sci. 10:647–659.

Nuimura T, Fujita K, Yamaguchi S, Sharma RR. 2012. Elevation changes of glaciers revealed by multitemporal digital elevation models calibrated by GPS survey in the Khumbu region, Nepal Himalaya, 1992–2008. J Glaciol. 58:648–656.

Osti R, Egashira S. 2009. Hydrodynamic characteristics of the Tam Pokhari Glacial Lake outburst flood in the Mt. Everest region, Nepal. Hydrol Process. 23:2943–2955.

Quincey DJ, Richardson SD, Luckman A, Lucas RM, Reynolds JM, Hambrey MJ, Glasser NF. 2007. Early recognition of glacial lake hazards in the Himalaya using remote sensing datasets. Glob Planet Change. 56:137–152.

Reynolds MJ. 2000. On the formation of supraglacial lakes on debris-covered glaciers. Paper presented at: Symposium at Seattle 2000 on Debris-covered Glaciers. IAHS Publication No. 264, p. 153–161; Seattle, WA.

Richardson SD, Reynolds JM. 2000a. An overview of glacial hazards in the Himalayas. Quat Int. 65/66:31–47.

Richardson SD, Reynolds JM. 2000b. Degradation of ice-cored moraine dams: implications for hazard development. Paper presented at: Symposium at Seattle 2000 on Debris-covered Glaciers. IAHS Publication No. 264, p. 187–197; Seattle, WA.

Sakai A, Fujita K. 2010. Formation conditions of supra-glacial lakes on debris-covered glaciers in the Himalaya. J Glaciol. 56:177–181.

Sawagaki T, Lamsal D, Byers AC, Watanabe T. 2012. Changes in surface morphology and glacial lake development of Chamlang South Glacier, eastern Nepal Himalaya, since 1964. Glob Environ Res. 16:83–94.

Vuichard D, Zimmermann M. 1987. The 1985 catastrophic drainage of a moraine dammed lake, Khumbu Himal, Nepal: cause and consequences. Mt Res Dev. 7:91–110.

Wang X, Ding Y, Liu S, Jiang L, Wu K, Jiang Z, Guo W. 2013. Changes of glacial lakes and implications in Tian Shan, central Asia, based on remote sensing data from 1990 to 2010. Environ Res Lett. 8: 044052.

Watanabe T, Ives JD, Hammond JE. 1994. Rapid growth of a glacial lake in Khumbu Himal, Himalaya: prospects for a catastrophic flood. Mt Res Dev. 14:329–340.

Watanabe T, Lamsal D, Ives JD. 2009. Evaluating the growth characteristics of a glacial lake and its degree of danger of outburst flooding: Imja Glacier, Khumbu Himal, Nepal. Nor Geogr Tidsskr. 63:255–267.

Yamada T. 1998. Glacier lakes and its outburst flood in the Nepal Himalaya. Monograph No. 1, Data Center for Glacier Research. Tokyo: Japanese Society of Snow and Ice.

12

Mapping disaster vulnerability in India using analytical hierarchy process

ANUSHEEMA CHAKRABORTY and P.K. JOSHI*
Department of Natural Resources, TERI University, New Delhi 110 070, India

Disasters are the coincidences between hazardous events, elements at risk, and conditions of vulnerability. Vulnerability integrates social and environmental systems to reduce the intensity and frequency of these risks. By categorizing regions according to their level of vulnerability, one can examine and assess the possible impacts of developmental and environmental degradation processes. This study is an attempt to map the sub-national areas (districts) in India that are vulnerable to natural and climate-induced disasters. The assessment is considered under the framework of Intergovernmental Panel on Climate Change definition of vulnerability. Using analytical hierarchy process as a multi-criteria decision-mapping method, vulnerability is measured in terms of exposure, sensitivity, and adaptive capacity. Based on this mapping assessment, districts in the state of Arunachal Pradesh, Assam, Bihar, Himachal Pradesh, Jharkhand, Manipur, Meghalaya, Mizoram, Uttar Pradesh, Uttarakhand, and West Bengal are the most vulnerable regions; while districts in the state of Punjab, Haryana, Gujarat, Tamil Nadu, Maharashtra, Goa, Andhra Pradesh, Kerala, and Karnataka are among the least vulnerable regions. The results of this study can serve as the basis for targeting prioritization efforts, emergency response measures, and policy interventions at district level for mitigating disaster vulnerability in the country.

1. Introduction

Vulnerability is defined as the susceptibility of human and environmental systems that are likely to experience harm or damage from the exposure to stresses and from the absence of capacity to adapt (IPCC 2001; Adger 2006). Despite the advances in scientific research and literature, the measurement and classification of vulnerability are highly uncertain (Kates et al. 2001; Kasperson et al. 2003; Adger 2006; Eakin & Luers 2006). The major challenge of developing vulnerability indices is the requirement of data-sets from various fields (Hinkel 2011). Many vulnerability indices are based on the assumption of equal importance of the indicators of vulnerability (Eakin & Bojorquez-Tapia 2008). The most common method for quantifying vulnerability in the global change scenario is the use of composite proxy indicators (Kelly & Adger 2000; Moss et al. 2001). More recently, the concept of multi-dimensionality has been applied for analysing vulnerability (IPCC 2001; Luers et al. 2003; Turner et al. 2003). It is typically described as a function of three overlapping elements: (1) *exposure* (the shocks and stresses experienced by the system), (2) *sensitivity* (the response of the system), and (3) *adaptive capacity* (the capacity of the system for any

*Corresponding author. Email: pkjoshi@teri.res.in

kind of adaptive action). While increase in sensitivity and exposure increases the vulnerability, an increase in adaptive capacity will reduce the vulnerability of the system (O'Brien et al. 2004; Adger 2006; Smit & Wandel 2006).

Disaster vulnerability lies under the circumstances affecting the social and economic structures of the society (Morrow 1999; de Chazal et al. 2008; Maskrey 1989). Natural and climate-induced disasters are considered as the coincidences between hazardous events (such as earthquake, cyclones, floods, droughts, and sea-level rise), elements at risk (such as environmental, financial, and political systems), and conditions of vulnerability of any system (Maskrey 1989; Birkmann 2011). The susceptibility to the risk of any disaster invariably becomes higher when one or more natural hazards occur in a vulnerable region (Wisner & Luce 1993). Numerous frameworks, theories, and modelling techniques have been developed for assessing disaster vulnerability (Adger 2006; Eakin & Luers 2006; Gallopín 2006; Füssel 2007). The theoretical conceptualization of vulnerability assessments to identify disaster risk zones have often been used as the basis for pre-impact studies and mitigation planning strategies (O'Brien et al. 2004; Brooks et al. 2005; Cutter et al. 2008). Various sectors, such as agriculture (Lin et al. 2013), forests (Simonovich & Sharabi 2013), water resources (Propastin 2013), glaciers and snow covers (Wang et al. 2013), and coastal regions (Paul & Ismail 2013), have been projected to have several catastrophic impacts due to natural and climate-induced disasters. The severity of these impacts are further exacerbated by the current vulnerability of any region from single stressors (hazards) to multiple stressors (global change) (Helmer & Hilhorst 2006; O'Brien et al. 2006; Schipper & Pelling 2006; Van Aalst 2006). The distribution of these incidences, however, varies with the geography encompassed throughout the entire globe (Cutter et al. 2008). In addition to this, multi-dimensional assessments of such impacts also vary across at different levels and scales of assessments (O'Brien et al. 2004). Therefore, understanding the cumulative impacts at the sub-national and national level to natural and climate-induced disasters is essential to decipher the complex processes that address human rights and humanitarian development. The prior knowledge of vulnerability of the system and ways to increase its adaptive capacity would help in reducing the adverse risks due to disasters. In resource-constrained countries, disaster vulnerability assessments will help in identifying and enhancing the actions required for adaptation and mitigation strategies for the most vulnerable communities.

The long-term developmental processes in India would be greatly influenced by the frequency and magnitude of the occurrence of natural and climate-induced disasters. With a wide variety of physical and climatic features, the vast diversity in India leads to different disaster-prone scenarios in the country. Despite this fact, India is often criticized for its lack of disaster preparedness and response strategies (Özerdem & Jacoby 2006). One of the major reasons for such a conclusion is the inadequate database and community or sector-specific adaptation and mitigation strategies (Guha-Sapir et al. 2013). The identification and characterization of the vulnerable communities and sectors are essential for targeting and planning any policy intervention.

The analytic hierarchy process (AHP) has been widely used to solve a broad range of multi-criteria decision-making problems, with the pairwise comparison matrix calculating the weights for each criterion considered (Ying et al. 2007; Aghataher et al. 2008; Pourghasemi et al. 2012; Cozannet et al. 2013). The flexibility of using AHP over multi-criteria methods brings intuitive appeal to decide solutions for integrated processes, with the enhanced ability to check inconsistencies of each criterion considered to study environmental impacts (Ramanathan 2001). The uniqueness of applying AHP in different studies helps in modelling situations of uncertainty without losing subjectivity and objectivity of any evaluation measure. Such procedure-oriented methodology can easily incorporate tangible and non-tangible factors in making major policy decisions.

With the aim of examining disasters under the concept of vulnerability, this study is an attempt to map the sub-national areas (districts) in India that are vulnerable to natural and climate-induced disasters. Using AHP as a multi-criteria decision-mapping method, vulnerability measured in terms of exposure, sensitivity, and adaptive capacity was assessed for entire country at the sub-national level. The results of this study can serve as the basis for targeting prioritization efforts, emergency response measures, channelizing funds, and raising environmental concern and policy interventions at district level for mitigating disaster vulnerability in the country.

2. Study area

The disaster vulnerability was assessed for the entire region of India. Officially known as the Republic of India, India is a country in South-Asia, situated between 6°44′ and 37°30′ north latitude and 68°7′ and 97°25′ east longitude, covering an area of 3.3 million sq. km. The country has a coastline measuring 7517 km in length; of this distance, 5423 km belongs to peninsular India and 2094 km to the Andaman, Nicobar, and Lakshadweep chain of islands. Figure 1 shows the vulnerable regions prone to natural and climate-induced disasters in the country.

3. Data and methodology

3.1. Data

The AHP was used to measure the vulnerability in terms of exposure, sensitivity, and adaptive capacity for the districts (Zilā) in India. As per the data available for the Census 2001, 594 districts in India were considered in the study. Five indicators were considered for each element. The disaster vulnerability assessment is based on the rankings across a number of composite indicators; each constructed using a different set of weights. The indicators considered for each element are as follows: (1) exposure (earthquake, cyclones, floods, droughts, and sea-level rise); (2) sensitivity (population density, marginal workers, forest cover, protected area, and net sown area); and (3) adaptive capacity (literacy rate, electricity availability, road facilities, communication facilities, and medical facilities). Since time immemorial, the Indian sub-continent has been exposed to many natural and climate-induced disasters. Around 57% of the Indian landmass is vulnerable to earthquakes (high seismic zones III–V), 28% is vulnerable to droughts, 12% is vulnerable to floods, and 8% is vulnerable to cyclones (Sinha 2003). Adding to this, the recent events of increase in the sea-level rise have damaged the human, social, and ecological resources in huge numbers. Major natural hazard and disaster, namely, earthquake, along with four climate-related risks, namely, cyclones, floods, droughts, and sea-level rise were considered as the best available proxies to predict future drastic disasters events. For computing sensitivity index, population sensitivity (indicators including marginal workers and population density) and ecological sensitivity (indicators including net sown area, forest cover, and protected areas) were considered to examine responsiveness to these disasters.

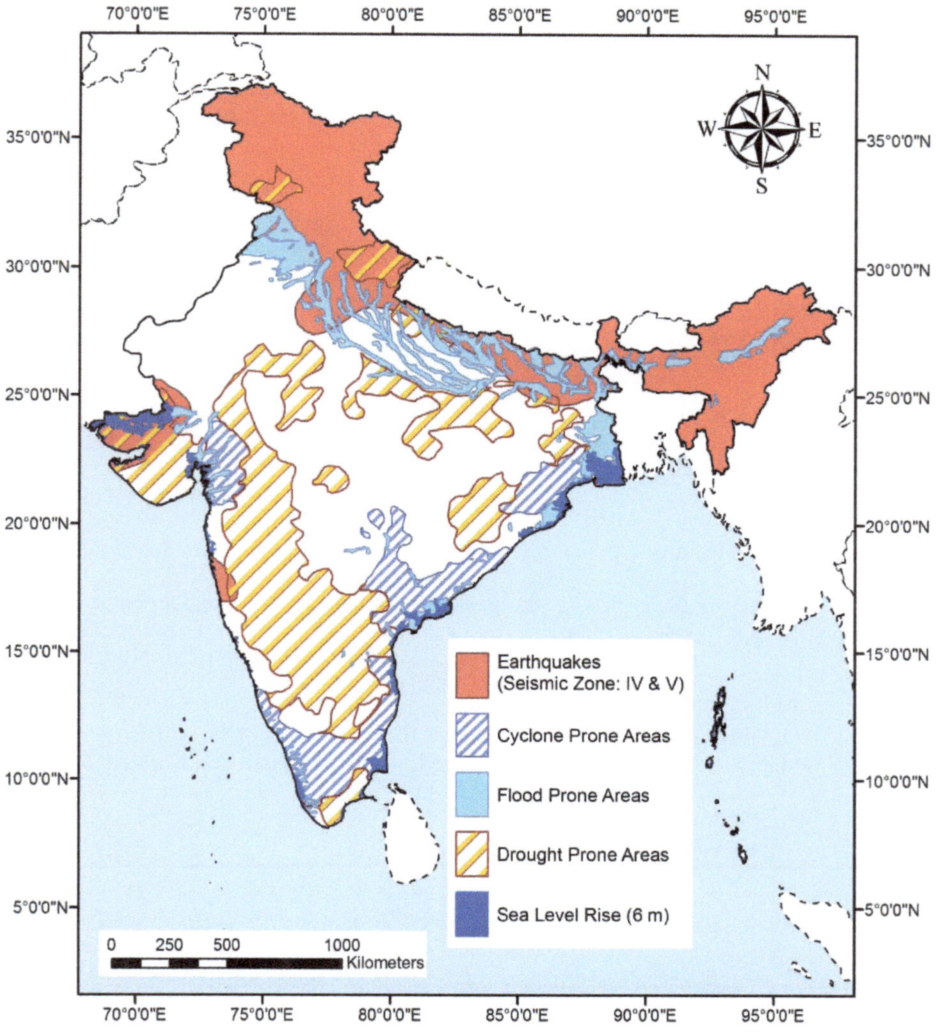

Figure 1. Location of the study area showing the natural and climate-induced disaster in the country.

To measure the adaptive capacity, indicators such as literacy rate, electricity availability, and facilities of health, communication, and accessibility (road networks) were considered.

The construction of the vulnerability profile requires explicit selection of indicators to assess vulnerability for each district in the country. In the area of vulnerability research, categorizing vulnerability of natural and climate-induced disasters requires an overall index from a suite of indicators to comprehend the underlying major cause of disaster vulnerability. The selection of indicators and the weights for each sub-indicator is, however, difficult considering the complex nature of vulnerability. Depending on the major causes of disaster vulnerability and availability of data, suitable indicators were selected. For this study, natural and climate-induced disasters (namely, earthquake, cyclones, floods, droughts, and sea-level rise), population and ecological sensitivity (namely, marginal workers, population density, net sown area,

Table 1. Vulnerability indicators used in the study.

Category	Indicator	Data source
Exposure	Earthquake	National Disaster Management Authority (NDMA)
	Cyclones	
	Floods	
	Droughts	Drought management in India: Department of Agriculture and Co-operation (Drought Management Division), Ministry of Agriculture, Government of India (GoI)
	Sea-level rise	Centers for the Remote Sensing of Ice Sheets (CReSIS), Haskell Indian Nations University
Sensitivity	Population density Marginal workers	Census of India, Office of The Registrar General and Census Commissioner, Ministry of Home Affairs, GoI
	Forest cover	Forest Survey of India, Ministry of Environment and Forest (MoEF), GoI
	Protected area	ENVIS Centre for Wildlife and Protected Areas, MoEF, GoI
	Net sown area	Indian Agricultural Statistics, Directorate of Economics and Statistics, Department of Agriculture and Co-operation, Ministry of Agriculture, GoI
Adaptive capacity	Literacy rate Electricity availability Road facilities Communication facilities Medical facilities	Census of India, Office of The Registrar General and Census Commissioner, Ministry of Home Affairs, GoI

forest cover, and protected areas), and adaptive capacity measured in terms of literacy, electricity, health, and communication facilities, and accessibility to road networks were considered as indicators for exposure, sensitivity, and adaptive capacity, respectively. The details of the indicators considered for mapping disaster vulnerability for the associated database were acquired from the respective responsible agencies (table 1).

3.2. Methodology

The approach of the study requires explicit selection of the indicator variables depicting the best available proxies of disastrous events. The indicators selected for each of these elements are essential variables for any kind of comprehensive vulnerability assessment. The composite vulnerability index approach was used to map the districts in India vulnerable to natural and climate-induced disasters. Vulnerability indices were calculated using data for each district in India. By aggregating the relative indices of exposure, sensitivity, and adaptive capacity, the cumulative composite vulnerability index has been calculated to categorize the highest and lowest vulnerable regions in India at the sub-national level (districts).

Using the indicators described in table 1, the disaster vulnerability in India was assessed. Apart from the indicators selected for *exposure*, the spatial database was created for *sensitivity* and *adaptive capacity* for each district. The details of the sources of the data are mentioned in table 1. Since each indicator was measured on different scales, dimension indices for each of the indicators for a district were calculated by normalizing the data using the following equation:

$$\text{Index} = \frac{X_a - X_{min}}{X_{max} - X_{min}} \quad (1)$$

where Index is the normalized value of an indicator, X_a is the actual value of the same indicator, and X_{min} and X_{max} are the minimum and maximum values, respectively, of the same indicator.

After normalizing all the indicators, each indicator of the respective element was assigned weights using the AHP. The AHP was used as a multi-criteria decision-making tool to develop priorities that best represent the respective element (exposure, sensitivity, and adaptive capacity). The AHP framework uses hierarchical structures to represent a problem and judgement alternatives for users by providing methodology to calibrate numeric scale for measuring the qualitative performances (Saaty 1980). The main steps involved in the AHP process are as follows: (1) defining the unstructured problem and developing the AHP hierarchy; (2) pairwise comparison of the criteria using Saaty's scale; (3) estimating the relative weights; (4) checking the consistency ratio (required threshold value of less than 0.1); and (5) generation of the resulting overall AHP indices. The pairwise comparison matrix for exposure, sensitivity, and adaptive capacity using the Saaty's scale of measurement (Saaty 2008) is shown in table 2 (a)–(c), respectively.

The quantification of the disaster vulnerability was carried out with the method of multi-level weighted sum after the normalization of indicators representing each element. The normalized weighted values of the indicators were added to generate the three sub-indices of exposure, sensitivity, and adaptive capacity. In other words, the following formulations were used to generate the sub-indices (exposure, sensitivity, and adaptive capacity) of disaster vulnerability:

$$\text{Sub-index} = \sum_{i=1}^{n} w_i \times u_i \quad (2)$$

where w is the weight of each indicator, u is the value of each indicator, and n is the total number of indicators, $i = 1, 2, 3,\ldots,n$.

To obtain the final composite vulnerability index, these sub-indices were combined using the additive (averaging) approach. The equation for calculating the disaster vulnerability was as follows:

$$V = \frac{[E + S + (1 - AC)]}{3} \quad (3)$$

where V, E, S, and AC represent vulnerability, exposure, sensitivity, and adaptive capacity, respectively.

Table 2. Pairwise comparison matrix using Saaty's scale of measurement (Saaty 2008).

(a) Exposure

Indicators	Cyclones	Floods	Droughts	Earthquake	Sea-level rise
Cyclones	1	3	5	1/3	4
Floods	1/3	1	3	1/4	3
Droughts	1/5	1/3	1	1/6	1/3
Earthquake	3	4	6	1	5
Sea-level rise	4	3	3	1/5	1

(b) Sensitivity

Indicators	Population density	Marginal workers	Forest cover	Protected area	Net sown area
Population density	1	1/3	5	6	3
Marginal workers	3	1	7	9	4
Forest cover	1/5	1/7	1	3	1/4
Protected area	1/6	1/9	1/3	1	1/5
Net sown area	1/3	1/4	4	5	1

(c) Adaptive capacity

Indicators	Medical facilities	Literacy rate	Electricity facilities	Communication facilities	Road facilities
Medical facilities	1	5	3	1/3	1/6
Literacy rate	1/5	1	1/3	1/7	1/9
Electricity facilities	1/3	3	1	1/5	1/8
Communication facilities	3	7	5	1	1/3
Road facilities	6	9	8	3	1

The summarized sequential method used in this study for evaluating disaster vulnerability in India has been shown in figure 2.

4. Results

The indicators used in this study measure the vulnerability parameter intended to characterize the system. For this study, the system of interest includes the entire heterogeneous landscape of India. The scale of the study is at the sub-national level, which is represented by the districts in the country. The data on the spatial distribution of the indicators were used to explain the exposure, sensitivity, and adaptive capacity collectively for each district.

By combining the natural and climate-induced disasters, the overall exposure has been mapped for India. Figure 3(*a*) depicts the inclusive exposure map showing the combination of natural (earthquakes) and climate-induced disasters (cyclones, floods, droughts, and sea-level rise). The map shows higher degrees of exposure in the districts located especially in the north-eastern part of the country. Many districts in other states including Arunachal Pradesh, Manipur, Meghalaya, Mizoram, and Nagaland in the north-eastern belt are also prone to extreme seismic events and flood-prone disasters. Along the northern and central part of the country, Jammu and Kashmir, Himachal Pradesh, Uttarakhand, parts of Punjab and Uttar Pradesh,

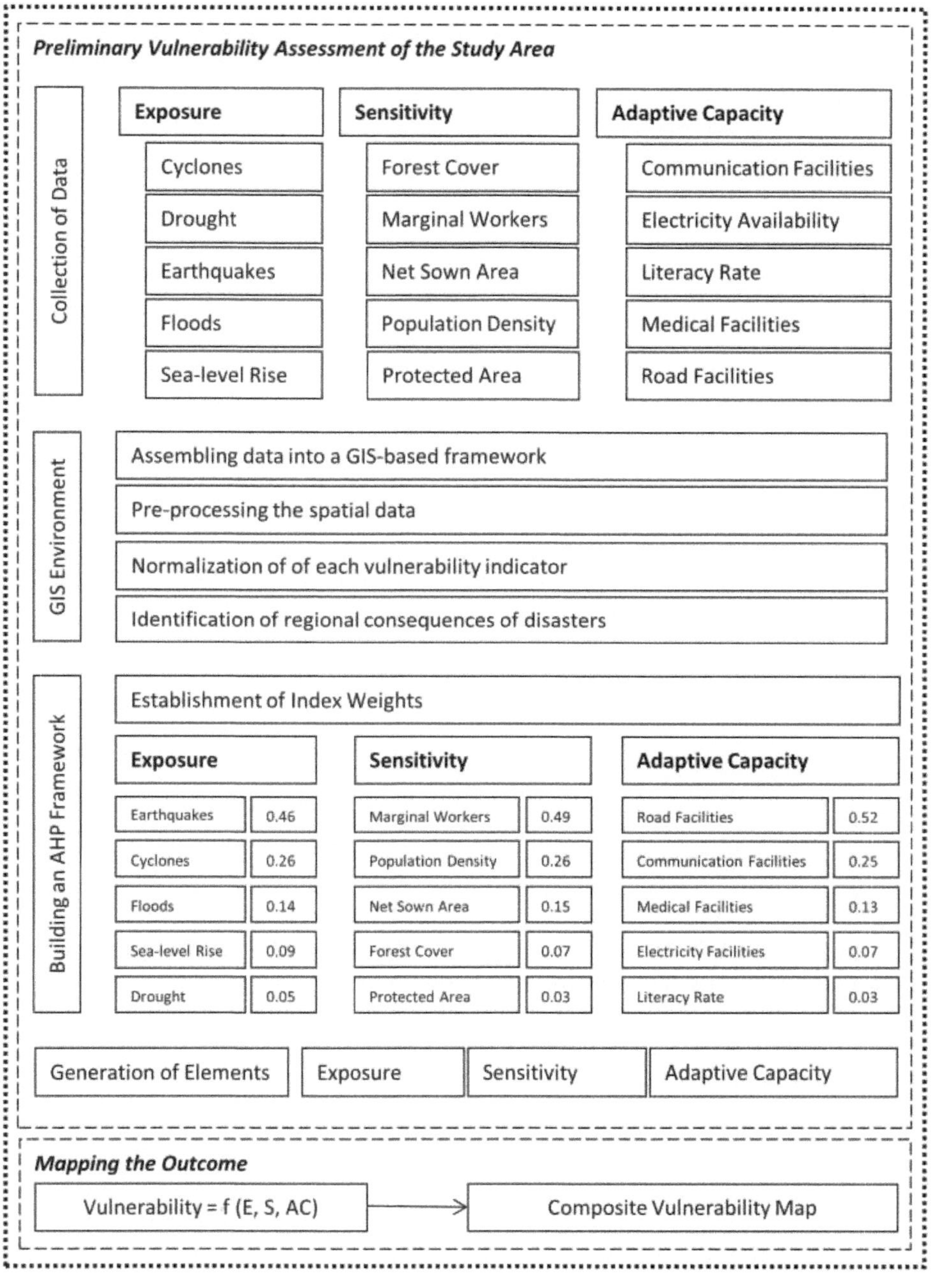

Figure 2. The sequential process in this study to map the disaster vulnerability.

the National Capital Territory (NCT) of Delhi, Bihar, West Bengal, and Kachchh district in Gujarat are more prone to events of incurring disasters, including the earthquake, drought, and flood-prone areas. Such instances of high exposure can be explained by the prevalence of earthquake as major natural hazard in these regions. The higher exposure could also be explained by the frequent climate-related occurrence of floods and droughts prolonging in these regions, especially in the

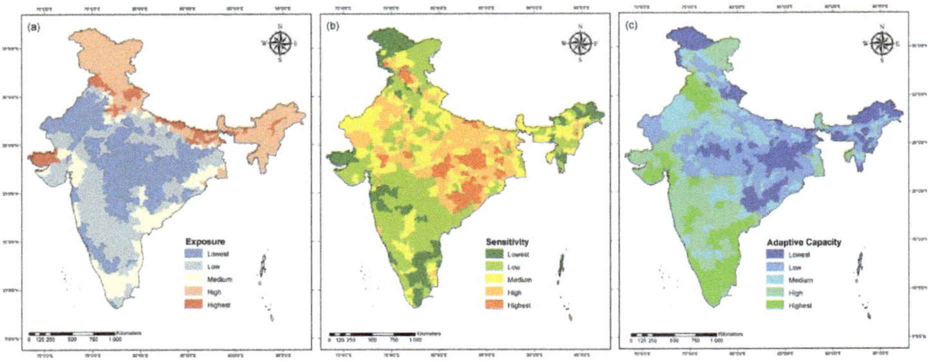

Figure 3. District-level mapping to disasters in India: (*a*) exposure, (*b*) sensitivity, and (*c*) adaptive capacity.

Indo-Gangetic river basin. The relatively less-exposed districts lie mostly in the coastal region, in states including Gujarat, Maharashtra, Kerala, Tamil Nadu, Andhra Pradesh, Orissa, and West Bengal, with the lesser probability of natural and climate-induced disaster. The state of Orissa (districts including Baleshwar, Bhadrak, Ganjam, Jagatsinghapur, Kendrapara, and Puri) has suffered momentous losses over the decades to tropical cyclones; these vulnerable regions are clearly highlighted in the exposure map (figure 3(*a*)). Districts in West Bengal and Gujarat have the higher probabilities of area inundated as a result of rising sea-water level. The situation has been aggravated due to anthropogenic influences disturbing the ecological systems in these exposed regions. The district-level map shows comparatively lower exposure to disasters in the interior portions of the country, particularly in the states of Madhya Pradesh, Chhattisgarh, Jharkhand, Rajasthan, and parts of the Maharashtra, Karnataka, Punjab, and Haryana to natural and climate-induced disasters considered in this study.

To measure sensitivity of the districts to the exposure of natural and climate-induced disasters, population sensitivity (indicators including marginal workers and population density), and ecological sensitivity (indicators best explaining the ecological sensitiveness including net sown area, forest cover, and protected areas) were considered to compute the sensitivity index. The presence of aggregated clusters of population, with large number of marginal workers, and the area covered by forests, or declared as a protected area, or the presence of irrigated/cultivated area, have been considered to measure the sensitivity of the districts in India to any possible natural and climate-induced disaster. Figure 3(*b*) shows the sensitivity of the districts to external stresses and shocks. The regions more sensitive owing to high population ratios and ecologically fragile environments make the states Himachal Pradesh (districts including Bilaspur, Chamba, Hamirpur, Kangra, and Mandi), Chhattisgarh (districts including Baster, Jangir-Champa, Jashpur, Koriya, and Surguja), Jharkhand (districts including Garhwa, Giridih, Godda, Gumla, Lohardada, palamu, and Pashchimi Singhbhum), Madhya Pradesh (districts including Balaghat, Dindori, Mandla, Rajgarh, Seoni, and Umaria), Orissa (districts including Balangir, Bargarh, Baudh, Debagarh, Gajapati, Kalahandi, Kandhamal, Koraput, Malkangiri, Mayurbhanj, Nabarangapur, Naupada, and Rayagada), Uttar Pradesh (districts including Kushinagar, Maharajganj, Siddharthnagar), and West Bengal (majorly in Puruliya district) more sensitive to extreme events (disasters). Few districts in Nagaland, Gujarat, Rajasthan, and Jammu and Kashmir are also sensitive to exposed stresses of disasters. The foremost reason for such higher sensitivity would be the continuous increase in the number of marginal workers with rising population in most of these states. The economy is inextricably tied to the climate-sensitive sector, primarily agriculture, which is the most prevalent source of income in these regions. Any drastic event such as cyclones, floods, and droughts would hamper the agricultural productivity. This would directly or indirectly endanger livelihoods of communities dependent on it and the food security of the states. Interlinkages between different sectors would have a ripple effect on the entire economy. The impacts will be most felt on the vulnerable and poor sections of the society. The state of Tamil Nadu (except the districts sharing the coastal boundary) is comparatively less sensitive to changes in the system. The extreme north-western district in the state of Jammu and Kashmir also shows lower sensitivity; however, this might be due to no available data for this district in the state.

Based on the indicators that explain the ability of the districts to accommodate the changes due to natural and climate-induced disasters, figure 3(*c*) depicts the adaptive capacity across the districts in India. The indicators (road facilities, communication facilities, medical facilities, electricity availability, and literacy rate) help in developing the index for measuring the adaptive capacity at the sub-national level in the country. Although the exposure and sensitivity increase the vulnerability, showing a negative relationship; higher value and increase in adaptive capacity reduce the vulnerability of any system, indicating a positive relationship. The study indicates highest degree of adaptive capacity in all the districts in the state of Kerala. Other than this region, particularly in the Deccan Plateau, the states including Tamil Nadu, parts of Andhra Pradesh, Maharashtra, and Gujarat, along with districts in Punjab and Haryana are the least affected in terms of their capacity to adapt to external stresses. Lower adaptive capacity can be clearly seen in states in the central and eastern part of the country. States, such as Arunachal Pradesh (districts including Lower and Upper Subansiri), Jharkhand (districts including Chatra, Deoghar, Dumka, Garhwa, Giridih, Godda, Kodarma, Pakuar, Pashchiimi, and Purbi Singhbhum), Orissa (districts including Kandhamal and Malkangiri), Bihar (districts including Banka and Jamui), and Uttarakhand (districts including Garhwal, Tehri Garwal, Chamoli, Bageshwar, and Almora), have the least adaptive capacity to natural and climate-induced disasters. Few districts in the north-eastern part of the country also have lesser capacity to adapt disastrous events. The district-based accessibility due to paved roads and availability of facilities for transport, health, communication, electricity, and education majorly contribute in determining the adaptive capacity for each district. The social, technological, and infrastructural indicators would affect the sensitivity of the districts, either positively or negatively.

The disaster vulnerability mapping at the sub-national level (districts) was measured as a composite of exposure, sensitivity, and adaptive capacity across the entire regions in India. Districts are ranked and represented as quantiles of high and low vulnerability profiles. Figure 4 reflects the vulnerable regions in the country to natural and climate-induced disasters. Districts in the states of Arunachal Pradesh (Lower Subansiri, Upper Subansiri, and West Kameng), Assam (Dhemaji, Dibrugarh, and Karbi Anglong), Bihar (Araria, Banka, Darbhanga, Gopalganj, Jamui, Katihar, Kishanganj, Lakhisarai, Nalanda, Pashchim Champaran, Purnia, Saran, Siwan, Sitamarhi, Saarsa, and Supaul), Himachal Pradesh (Chamba, Mandi, and

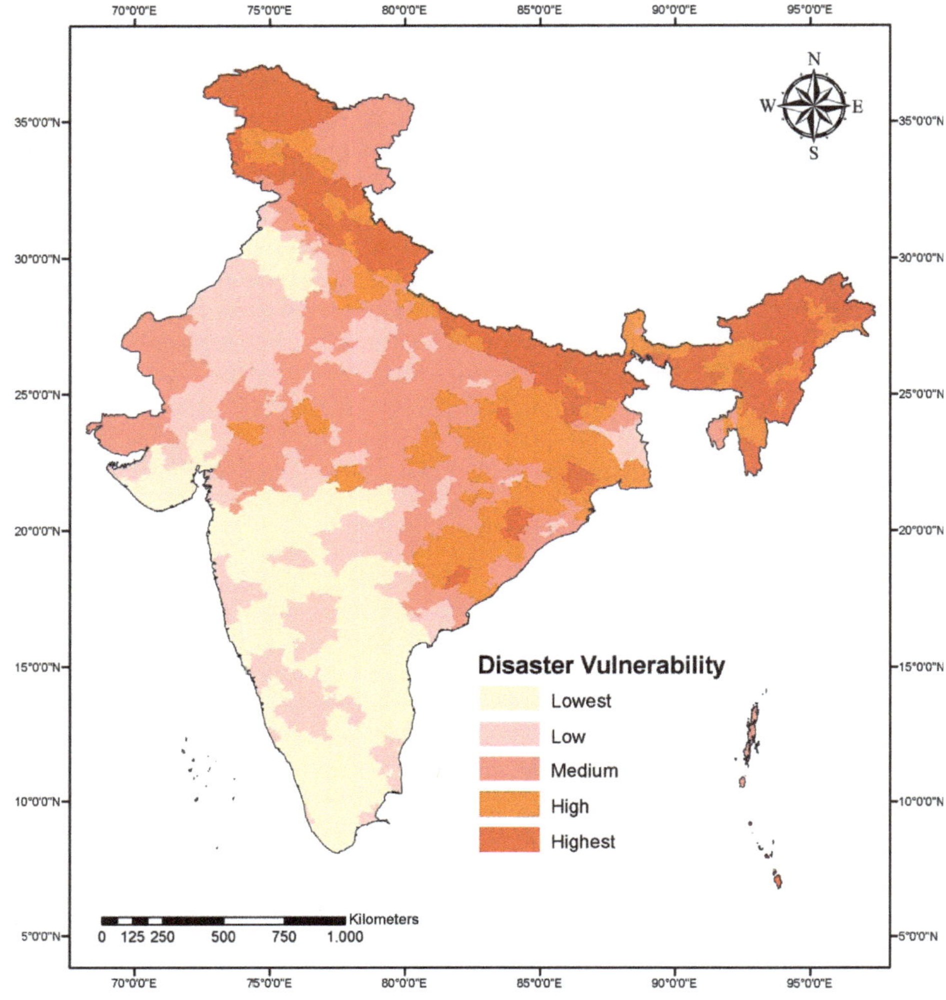

Figure 4. District-level mapping of disaster vulnerability in India, measured as a composite of exposure, sensitivity, and adaptive capacity indices.

Solan), Jharkhand (Godda and Chandel), Manipur (Chandel and Churachandpur), Meghalaya (all the districts), Mizoram (Lawngtlai district), Uttar Pradesh (Balrampur, Kushinagar, Maharajganj, Shrawasti, and Siddarthnagar), Uttarakhand (districts including Bageshwar, Chamoli, Champawat, Garhwal, Tehri Garhwal, and Pithoragarh), and West Bengal (Dakshin Dinajpur district) are among the most vulnerable regions in India. While districts in the state of Punjab, Haryana, Gujarat, Tamil Nadu, Maharashtra, Goa, Andhra Pradesh, Kerala, and Karnataka are among the least vulnerable regions in the country. The districts in the state of Rajasthan, Madhya Pradesh, the union territories (NCT of Delhi, Chandigarh, Pondicherry, Daman and Diu, Dadra and Nagar Haveli), along with the chain of islands in Lakshadweep and Andaman and Nicobar show comparatively medium vulnerability to disasters and risk hazards in the country. Table 3 shows the vulnerability of the districts in India to natural and climate-induced disasters.

Table 3. District-wise disaster vulnerability in India.

Disaster	High vulnerability	Low vulnerability
Cyclones	Mayurbhanj, Malkangiri, Kandhamal, Koraput, Visakhapatnam, Baleshwar, Bhadrak, Kendrapara, Jajapur, South Twenty Four Parganas	Alappuzha, Idukki, Kottayam, Kozhikode, Kanchipuram, Palakkad, Ernakulam, Thrissur, Pathanamthitta, Kollam
Floods	Patiala, Kapurthala, Amritsar, Faridkot, Bhatinda, Monga, Gurdaspur, Sangrur, Hoshiarpur, Mansa, East Godavari, West Godavari, Krishna, Cuttack, Jagatsinghpur, Kendrapara, Birbhum, Nadia, Murshidabad, Barddhaman, North and South Twenty Four Parganas, Banka, Samstipur, Sitamarhi, Araria, Kishanganj, Farukhabad, Hardoi, Gorakhpur, Chandoli, Bareli, Golaghat, Nalbari	Tinksukia, Morigaon, Shimla, Kangra, Azamgarh, Jaunpur, Mirzapur, Varanasi, Fatehgarh Sahib, Panchkula, Rupnagar
Droughts	Kutch, Bid, Satara, Bijapur, Gulbarga, Dharwad, Chitradurga, Tumkur, Bangalore Rural, Ramanathapuram, Phulbani, Kalahandi, Bolangir, Bankura, Midnapore, Purulia, Allahabad, Nawada, Garhwa, Jhalawar, Rajgarh, Banswara, Dungarpur, Betul, Doda, Udhampur	Nashik, Aurangabad, Ahmadnagar, Bgalakot, Belgaon, Sangli, Cuddapah, Anantapur, Kurnool, Chittoor
Earthquake	Chamba, Doda, Lahul & Spiti, Udhampur, Kangra, Mandi, Shimla, Sirmaur, Solan, Chamoli, Rudraprayag, Tehri Garhwal, Uttarkashi, Almora, Bageshwar, Pitthoragarh, East Kamend, West Kameng, Dibang Valley, Gaolpara, Cachar, Lohit, East Khasi Hills, West Khasi Hills, Lower Upper Subansiri, Upper Subansiri, Kutch	Leh (Ladakh), Amritsar, NCT of Delhi, Hoshiarpur, Jalandhar, Kapurthala, Nawanshahr, Patan, Banas Kantha, Jalor, Rajkot, Surendernagar, Ratnagiri, Raigarh, Satara
Sea-level rise	Kutch, Anand,South Twenty Four Parganas, Howrah, Bhadrak, Kendrapara, Puri, West Godavari, East Godavari, Nagapattinam, Thiruvarur, Thanjavur, Ramanathapuram	Mednipur, Baleshwar, Ganjam, Guntur, Prakassam, Nellore, Thoothukkudi, Alappuzha

5. Discussion

In a country, such as India, with the rising population to about one billion people, declining economic scenario, and changing climatic conditions, urgent necessity for better understanding of natural and climate-induced disasters and developmental processes would require an integrated approach to comprehensively assess the vulnerability of any region. The availability and accessibility in the database in terms of presence of heterogeneity in the scales of assessment make vulnerability mapping even more complicated (Preston et al. 2011). Despite such inconsistencies, vulnerability has been often studied for many fields including natural hazards, extreme rainfall, economic globalization, and climate change (Janssen et al. 2006).

Using multi-criteria decision-mapping methods, considerable research has been attempted to evaluate the risk of vulnerability in different systems. There are numerous examples of the use of AHP in mapping natural and climate-induced disasters. The AHP has been used for landslide hazard and susceptibility (Yoshimatsu & Abe 2006; Pourghasemi et al. 2012), soil erosion hazard (Rahman et al. 2009), earthquake hazard (Aghataher et al. 2008), flood hazard (Sinha et al. 2008), and coastal risks (Hu et al. 2009) mapping.

Many studies in India have attempted to map vulnerability across smaller regions in the country (O'Hare 2001; Roy & Venema 2002; Gaiha & Imai 2004; Brenkert & Malone 2005; Sanyal & Lu 2005; Kelkar et al. 2008; Rahman 2008; Singh et al. 2009). However, large-scale study for the entire nation was first attempted by O'Brien et al. (2004) to map regional vulnerability to climate change primarily for the agriculture sector. The inclusion of disasters using multi-criteria decision-mapping method under the concept of vulnerability has not been yet attempted for India. In this study, the first attempts to map the vulnerability to natural and climate-induced disasters for the entire country at the sub-national (district) level are provided.

According to this study, the north-eastern part of the country is highly vulnerable to natural and climate-induced disasters. This region has always experienced many earthquakes; the north-eastern India is one among the major seismically active intra-continental regions of the world. Some of the examples include the Cachar earthquake in Assam and subsequent tremors in the vicinity of the district, and seismic activities near the Indo-Burma border showing consistent exposure to earthquakes (Gupta 1989). Earthquake incidences in Latur district in Maharashtra (Valdiya 1993) and Bhuj in the Kutch district of Gujarat (Rajendran et al. 2001) have also caused heavy losses to life and property in the western part of the country. The north-eastern India is also prone to frequent flooding in the Brahmaputra river basin. The river along with its tributaries causes recurrent devastating floods in the state of Assam almost every year. Nagaon district in Assam is the worst affected flood area suffering major losses and damages in the many villages near the Kopili and Brahmaputra rivers (Sharma et al. 2012). Many flood-prone and drought-prone areas have also been identified in the Indo-Gangetic basin. The recent and past events of flash floods and earthquakes in Uttarkashi and Garhwal district, Gaurikund, Rambara, and Kedarnath in Rudraprayag district, Alaknanda and Birahi valley in Chamoli districts of Uttarakhand (Dobhal et al. 2013; Gupta et al. 2013), flooding in Satluj catchment in Kinnaur district in Himachal Pradesh, Ludhiana, and Firozpur districts in Punjab (Sharma et al. 1996; Gupta & Sah 2008), the flood plains of Ganga and Brahmaputra river basins (Singh et al. 2009), Bhagmati and Kosi river basin flooding in the alluvial plains of north Bihar (Chandran et al. 2006), intense rainfall resulting in sudden flooding in many parts of Mumbai (Prasad & Singh 2005), high danger seismic-risk zones in northern India, particularly in the Himalaya belt (Valdiya 1993), some incidents of extreme seismicity in the Kangra district of Himachal Pradesh (Arya 1992) and Munger and Muzaffarpur districts of Bihar (Krishnan 1934) are examples showing the extreme vulnerability of these regions to natural and climate-induced disasters. The submerging coasts due to increase in the sea levels and cyclones in the Bay of Bengal have affected the eastern coasts of India and continue to rise from the past few decades (Tripati & Unnikrishnan 2011; Unnikrishnan et al. 2011). The eastern coast is prone to increase in the sea-water level, especially, the states including West Bengal, Odisha, and Andhra Pradesh show much area inundated due to the rising water level (Naik et al. 2013). With the increasing impacts of climate change, the rising sea levels would pose much threat to the coastal vulnerable areas in India.

With the help of using AHP as a tool for assessing disaster vulnerability, information from qualitative assessments can be converted to quantitative information by defining the weightages of various indicators in multi-criteria analysis. The consequences of identifying vulnerable regions to natural and climate-induced disasters by generating such a database would aid in pre-impact studies, contingency planning, and annual resource allocations with inclusion of ecological sensitivities, developing innovative projects including environmental impact assessments, planning and decision making for disaster management systems, and increasing skills sets of communities and various sectors to increase the adaptive capacity by infrastructure and urban planning (Clark et al. 2000; Turner et al. 2003; O'Brien et al. 2004; Brooks et al. 2005; IPCC 2007; Cutter et al. 2008).

6. Conclusion

From time immemorial, the Indian sub-continent has been exposed to natural disasters and more recently, higher magnitudes and frequencies of climate-induced disasters. This drastic increase in recent years poses serious threats to the overall development of the country. However, the driving forces behind these developmental processes in itself have contributed to the increase in susceptibility of any region to such extreme events. Lack of information and communication channel challenges in the progress of the entire country as a whole.

No single methodology will be able to capture completely the multi-scale and multi-faceted dimensions of vulnerability. Ultimately, the framework for analysing vulnerability would require various set of methods and modelling techniques to explain vulnerability characteristics within and between systems. Perhaps, the best probable results of vulnerability can be expected in scenarios when wide range of indicators are representing the system in consideration.

This study identifies the districts in India vulnerable to natural and climate-induced disasters. Since the districts have different levels of exposure, highest sensitivity does not always lead to highest disaster vulnerability, and the highest adaptive capacity does not always necessarily result in the lowest disaster vulnerability. The combined influence of the three elements of vulnerability (exposure, sensitivity, and adaptive capacity) was used to generate the composite vulnerability index for the country. The districts in the state of Arunachal Pradesh, Assam, Bihar, Himachal Pradesh, Jharkhand, Manipur, Meghalaya, Mizoram, Uttar Pradesh, Uttarakhand, and West Bengal are the most vulnerable regions in India. While districts in the state of Punjab, Haryana, Gujarat, Tamil Nadu, Maharashtra, Goa, Andhra Pradesh, Kerala, and Karnataka are among the least vulnerable regions in the country. The districts in the state of Rajasthan, Madhya Pradesh, the union territories (NCT of Delhi, Chandigarh, Pondicherry, Daman and Diu, and Dadra and Nagar Haveli), along with the chain of islands in Lakshadweep and Andaman and Nicobar show comparatively medium vulnerability to disasters and risk hazards in the country.

Such a database would help in creating policy intervention for the most critical regions in the country. The impacts of natural and climate-induced disasters continue to rise across the globe. Instead of dealing with disaster risks in isolation, management practices should integrate the social and environmental systems to reduce the intensity and frequency of these risks. The results of this study can serve as the basis for targeting prioritization efforts, emergency response measures, channelizing funds, and raising environmental concern and policy interventions at district level for mitigating disaster vulnerability in the country.

Acknowledgements

Authors are thankful to the anonymous reviewers for valuable suggestions. A. Chakraborty would like to acknowledge HSBC Climate Scholarship of TERI University for funding PhD research.

References

Adger WN. 2006. Vulnerability. Global Environ Change. 16:268–281.

Aghataher R, Delavar MR, Nami MH, Samnay N. 2008. A fuzzy-AHP decision support system for evaluation of cities vulnerability against earthquakes. World Appl Sci J. 3:66–72.

Arya AS. 1992. Possible effects of a major earthquake in Kangra region of Himachal Pradesh. Curr Sci. 62:251–256.

Birkmann J. 2011. First-and second-order adaptation to natural hazards and extreme events in the context of climate change. Nat Hazard. 58:811–840.

Brenkert AL, Malone EL. 2005. Modeling vulnerability and resilience to climate change: a case study of India and Indian states. Clim Change. 72:57–102.

Brooks N, Neil Adger W, Mick Kelly P. 2005. The determinants of vulnerability and adaptive capacity at the national level and the implications for adaptation. Global Environ Change. 15:151–163.

Chandran RV, Ramakrishnan D, Chowdary VM, Jeyaram A, Jha AM. 2006. Flood mapping and analysis using air-borne synthetic aperture radar: a case study of July 2004 flood in Baghmati river basin, Bihar. Curr Sci. 90:249–256.

Clark WC, Jäger J, Corell R, Kasperson R, McCarthy JJ, Cash D, Cohen SJ, Desanker P, Dickson NM, Epstein P et al. 2000. Assessing vulnerability to global environmental risks. Discussion paper 2000–12, Environment and Natural Resources Program; Belfer Center for Science and International Affairs, Harvard Kennedy School, Cambridge (MA).

Cozannet GL, Garcin M, Bulteau T, Mirgon C, Yates ML, Méndez M, Baills A, Idier D, Oliveros C. 2013. An AHP-derived method for mapping the physical vulnerability of coastal areas at regional scales. Nat Hazard Earth Syst Sci. 13:1209–1227.

Cutter SL, Barnes L, Berry M, Burton C, Evans E, Tate E, Webb J. 2008. A place-based model for understanding community resilience to natural disasters. Global Environ Change. 18:598–606.

De Chazal J, Quétier F, Lavorel S, Van Doorn A. 2008. Including multiple differing stakeholder values into vulnerability assessments of socio-ecological systems. Global Environ Change. 18: 508–520.

Dobhal DP, Gupta AK, Mehta M, Khandelwal DD. 2013. Kedarnath disaster: facts and plausible causes. Curr Sci. 105: 171–174.

Eakin H, Bojorquez-Tapia LA. 2008. Insights into the composition of household vulnerability from multicriteria decision analysis. Global Environ Change. 18:112–127.
Eakin H, Luers AL. 2006. Assessing the vulnerability of social-environmental systems. Ann Rev EnvironResour. 31:365.
Füssel HM. 2007. Vulnerability: a generally applicable conceptual framework for climate change research. Global Environ Change. 17:155–167.
Gaiha R, Imai K. 2004. Vulnerability, shocks and persistence of poverty: estimates for semi-arid rural South India. Oxford Dev Stud. 32:261–281.
Gallopín GC. 2006. Linkages between vulnerability, resilience, and adaptive capacity. Global Environ Change. 16:293–303.
Guha-Sapir D, Santos I, Borde A. 2013. The economic impacts of natural disasters. New York (NY): Oxford University Press.
Gupta HK. 1989. Patterns preceding major earthquakes in north-east India. Curr Sci. 64:889–893.
Gupta V, Dobhal DP, Vaideswaran SC. 2013. August 2012 cloudburst and subsequent flash flood in the Asi Ganga, a tributary of the Bhagirathi river, Garhwal Himalaya, India. Curr Sci. 105, 249–253.
Gupta V, Sah MP. 2008. Impact of the trans-Himalayan landslide lake outburst flood (LLOF) in the Satluj catchment, Himachal Pradesh, India. Nat Hazard. 45:379–390.
Helmer M, Hilhorst D. 2006. Natural disasters and climate change. Disasters. 30:1–4.
Hinkel J. 2011. Indicators of vulnerability and adaptive capacity: towards a clarification of the science-policy interface. Global Environ Change. 21:198–208.
Hu B, Zhou J, Wang J, Chen Z, Wang D, Xu S. 2009. Risk assessment of land subsidence at Tianjin coastal area in China. Environ Earth Sci. 59:269–276.
[IPCC] Intergovernmental Panel on Climate Change. 2001. Chapter 1, Overview of impacts, adaptation, and vulnerability to climate change. In: McCarthy JJ, Canziani OF, Leary NA, Dokken DJ, White KS, editors. Climate change 2001: impacts, adaptation & vulnerability: contribution of working group II to the third assessment report of the intergovernmental panel on climate change. Cambridge (UK): Cambridge University Press; p. 75–104.
[IPCC] Intergovernmental Panel on Climate Change. 2007. Chapter 19, Assessing key vulnerabilities and the risk from climate change. In: Parry ML, Canziani OF, Palutikof JP, van der Linden PJ, Hanson CE, editors. Climate change 2007: impacts, adaptation and vulnerability. Contribution of working group II to the fourth assessment report of the intergovernmental panel on climate change. Cambridge (UK): Cambridge University Press; p. 779–810.
Janssen MA, Schoon ML, Ke W, Börner K. 2006. Scholary networks on resilience, vulnerability and adaptation within the human dimensions of global environmental change. Global Environ Change. 16:240–252.
Kasperson JX, Kasperson RE, Turner II BL, Schiller A, Hsieh WH. 2003. Vulnerability to global environmental change. In: Diekmann A, Dietz T, Jaeger C, Rosa ES, Editors. The human dimensions of global environmental change. Cambridge (MA): MIT.
Kates RW, Clark WC, Corell R, Hall JM, Jaeger CC, Lowe I, McCarthy JJ, Schellnhuber HJ, Bolin B, Dickson NM, et al. 2001. Sustainability science. Science 292:641–642.
Kelkar U, Narula KK, Sharma VP, Chandna U. 2008. Vulnerability and adaptation to climate variability and water stress in Uttarakhand State, India. Global Environ Change. 18: 564–574.
Kelly PM, Adger WN. 2000. Theory and practice in assessing vulnerability to climate change and facilitating adaptation. Clim. Change. 47:325–352.
Krishnan MS. 1934. The north Bihar Earthquake of the 15th January, 1934. Curr Sci. 2:323–326.
Lin Y, Deng X, Jin Q. 2013. Economic effects of drought on agriculture in North China. Int J Disaster Risk Sci. 4:59–67.
Luers AL, Lobell DB, Sklar LS, Addams CL, Matson PA. 2003. A method for quantifying vulnerability, applied to the agricultural system of the Yaqui Valley, Mexico. Global Environ Change. 13:255–267.

Maskrey A. 1989. Disaster mitigation: a community based approach. Dev Guidel. 3:1–4.

Morrow BH. 1999. Identifying and mapping community vulnerability. Disasters. 23:1–18.

Moss RH, Brenkert AL, Malone EL. 2001. Vulnerability to climate change: a quantitative approach. Prepared for the US Department of Energy, Pacific Northwest National Laboratory (PNNL-SA-33642). Washington (DC): Battelle Institute.

Naik BS, Panda RK, Jakhar P, Dass A, Madhu M. 2013. Appraisal of the erosion status in the eastern coastal region of India. Indian J Soil Conservation. 41:121–126.

O'Brien G, O'Keefe P, Rose J, Wisner B. 2006. Climate change and disaster management. Disasters. 30:64–80.

O'Brien K, Leichenko R, Kelkar U, Venema H, Aandahl G, Tompkins H, Javed A, Bhadwal S, Barg S, Nygaard L, West J. 2004. Mapping vulnerability to multiple stressors: climate change and globalization in India. Global Environ Change. 14:303–313.

O'Hare G. 2001. Hurricane 07B in the Godavari Delta, Andhra Pradesh, India: vulnerability, mitigation and the spatial impact. Geogr J. 167:23–38.

Özerdem A, Jacoby T. 2006. Disaster management and civil society: earthquake relief in Japan, Turkey and India. London: IB Tauris.

Paul GC, Ismail AIM. 2013. Contribution of offshore islands in the prediction of water levels due to tide-surge interaction for the coastal region of Bangladesh. Nat Hazard. 65:13–25.

Pourghasemi HR, Pradhan B, Gokceoglu C. 2012. Application of fuzzy logic and analytical hierarchy process (AHP) to landslide susceptibility mapping at Haraz watershed, Iran. Nat Hazard. 63:965–996.

Prasad AK, Singh RP. 2005. Extreme rainfall event of July 25–27, 2005 over Mumbai, West Coast, India. J Indian Soc Remote Sensing. 33:365–370.

Preston BL, Yuen EJ, Westaway RM. 2011. Putting vulnerability to climate change on the map: a review of approaches, benefits, and risks. Sustainability Sci. 6:177–202.

Propastin P. 2013. Assessment of climate and human induced disaster risk over shared water resources in the Balkhash Lake drainage basin. In: Climate change and disaster risk management. Springer: Heidelberg; p. 41–54.

Rahman A. 2008. A GIS based DRASTIC model for assessing groundwater vulnerability in shallow aquifer in Aligarh, India. Appl Geogr. 28:32–53.

Rahman MR, Shi ZH, Chongfa C. 2009. Soil erosion hazard evaluation – an integrated use of remote sensing, GIS and statistical approaches with biophysical parameters towards management strategies. Ecol Modelling. 220:1724–1734.

Rajendran K, Rajendran CP, Thakkar M, Tuttle MP. 2001. The 2001 Kutch (Bhuj) earthquake: coseismic surface features and their significance. Curr Sci. 80:1397–1405.

Ramanathan R. 2001. A note on the use of the analytic hierarchy process for environmental impact assessment. J Environ Manage. 63:27–35.

Roy M, Venema HD. 2002. Reducing risk and vulnerability to climate change in India: the capabilities approach. Gend Dev. 10:78–83.

Saaty TL. 1980. The analytic hierarchy process: planning, priority setting, resources allocation. New York (NY): McGraw-Hill.

Saaty TL. 2008. Decision making with the analytic hierarchy process. Int J Serv Sci. 1:83–98.

Sanyal J, Lu XX. 2005. Remote sensing and GIS-based flood vulnerability assessment of human settlements: a case study of Gangetic West Bengal, India. Hydrol Process. 19:3699–3716.

Schipper L, Pelling M. 2006. Disaster risk, climate change and international development: scope for, and challenges to, integration. Disasters. 30:19–38.

Sharma PK, Chopra R, Verma VK, Thomas A. 1996. Technical note flood management using remote sensing technology: the Punjab (India) experience. Int J Remote Sensing. 17:3511–3521.

Sharma SVSP, Rao GS, Bhanumurthy V. 2012. Development of village-wise flood risk index map using multi-temporal satellite data: a study of Nagaon district, Assam, India. Curr Sci. 103:705–711.

Simonovich J, Sharabi M. 2013. Dealing with environmental disaster: the intervention of Community Emergency Teams (CET) in the 2010 Israeli forest fire disaster. J Sustainable Dev. 6:86–93.

Singh RP, Kumar R, Tare V. 2009. Variability of soil wetness and its relation with floods over the Indian subcontinent. Can J Remote Sensing. 35:85–97.

Sinha AK. 2003. Development of an integrated disaster management system in India: importance of reliable information. Paper presented at: The International Conference on Total Disaster Risk Management; Kobe. p. 2–4.

Sinha R, Bapalu GV, Singh LK, Rath B. 2008. Flood risk analysis in the Kosi river basin, north Bihar using multi-parametric approach of analytical hierarchy process (AHP). J Indian Soc Remote Sensing. 36:335–349.

Smit B, Wandel J. 2006. Adaptation, adaptive capacity and vulnerability. Global Environ Change. 16:282–292.

Tripati S, Unnikrishnan AS. 2011. Role of cyclones and other factors in the decline of the ports of northern Orissa. Curr Sci. 100:305–312.

Turner BL, Kasperson RE, Matson PA, McCarthy JJ, Corell RW, Christensen L, Eckley N, Kasperson JX, Luers A, Martello ML, et al. 2003. A framework for vulnerability analysis in sustainability science. Proc Nat Acad Sci. 100:8074–8079.

Unnikrishnan AS, RameshKumar MR, Sindhu B. 2011. Tropical cyclones in the Bay of Bengal and extreme sea-level projections along the east coast of India in a future climate scenario. Curr Sci. 101:327–331.

Valdiya KS. 1993. Latur earthquake of 30 September 1993: implications and planning for hazard-preparedness. Curr Sci. 65:515–517.

Van Aalst MK. 2006. The impacts of climate change on the risk of natural disasters. Disasters. 30:5–18.

Wang W, Liang T, Huang X, Feng Q, Xie H, Liu X, Chen M, Wang X. 2013. Early warning of snow-caused disasters in pastoral areas on the Tibetan Plateau. Nat Hazard Earth Sys Sci. 13:1411–1425.

Wisner B, Luce HR. 1993. Disaster vulnerability: scale, power and daily life. GeoJournal. 30:127–140.

Ying X, Zeng GM, Chen GQ, Tang L, Wang KL, Huang DY. 2007. Combining AHP with GIS in synthetic evaluation of eco-environment quality – a case study of Hunan Province, China. Ecol Modelling. 209:97–109.

Yoshimatsu H, Abe S. 2006. A review of landslide hazards in Japan and assessment of their susceptibility using an analytical hierarchic process (AHP) method. Landslides. 3:149–158.

Permissions

All chapters in this book were first published in GNHR, by Taylor & Francis Online; hereby published with permission under the Creative Commons Attribution License or equivalent. Every chapter published in this book has been scrutinized by our experts. Their significance has been extensively debated. The topics covered herein carry significant findings which will fuel the growth of the discipline. They may even be implemented as practical applications or may be referred to as a beginning point for another development.

The contributors of this book come from diverse backgrounds, making this book a truly international effort. This book will bring forth new frontiers with its revolutionizing research information and detailed analysis of the nascent developments around the world.

We would like to thank all the contributing authors for lending their expertise to make the book truly unique. They have played a crucial role in the development of this book. Without their invaluable contributions this book wouldn't have been possible. They have made vital efforts to compile up to date information on the varied aspects of this subject to make this book a valuable addition to the collection of many professionals and students.

This book was conceptualized with the vision of imparting up-to-date information and advanced data in this field. To ensure the same, a matchless editorial board was set up. Every individual on the board went through rigorous rounds of assessment to prove their worth. After which they invested a large part of their time researching and compiling the most relevant data for our readers.

The editorial board has been involved in producing this book since its inception. They have spent rigorous hours researching and exploring the diverse topics which have resulted in the successful publishing of this book. They have passed on their knowledge of decades through this book. To expedite this challenging task, the publisher supported the team at every step. A small team of assistant editors was also appointed to further simplify the editing procedure and attain best results for the readers.

Apart from the editorial board, the designing team has also invested a significant amount of their time in understanding the subject and creating the most relevant covers. They scrutinized every image to scout for the most suitable representation of the subject and create an appropriate cover for the book.

The publishing team has been an ardent support to the editorial, designing and production team. Their endless efforts to recruit the best for this project, has resulted in the accomplishment of this book. They are a veteran in the field of academics and their pool of knowledge is as vast as their experience in printing. Their expertise and guidance has proved useful at every step. Their uncompromising quality standards have made this book an exceptional effort. Their encouragement from time to time has been an inspiration for everyone.

The publisher and the editorial board hope that this book will prove to be a valuable piece of knowledge for researchers, students, practitioners and scholars across the globe.

List of Contributors

Samy Ismail Elmahd
Civil and Environmental Engineering Department, United Arab Emirates University,Al-Ain P.O. Box 15551, United Arab Emirates

Maged Mahmoud Marghan
Faculty of Geoinformation and Real Estate, Institute of Geospatial Science and Technology (INSTeG), Universiti Teknologi Malaysia (UTM), Johor 81310, Malaysia

Mohamed Mostafa Mohamed
Civil and Environmental Engineering Department, United Arab Emirates University,Al-Ain P.O. Box 15551, United Arab Emirates Faculty of Engineering, Irrigation and Hydraulics Department, Cairo University, Giza
P.O. Box 12211, Egypt

Mahboob Alam
Department of Meteorology, COMSATS Institute of Information Technology,Islamabad 44000, Pakistan

Raja Rizwan Hussain
Civil Engineering Department, College of Engineering, King Saud University, Riyadh 11421, Saudi Arabia

A.B.M. Saiful Islam
Department of Civil Engineering, Faculty of Engineering, University of Malaya,Kuala Lumpur 50603, Malaysia

Ahmed M. Youssef
Geology Department, Faculty of Science, Sohag University, Sohag 82524, Egypt
Geological Hazards Department, Applied Geology Sector, Saudi Geological Survey,Jeddah 21514, Kingdom of Saudi Arabia

Mohamed Al-Kathery and Turki El-Sahly
Geological Hazards Department, Applied Geology Sector, Saudi Geological Survey,Jeddah 21514, Kingdom of Saudi Arabia

Biswajeet Pradhan
Department of Civil Engineering, Faculty of Engineering, Geospatial Information Science Research Center (GISRC), University Putra Malaysia (UPM), Serdang 43400, Malaysia

Kiran Prasad Acharya
Graduate School of Science and Engineering, Ehime University, 3 Bunkyo-cho,Matsuyama 790-8577, Japan

Ryuichi Yatabe and Netra Prakash Bhandary
Department of Civil and Environmental Engineering, Graduate School of Science and Engineering, Ehime University, 3 Bunkyo-cho, Matsuyama 790-8577, Japan

Ranjan Kumar Dahal
Department of Geology, Trichandra Campus, Tribhuvan University, Ghantaghar,Kathmandu, Nepal

Cemal Ozer Yigit
Gebze Institute of Technology, Department of Geodetic and Photogrammetric Engineering, Gebze 41400, Kocaeli, Turkey

Mosbeh R. Kaloop
Department of Public Works and Civil Engineering, Faculty of Engineering, Mansoura University, El-Mansoura 35516, Egypt

Dookie Kim
Department of Civil Engineering, Kunsan National University, Kunsan 573-701,Republic of Korea

M.J. Lee
Korea Adaptation Center for Climate Change, Korea Environment Institute (KEI),Seoul 122-706, Korea

I. Park
Department of Geoinformatics, University of Seoul, Seoul 130-743, Korea

J.S. Won
Department of Earth System Sciences, Yonsei University, Seoul 120-749, Korea

S. Lee
Geological Research Division, Korea Institute of Geoscience & Mineral Resources (KIGAM), Daejeon 305-350, Korea
Korea University of Science and Technology, Daejeon 305-350, Korea

V.M. Patel
Civil Engineering Department, K.D. Polytechnic, Patan 384265, Gujarat, India
z Kadi Sarva Vishwavidyalaya, Gandhinagar 382016, Gujarat, India

M. B. Dholakia
L.D. College of Engineering, Ahmedabad 380015, Gujarat, India

A. P. Singh
Institute of Seismological Research, Raisan, Gandhinagar 382009, Gujarat, India

W. Chen, K.Moriya, T. Sakai and L. Koyama
Department of Social Informatics, Biosphere Informatics Laboratory, Graduate School of Informatics, Kyoto University, Kyoto 606-8501, Japan

C.X. Cao
State Key Laboratory of Remote Sensing Science, Institute of Remote Sensing and Digital Earth, Chinese Academy of Sciences, Beijing 100101, P. R. China

Manoj Kumar
National Institute of Rock Mechanics, Kolar Gold Fields-563 117, Karnataka, India

Pijush Samui
Centre for Disaster Mitigation and Management, VIT University, Vellore-632014, India

Ajay Kumar Naithani
Department of Engineering Geology, National Institute of Rock Mechanics, P.O.- Champion Reefs, Kolar Gold Fields–563117, Karnataka, India

Damodar Lamsal
Department of Earth and Environmental Sciences, Graduate School of Environmental Studies, Nagoya University, Furo-cho, Chikusa-Ku, Nagoya 464-8601, Aichi, Japan
Division of Earth System Science, Graduate School of Environmental Science,Hokkaido University, Sapporo 060-0810, Hokkaido, Japan

Takanobu Sawagaki and Teiji Watanabe
Section of Integrated Environmental Science, Faculty of Environmental Earth Science, Hokkaido University, Sapporo 060-0810, Hokkaido, Japan

Alton C. Byers
The Mountain Institute, 108 LA, Elkins, WV 26241, USA

Anusheema Chakraborty and P.K. Joshi
Department of Natural Resources, TERI University, New Delhi 110 070, India

Index